BIODIVERSITY - THE DYNAMIC BALANCE OF THE PLANET

Edited by **Oscar Grillo**

Biodiversity - The Dynamic Balance of the Planet
http://dx.doi.org/10.5772/57015
Edited by Oscar Grillo

CBS, Edition **2016**

Contributors

Leslie Robert Brown, Louise Pretorius, George Bredenkamp, Cornie Van Huyssteen, Andrea Berruti, Roberto Borriello, Alberto Orgiazzi, Antonio Carlo Barbera, Erica Lumini, Valeria Bianciotto, Ronaldo Barthem, Marcio Candido da Costa, Fernanda Cassemiro, Rosseval Galdino Leite, Anita Srivastava, Vivek Khandekar, Johan Pieter Johannes Du Preez, Enrique Javier Javier Carvajal Barriga, Patricia Portero Barahona, Carolina Tufiño, Bernardo Bastidas, Larissa Freitas, Carlos Rosa, Cristina Guamán-Burneo, Emily Weeks, Sergey Alexandrovich Lashin, Yuri Matushkin, Nikolay Kolchanov, Alexandra Klimenko, Valentin Suslov, Waleed Hamza, Humood Naser, Stephane Christian La Barre, Michael Danquah, Indira Sarethy, Sharadwata Pan, Nagaraja B.C., Somashekar R.K., Sunil C, Jose Schoereder

Published by InTech

Janeza Trdine 9, 51000 Rijeka, Croatia
Individual chapters are under their authors' copyright and distributed under the Creative Commons Attribution 3.0 license, which allows users to download, copy and build upon published chapters provided that the authors and publisher are properly credited, which ensures maximum dissemination and a wider impact of the authors' work. Any republication, referencing or personal use of the individual chapters or any of their contents must explicitly identify the original source.

Permission for commercial use of the book as a whole, such as (but not limited to) reprint rights, republication, distribution, sales, translation, and reproduction in any and all forms of media, must always be obtained from InTech.

Publishing Process Manager Iva Lipovic

Technical Editor InTech DTP team

Cover InTech Design team

Additional hard copies can be obtained from orders@intechopen.com

Biodiversity - The Dynamic Balance of the Planet, Edited by Oscar Grillo
p. cm.
ISBN 978-953-51-1315-7

Contents

Preface

Since the first unicellular organisms came to life on our planet, they evolved to colonize ever more hostile territories, developing, for this purpose, special adaptation features. During the whole evolutionary process of the Earth, these capabilities have enabled them to respond to environmental changes that continuously follow each other and that are at the bottom of the current biodiversity.

Since the first unicellular organisms came to life on our planet, they evolved to colonize ever more hostile territories, developing, for this purpose, special adaptation features. During the whole evolutionary process of the Earth, these capabilities have enabled them to respond to environmental changes that continuously follow each other and that are at the bottom of the current biodiversity.

Since the first unicellular organisms came to life on our planet, they evolved to colonize ever more hostile territories, developing, for this purpose, special adaptation features. During the whole evolutionary process of the Earth, these capabilities have enabled them to respond to environmental changes that continuously follow each other and that are at the bottom of the current biodiversity.

In fact, the complex interactions of macro and microscopic species of plants and animals, together with the rich arrays of symbiotic fungi and lichens that are at the basis of every ecosystem, indiscriminately suffer the effects of these changes. The ecosystems are not only basin of very high biodiversity, they are extremely productive areas, providing a lot of benefits to mankind, removing CO_2 from the atmosphere, maintaining the water quality and much more. Consequently, their progressive decline may accelerate climate change, influencing flora and fauna composition and distribution, also resulting in loss of productivity and human lifestyle quality.

Many ecological studies, environmental evaluations and monitoring, as well as new models and methods to assess and preserve the richness of life forms on Earth have been conducted in the last decades, highlighting the current condition in which our planet is, and the future perspectives.

Biodiversity - The Dynamic Balance of the Planet presents comprehensive overviews and original studies focused on biological diversity and conservation of various ecosystems. This volume contains 14 chapters written by international experts, presenting thorough research results and critical reviews of the most relevant aspects and most ecologically interesting areas of the Earth. Topics of the book are ecological and ecosystem functioning studies, hazards and conservation management, assessment of environmental variables affecting species diversity, also considering species richness and distribution.

Oscar Grillo
Stazione Sperimentale di Granicoltura per la Sicilia
Italy

Centro Conservazione Biodiversità
Dipartimento di Scienze della Vita e dell'Ambiente
Università degli Studi di Cagliari
Italy

Prioritising Land-Use Decisions for the Optimal Delivery of Ecosystem Services and Biodiversity Protection in Productive Landscapes

Emily S. Weeks, Norm Mason,
Anne-Gaelle E. Ausseil and Alexander Herzig

Additional information is available at the end of the chapter

http://dx.doi.org/10.5772/58255

1. Introduction

Over the past 50 years, ecosystems have changed more rapidly than at any other period of human history [1]. Considerable portions of the world's thirteen terrestrial biomes are being converted to less ecologically diverse ecosystems [2]. Such a high degree of conversion is leading to extensive changes in biodiversity composition and ecological processes, which results in the diminishing of the ecosystem services that help sustain biological diversity and human populations [3].

Estimates of current extinction rates are several magnitudes above average extinction rates through geological time [4]. Some biologists suggest that a sixth mass extinction is underway, but there is large uncertainty in estimates of global extinction rates [5]. Recently, however, there has been considerable evidence for widespread loss of species at the local and regional level. Studies have shown that the loss of biodiversity at this level has led to the simplification of ecosystem function and resilience [6], and is altering key process important to productivity and sustainability of Earth's ecosystems.

Biodiversity is considered to provide a range of services of varying values to humanity [7] associated with the normal functioning of both their individual components and different combinations of these components in integrated functional ecological systems (Figure 1).The type and level of service inevitably varies among ecosystems but each one can contribute significantly depending on type and their degree of intactness. As the human population increases so do the demands on most ecosystem services provided by indigenous ecosystems, but their ability to provide these services generally decreases with increasing degradation of

the system. This is generally associated with increased human population pressure. Understanding this complex relationship in a particular ecosystem or related ecosystems is most important and should be an integral component of the planning of ecosystem management.

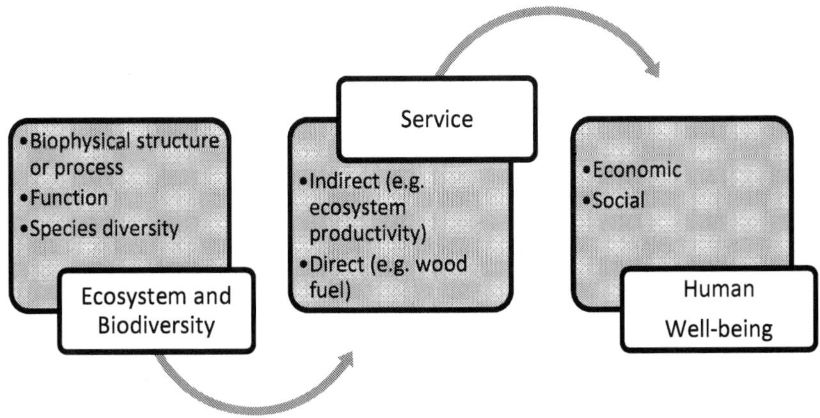

Figure 1. The links between natural ecosystems and human well-being adapted from Haines-Young et al. [7].

A global synthesis reveals that biodiversity loss is a major driver of ecosystem change [6, 8]. There are numbers of causes of biodiversity loss and differences across ecosystems, including: land-use change, climate change, nitrogen deposition, biotic exchanges, and atmospheric carbon dioxide changes [9]. Land-use change has been shown to be one of the leading causes of biodiversity loss in terrestrial ecosystems [3, 10, 11]. An increasing global population and greater demand for food, fodder, fibre and fuel has led to rapid changes in land-use patterns. Areas of low production value, once considered impervious to human activity, have increasingly become susceptible to intensive land-use changes [2, 12, 13].

There is growing evidence of disconnection or opposition between environmental conservation and socio-economic development. In 2005, The Millennium Assessment showed that changes to ecosystems have contributed to human well-being and economic development, but this has been achieved at the expense of many ecosystem services, and increased poverty for some groups of people [1]. Furthermore, the degradation of ecosystem services could escalate during the first half of this century.

A global study, The Economics of Ecosystems and Biodiversity (TEEB, more information at http://www.teebweb.org/), recently revealed the global economic benefit of biodiversity and made the case for better natural resources management [10]. As part of the process, an Intergovernmental Science-Policy Platform on Biodiversity and Ecosystem Services (IPBES, www.ipbes.net) has been established as a follow-up initiative to the Millenium Ecosystem Assessment [1]. Since then, there has been significant interest in converting the concept of ecosystem services into practice, both as a rationale for conservation of biological diversity and as a method to design policies that maximize benefits from the sustainable management of ecosystems. Yet despite this level of attention, understanding of the relationship between biodiversity and ecosystem function is far from complete and is biased towards a small number of ecosystems [14, 15].

The following chapter provides an overview of some the research completed over the last decade for ecosystem services and biodiversity protection. It discusses the general question of how best to allocate land use (or prioritise decisions) to ensure optimal delivery of both ecosystem services and biodiversity protection in productive landscapes. The first section explores the links between biodiversity and ecosystem function and then discusses the role of biodiversity in productive landscapes. This is followed by a brief overview of the different tools currently available that aim to prioritise land-use decisions based on the optimal delivery of ecosystem services and/or biodiversity protection within productive landscapes. The last section highlights challenges and opportunities for multi-objective land-use planning and concludes with recommendations for future research.

2. Linking biodiversity to ecosystem services

Although there are good estimates of society's willingness to pay for various non-marketed ecosystem services, as yet there is no universally accepted framework for assigning values to biological diversity. One of the most useful frameworks to value biodiversity divides direct and indirect use values. Direct-use values can be readily calculated by observing the activities of representative groups of people, by monitoring collection points for natural production, and by examining import and export statistics. Indirect-use values are harder to measure. They are based on the indirect benefits people gain from biological diversity such as ecosystem productivity, water purification, climate regulation, eco-tourism, and recreation [16].

The role of biodiversity in ecosystem function and services is widely debated. Most theoretical and empirical work on measuring this relationship has focused on the congruence between species richness and ecological function. Some studies have found a strong link between the two [8, 17], while others have found little evidence supporting these findings [19]. A review of existing studies illustrates that only a few empirical studies demonstrate improved function with high level of species richness [19]. This suggests changes in species evenness (relative abundance) may deserve greater attention than species richness [20]. Species evenness has a more immediate impact on ecosystems and in most services few dominant species play a major role [6]. In most cases many species are critical for a range of ecosystem functions.

Recent studies have found that certain biodiversity facets co-occur with ecosystem services [21–25]. Large ecosystems, e.g. grasslands, show a high overlap along with ecosystems that provide a range of services that could be used to justify conservation action [21]. There is also a positive relationship between many ecosystem services and high species richness illustrated in several studies that reveal a positive relationship between species richness and productivity [17, 25–27]. However, with some services, such as soil services, the relationship with species diversity is not clear, indicating no single biodiversity measure could be used as a substitute for ecosystem services, and vice versa.

Global efforts to conserve biological diversity have the potential to deliver economic benefits to people [28]. Whether biodiversity conservation could be justified on economic grounds depends on the scale over which benefits are measured [29] or the policy context [30]. A recent study [29] found that at local and global scales conservation benefits outweigh the benefits offered by development, but the balance changes when assessing economic benefits at the national level. For example, the strong positive correlations between species richness and productivity might be interpreted as a win-win for economic growth and the protection of species; however, the economic benefit of conserving high value land at the national scale is limited, and conservation is more likely through smaller scale reserve selection, which would likely benefit only a few species.

Greater diversity in terms of numbers of species may be linked to greater system productivity [19]. For example, decreased local diversity can lead to lower ecosystem productivity, lower use of limiting resources, and lower temporal stability. This is of potential relevance to economists because it provides evidence of the direct value of biodiversity to ecosystem function and the services provided to society by ecosystems. However, there are a number of ecosystem services that are not adequately represented by productivity measures, such as pollination and control of pests. This presents an on-going joint challenge between economists and ecologists – how to quantify these services in a way that can be valued by tools of economists.

3. Scheduling conservation action into productive landscapes

3.1. Potential conservation value of productive landscapes

Production ecology and conservation biology have long been viewed as two opposing approaches to managing ecological systems. With increasing global food demand projected in the next 50 years and decreasing biological diversity that is vital to future productivity, separating the two fields seems counter-productive. A study by Broussard *et al.* [31] evaluated a range of approaches to agricultural management and found that by integrating an ecological approach into environmental management it is possible to feed growing populations without further encroaching into natural systems. To do so, however, may require adopting a landscape management approach to intensification, biodiversity, and ecosystem service protection [32].

There is growing evidence that productive land use can contribute to the conservation of biological diversity [33–36]. Some studies highlight the importance of population exchanges among areas of different disturbance regimes and among early and late successional habitats [37, 38]. Though intensified land use is undeniably the main cause of biodiversity loss, there are opportunities for low intensity land-use systems or those in the form of polycultures and/ or agroforestry patterns to play important roles in large-scale biodiversity conservation. Moreover, pockets of native vegetation found in productive landscapes are refuges for native

flora and fauna [39–41] and provide land corridors for a range of wildlife [42]. These fragmented landscapes are also of great importance for the establishment of studies related both
to species preservation in the long term (including the re-introduction and translocation of
species) and to the genetic health of isolated populations. Remaining riparian habitats can also
play a major role for both humans and nature in productive landscapes [35] by providing
habitat for wildlife and maintaining important ecosystem services (such as clean water) that
are important for productivity.

In addition to contributing to biodiversity protection, productive landscapes are interrelated
to a range of ecosystems services that are associated with biological conservation. Such
landscapes receive services such as pollination, soil fertility, and water retention from surrounding natural systems but also contribute to services such as soil retentions and food
production. The approach adopted for managing productive landscapes can have significant
impact on the services on which it depends or provides. Water quality, pollination and nutrient
cycling, soil retention, carbon sequestration, and biodiversity conservation are all highly
vulnerable to changes in management practices. Some relationships are easy to identify, while
others are much more difficult to measure [43]. For example, the relationship between the
number of pollinators, crop yields, and the use of pesticides is easy to identity (pollinators will
directly increase with crop yield and decrease with increased use of insecticides), while the
benefits of wetlands are much more indirect (wetlands reduce the load of nitrogen in surface
water resulting from agricultural fields).

Another important ecosystem service that has been associated with biodiversity is natural pest
control [32, 44, 45]. Natural pest control provides environmental and economic benefits.
Although productive landscapes with networks of natural habitat can provide refuge for a
range of pests [46], there is evidence that multiple non-productive habitat types may also
favour natural pest control (e.g. grassland, herbaceous wooden habitats and wetlands) [44, 47,
48]. Spatial scale and the distribution of natural habitat may influence the natural pest control
function. For example, diverse small-scale landscapes provide better conditions for natural
pest control than do large-scale landscapes [49]. Overall, there is a need for more studies to
quantify the effects of landscape composition on natural pest control, and further investigation
into the benefits biodiversity restoration programmes may offer to productive landscapes.

3.2. Integrated conservation planning in productive landscapes

There is an increasing expectation that productive (i.e. agricultural) landscapes should be
managed to preserve or enhance biodiversity (e.g. [50]). Often, the impacts of pressures
associated with productive landscapes (and management interventions aimed at mitigating
them) are assessed using local measures, such as native species richness or dominance.
However, it is questionable how relevant such measures are for national-scale conservation
priorities, since they may merely reflect changes in the occurrence and abundance of common,
unthreatened indigenous species [50]. Ideally, any attempts to enhance biodiversity in
productive landscapes should contribute to national conservation objectives. Integrated
conservation planning [51] provides an obvious means for achieving this.

Generally, two independent strands contribute to integrated conservation planning – ecosystem-centred and species-centred prioritisation [52]. An ecosystem-centred approach prioritises efforts that increase the representation of indigenous biodiversity across the full range of environment, ecosystem, and habitat types by enhancing or protection highly modified ecosystem types (thus enhancing or protecting Environmental Representation). A species-centred approach prioritises species based on their conservation status, or some measure of current vs potential distribution – with conservation efforts benefiting the most severely threatened species receiving the highest priority. Some frameworks also consider existing conservation efforts in prioritising new efforts. For instance, threatened species or environment types that already receive a high degree of protection may be assigned lower priority than those that receive little or no protection.

Clearance of indigenous vegetation for agriculture and land-use intensification has severely reduced indigenous biodiversity representation within productive lowland ecosystems (i.e. has reduced environmental representation), so that there is often little or no remnant habitat available for conservation (e.g. [53, 54]. Consequently, ecological restoration is necessary to ensure representation of these ecosystems in conservation networks [53, 55, 56].

In many countries, clearance of indigenous vegetation has been especially severe in environments of limited geographic extent, such as coastal and riparian habitats (e.g. [57]), or ecosystems on unusual substrates (e.g. [58, 59]). Thus areas providing very high gains in environmental representation through restoration or protection will often occupy very small sites. This leads to a right-skewed distribution where most sites provide low environmental representation gains while a few sites provide very large gains. Because environmental representation gain will often be strongly right-skewed, it may be especially vulnerable to trade-offs in multi-objective optimisations of restoration effort. This arises because high values for environmental representation gain are unlikely to co-occur with high values for ecosystem service gains [60]. This means that when ecosystem service benefits are included as criteria for deciding where to apply restoration effort, environmental representation gains will often be much lower than if it were the only criterion. The environmental representation strand of integrated conservation planning thus reveals that a focus on non-biodiversity objectives in designing restoration programmes may result in drastically lower rates of biodiversity gain per unit of restoration effort.

Perhaps the most important implication of integrated conservation planning for biodiversity enhancement schemes is that programmes focussed on the farm scale will likely be very inefficient at contributing to national biodiversity objectives. Not all farms will contain significant areas of highly modified environment types. Hence the potential gain in environmental representation for many farms will often be quite low. Similarly, few farms are likely to contain any threatened species, or have the potential to provide suitable habitat for threatened species. Therefore, any scheme that operates primarily by incentivising individual landowners to manage for biodiversity will result in relatively low gains in national-level conservation priorities per unit effort. By contrast, schemes focussing on the landscape scale will be able to target resources to areas where the potential gains are highest.

This has obvious implications for designing funding models and legislative frameworks to enhance biodiversity in productive landscapes. For instance, if society decides that biodiversity enhancement in a requirement for agricultural industries to obtain a licence to operate, it may be inefficient to demand that every landowner embarks on a significant biodiversity enhancement programme. Rather, it will be more efficient for biodiversity enhancement to operate at the industry level, where industry bodies collect fees from landowners which are then used to fund management in areas of high potential biodiversity gain. Similarly, it would be inefficient to offer every landowner a subsidy in return for carrying out biodiversity enhancements. Rather, it would be better to target subsidies at landowners whose farms contain large areas of high potential biodiversity gain.

4. Land-use planning for ecosystem services and biodiversity protection in productive landscapes

4.1. Spatial optimisation of ecosystem services

Spatial optimisation is a powerful method to explore the potentials of a given area to improve the spatial coherence of land-use functions. It is suitable for identifying land-use configurations which optimally match with spatially varying ecosystem characteristics as well as stake-holder expectations.

Spatial optimisation models have been successfully used to address complex spatial planning problems [61–65] including forest management and timber harvest [66], agricultural issues [61, 65, 67], general issues of land-use change [68], and habitat suitability [69]. Modelling methodologies range from dynamic models based on difference equations of exponential growth [66, 69] to complex models based on systems of non-linear differential equations [70].

The complexity of an optimisation model depends on the complexity of the ecosystem (number of variables, degree of non-linearity, etc.) and the spatial complexity (size of the study area, grid cell size, number of spatially interacting processes). Within land-use planning linear optimisation methods are often not applicable because of the qualitative character of the relations and the large number of variables and/or relations to be optimised. In this case, heuristic methods such as Genetic Algorithms are applied, given that there are few restrictions regarding the formulation of the variables and their relations [61].

Using spatial optimisation tools that systematically consider a range of scenarios, objectives, constraints, and stakeholder or societal preferences helps decision-makers gain insight into the full spectrum of feasible solutions. The tools also allow them to explore opportunities creatively in relation to the imposed limits. However, such use also can result in a simplified representation of options and trade-offs. The accuracy of the result of a spatial optimisation exercise depends on the quality of the input data and the complexity of the model. The more complex the model and the more spatial relationships considered, the greater the uncertainty in the optimisation. Furthermore, a relatively stable land-use pattern indicates a larger degree of freedom in terms of planning alternatives, whereas a relatively unstable land-use pattern indicates there is little room for trade-offs without significantly changing the expectations (i.e. constraints) [71].

In recent years spatial optimisation tools have also been used to link supply of ecosystem services to land use, climate and soil information [71]. Spatial land-use optimisation techniques can help to raise awareness on trade-offs and understand how a landscape configuration could be optimally manage for ecosystem services. LUMASS, for example, is a freeware that has been specifically developed to address this situation [71]. It is a multi-objective decision making tool that can be used for spatial optimisation of ecosystem services. It uses linear programming to optimise the spatial allocation of resources to satisfy an objective (single or multiple), subject to some constraints. Objectives and constraints are specified with regard to a set of criteria representing indicators (such as sediment loss, nitrate leaching and carbon sequestration) of ecosystem services (Figure 2). The optimal allocation of (area of) land uses across the land scape is expressed by the decision variables.

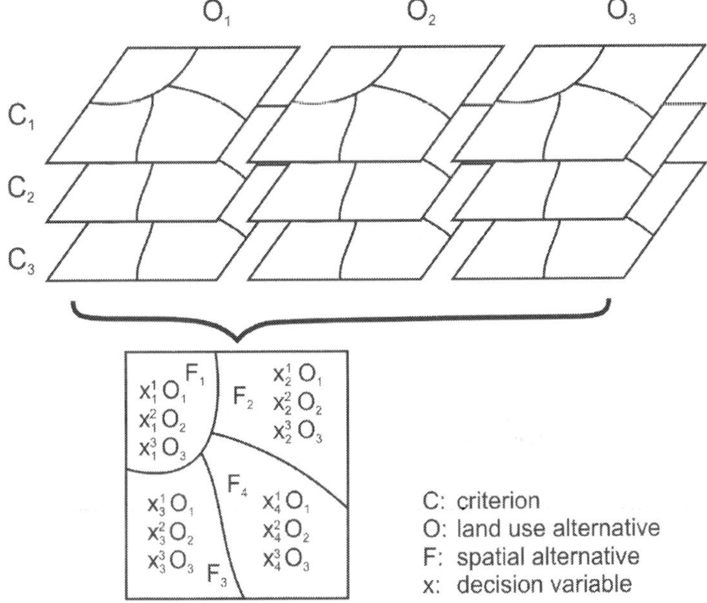

Figure 2. Diagram of the layers incorporated into a multi-objective decision making tool (adapted from71).

The accuracy of the result of a spatial optimisation model depends critically on the quality of the input data. Performance scores used as input data for an optimisation module can be derived from quantitative process-based landscape models or from expert empirical knowlededge. Little is known about the impact of uncertain input data in terms of performance scores and constraints on the produced land-use pattern. However, Herzig et al. [73] found that uncertainty associated with input performance scores may lead to an overestimation of the optimisation benefits.

4.2. Modelling patterns of land-use change

Modelling patterns of land-use change is a fundamental component of conservation planning in productive landscapes [74–78]. Land-use change models can provide a tool for capturing the essence of where land-use change is likely to take place and what is driving these patterns [75, 79] This information can then be used to assess the vulnerability of remaining indigenous habitats and help identify their relative urgency for protection.

Assessment of the vulnerability of species and habitats to imminent proximate threats such as land-use change is a fundamental component of conservation management and planning [76, 77, 80]. Spatially explicit models can provide a tool for capturing the essence of where land-use change is likely to take place and what is driving these patterns [79]. This offers the opportunity to develop an understanding of how land-use change responds to different changes in policy or land-management plans. They can also be used to assess the vulnerability of remaining indigenous habitat and help identify the relative urgency of their protection.

Over the past two decades a range of models of land-use change have been developed, including process-based and statistical models [81]. Statistical models typically rely on the implicit assumption that land-use change processes are stationary. Process-based models, on the other hand, are able to deal with temporal changes in driving forces or processes. While process-based models have sound theoretical basis, statistical models can be easier to implement [82]. As a result, most land-use change models have relied more on a statistical approach to land-use change modelling. These models include Markov, logistic function, and regression models [83–88].

Each land-use change model adopts different statistical techniques to capture present and future land-use patterns. These techniques are either regression or transition based. The regression-based approach is used to understand historical, current, and future land-use patterns by establishing a relationship between a wide range of environmental or socio-economic variables [82]. The influence of these locational factors on land-use change is modelled with distance decay functions, where influence decreases with increasing distance from some feature [89]. In comparison, the transition-based models are rooted in a stochastic Markov-chain technique, where the transition probability is calculated by determining how often the system moved from one state to another [83].

Most land-use change models are spatial transition models and are designed to be part of a decision support system aimed to capture present and future land-use patterns [81, 84, 85, 90]. These models have relied on user-supplied assumptions about how people actually used the land [91]. These assumptions are based on the 'maximum power principle', i.e. people will use the most economically productive land first [92]. However, recent observations of land-use conversion indicate these patterns are changing [93]. Economically viable land is becoming less available, although demand is rapidly increasing [94, 95]. This suggests current and future land-use trends are no longer following traditional decision-making processes supported by the maximum power principle.

Spatial transition models are useful for land-use change modelling because of their robustness but they do not account for temporal heterogeneity. Land-use decisions are often triggered by single events such as economic crises and/or are often remote in space and time and operate at a higher hierarchical level [93]. In these cases, process-based models or models using an economic framework provide better representations of the decision making process. However, for systematic conservation planning these different models may complement each other – logistic regression models effectively identifying areas vulnerable to change, while process-based models can be used to better understand the drivers of these changes.

Most models attempt to illustrate the temporal patterns of land-use change [12, 79, 88, 96]. Quantifying the extent of future land-use change is difficult in itself, given the numerous social, political, and economic factors that drive such change [84, 97–99]. As a result, decision makers often prefer to use a Boolean map (change/no change) to illustrate the temporal and spatial patterns of land-use change. Boolean maps are easier to interpret and decipher and allow decision makers to compare scenarios of land-use change and make management decisions based on the interpolations of the various scenarios.

Because land-use systems respond to a combination of proximate (biophysical) and ultimate (socio-economic) drivers, modelling land-use change ideally requires a multidisciplinary approach [81, 100, 101]. It is often useful to incorporate a wide range of socio-economic and environmental predictors of change; particularly because the individual importance of factors in explaining patterns of land use for a past period will not necessarily reflect their ability to predict a future landscape. The predictive strength of empirical (observed) patterns of land-use change can be enhanced or diminished based on the combination of different. This also underlies the importance of validation in the modelling process [75].

Because land-use change is a dynamic threat, it is important that practitioners keep abreast of change and regularly validate the utility of their vulnerability assumptions and models [80, 102]. A common weakness of land-use modelling is the use of the same data for both calibration (making the model as consistent as possible with the data from which the parameters were estimated) and validation (assessment of the predictive power of the model) [87]. Lack of consideration of model uncertainty through rigorous validation has been shown to result in inaccurate and over-confident predictions. Validation therefore requires testing the predictions of independent data (i.e. not those used in model parameterisation) to ensure the relationships inferred by a model are robust and the predictions are reliable [75].

Spatial statistical models provide a tool for predicting where land-use change is most likely to take place [75, 79, 91]. However, models based on patterns of past change will not necessarily provide reliable predictions of future change because over time exhaustion of formerly suitable areas and changes in global markets, technology, and crops can alter both the distribution and rate of habitat conversion [77, 103]. The reliability of land-use change predictions is therefore likely to decrease as they are projected further into the future, risking misallocation of scarce conservation resources [75]. However, though recent land-use change data will provide the best estimates of future land-use change, vulnerability estimates based on too narrow a time range may also provide less accurate forecasts, because they are based on a small sample of conversion events.

4.3. The application of conservation planning tools for ecosystem services

Systematic conservation planning is the process of identifying and configuring complementary actions required to achieve conservation goals [73, 76]. Since the 1980s, numerous spatial approaches have been developed for identifying priority areas for conservation [103]. These approaches rely on information on the distribution of biodiversity (e.g. [104]); the distribution and effects of threatening processes or 'pressures' on biodiversity (such as pest, weeds, pollution and habitat conversion) and consequent vulnerability (the likelihood or imminence of biodiversity loss to current or impending threatening processes, in the sense of Pressey et al [51] and Wilson et al. [77]; and the effects, and costs, of potential management of pressures [105, 106].

A number of software packages have been developed for conservation planning and resource allocation. Among these are Marxan [107], C-Plan [108], ResNet [109] and Zonation [112, 113]. The two most widely used tools for conservation planning that have integrated ecosystem series into their configuration are Marxan and Zonation.

Marxan is popular with conservation practitioners worldwide and has been applied to both terrestrial and marine ecosystems [107]. It is designed to identify a set of planning units that meets a number of targets for a minimum cost. It can be applied to a variety of conservation features considered in conservation planning and can incorporate ecological processes, site condition, or socio-political influences [110]. It addresses most objectives typically considered in conservation planning and also uses a flexible algorithm that has a variety of applications.

Chan et al. [111] were the first to publish the integration of ecosystem services into Marxan. They compared differences in service provisions between conservation and development in the central coast of California. Marxan has also been used to examine the spatial congruence between biodiversity and ecosystem services in South Africa [21]. In both cases, Marxan produces a map of reserve networks that captured all biodiversity and economic targets at an optimal cost. However, the integration of ecosystem services into the Marxan framework still requires further development and the application should be used with caution. The tool lacks several features that are required for ecosystem service planning [16, 111].

Zonation [112, 113] differs from Marxan and other conservation planning approaches in that it primarily produces problem ranking rather than meeting targets at a minimum cost. Zonation can produce a priority ranking across an entire landscape using large data sets, while also identifying fine-scale prioritisation of biodiversity. Zonation provides priority ranking that balances the needs of biodiversity and competing land uses [114]. This capability, along with its ecologically based model of conservation value, makes it possible for Zonation to incorporate the ecosystem services provision into the prioritisation process

Both Marxan and Zonation address the basic principles of conservation planning. However, neither account for the dynamic processes such as on-going habitat loss, site availability rates, changing or unknown acquisition costs, species-specific connectivity requirements, or temporally varying distributions of features [115]. Nor do they formally incorporate multiple conservation actions such as land acquisition, restoration or easements. Furthermore, both software packages allow for detailed non-linear process descriptions and/or account for sophisticated spatial neighbourhood relationships. However, they are predominantly focused on conservation biology and hence offer only limited flexibility to configure the number and type (i.e. minimisation or maximisation) of objective functions as well as the specification of constraints. It would therefore appear they are less applicable to general land-use pattern optimisation for maximising ecosystem services.

5. Challenges and opportunities for multi-objective land-use planning

5.1. Funding multiple objectives

Despite the mounting interest in focusing conservation efforts on ecosystem services, there is still much debate over the implications for the protection of biological diversity. There is growing evidence of disconnection or opposition between environmental conservation and socio-economic development. Although the maintenance of ecosystem services is often used to justify biodiversity conservation actions, it is still unclear how ecosystem services relate to different aspects of biodiversity and to what extent the conservation of biodiversity will ensure the provision of services.

Part of the difficulty of using biodiversity as a measure of success is that its link to value is unclear. Value of ecosystem services, on the other hand, is easier to define and provides a useful common metric and measure of success [116–118]. Budgets for biodiversity conservation are thinly stretched and thus measuring success is essential to ensuring that scarce funds are optimally used. This recent switch to ecosystem services rather than biodiversity as an organizing principal has led to a more trans-disciplinary approach to biodiversity planning. Though the concept of ecosystem services is anthropogenic (measurements of success are particularly focused on monetary gains or losses), tools such as Ecosystem Services Valuation aim to build and link economic and ecological/biological metrics and work out solutions that correspond to optimal social and ecological decisions. Providing a monetary value to ecosystem services also provides a mechanism to uncover which economic decisions affect biodiversity of an ecosystem and consequently resilience, productivity and value of services.

An ecosystem service approach to managing productive landscapes offers a way to align multiple objectives such as the protection of biodiversity and increased land productivity. Where traditional approaches focus on setting aside land for protection the ecosystem service approach aims to engage a wider range of stakeholders and integrate economic incentives into the planning framework. This is particularly important given that the majority of Earth's natural systems, (containing important biodiversity and provide key ecological services) rest outside protected areas and it is projected that human impacts on these systems are to continue to intensify. In recent years ecosystem services projects have attracted on average more than four times as much funding and are more likely to expand opportunities for conservation [119, 120]. Given that ecosystems services projects are engaging a wider set of funders and becoming increasingly popular around the globe, there is a great need to continue to build alignment between biodiversity protection, human well-being and the delivery of ecosystem services.

There is growing support for understanding the economic costs and benefits of conserving ecosystems, particularly if it will help allocate scarce dollars more efficiently. Investments in biodiversity conservation may be strategically aligned to ecological services of high economic value, and vice versa. By explicitly valuing the costs and benefits associated with services, it may be possible to achieve meaningful biodiversity conservation at lower costs with greater co-benefits [111]. Cost-benefit analyses are widely used in other fields to inform policy decision making (e.g. health, safety, and transport), however conservation planning tools have been slow to integrate this into their framework [121–123]. Spatial cost-benefit analysis could prove invaluable for informing conservation planning, even when relevant data may be limited [122]. There is increased awareness of the economic value of ecosystem services (including biodiversity) and quantifying these values can help decision makers best allocate scarce resources to various policy objectives.

5.2. Thinking beyond habitat provision

Provision of habitat is a necessary but not sufficient condition for threatened species population increases in productive landscapes. Threatened species may be completely absent from the landscape so that translocations will be required for them to occupy habitat made available through restoration or preserved through protection. Further management actions may be required, such as exclusion of domestic livestock, control of invasive predators, herbivores, and weeds (e.g. 124, 125). Consequently, ecological restoration activities required to improve ecosystem services such as water quality or carbon storage may often be insufficient to enhance biodiversity (Figure 3;60).

Obviously, any attempt to enhance threatened species requires an understanding of the primary factors limiting threatened species populations in productive landscapes. We also need to know whether or not threatened species populations are likely to increase in response to management interventions before they are applied on large scales. Obtaining this information for all groups of the indigenous biota is challenging. It may be impracticable to document responses of all high priority species to the pressures imposed by productive landscapes. Similarly, it may not possible to document the response of all high priority species to management interventions aimed at mitigating pressures. Indeed, many studies focus on demonstrating the effect of pressures and management interventions on community-level changes in species composition, without considering implications for high-priority species. This is understandable since rare species are often poorly captured by objective sampling designs. A shift towards studies focussed on capturing variation in high-priority species might help improve our understanding of how pressures and management interventions harm and benefit national conservation goals. However, it might be more efficient to find a way of using existing studies on changes in species composition to predict responses of high-priority species to pressures and management interventions. Functional traits provide such a means of

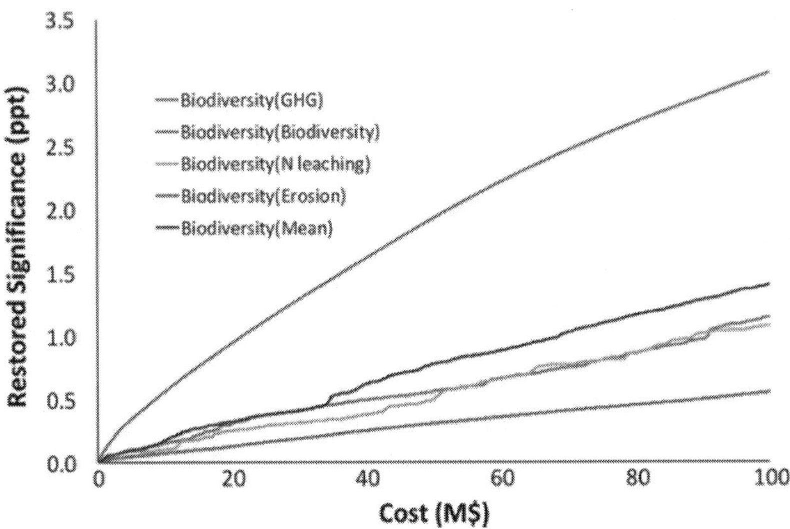

Figure 3. Example of the cumulative averted impacts against restoration effort (millions of New Zealand dollars biodiversity (i.e. restored significance) gain, the cost–benefit optimisations. The variable being optimised is given in brackets. For example, Erosion (Biodiversity) indicates the cumulative erosion reduction achieved when allocating restoration effort to maximise biodiversity gains. Mean indicates that the mean cost–benefit ratio across all variables was optimised. For each optimisation, land was selected in steps of NZ$100, 000 in order of decreasing benefit:cost ratio. In each panel, the vertical distances between the uppermost curve and the other curves indicate the size of trade-offs [60].

generalising results obtained from detailed studies on small subsets of the total indigenous biota to predict responses for threatened species.

Advanced statistical methods using traits to predict species distributions along environmental and disturbance gradients [126–128] could also be used to develop trait-based responses to pressures and management interventions. There is a large literature demonstrating relationships between functional traits to pressures such as anthropogenic disturbance or herbivory from domestic livestock (e.g. 129–133). These relationships could be used to predict threatened species vulnerabilities to such pressures. A smaller number of studies have demonstrated the influence of traits on species responses to the removal or reduction of pressures, such as exclusion of invasive herbivores [59], and post-agricultural forest spread in heavily fragmented landscapes [134]. Thus there appears to be potential for traits to fill in the knowledge gaps that might hinder integrated conservation planning from being applied to productive landscapes, especially in terms of responses to pressure and management. However, we are unaware of any studies testing the ability of trait-based response models to predict accurately responses on independent data or for a different set of species. There is a real need to test whether trait relationships demonstrated in a small-scale study on a subset of species can really be used to extrapolate outcomes for a larger set of species.

5.3. Land-use planning when data are scarce

In most practical land-use and biodiversity planning situations, the information available falls short of that required for informed prioritisation of resource allocation or conservation actions. Many organisations therefore invest resources into gathering and developing data instead of other environmental management activities. There is a growing literature on the cost-effectiveness and optimality of data gathering for guiding land-use planning [103, 131, 135, 136]. A key message from this work is that diminishing returns are inherent in data gathering for conservation planning; at some point, it becomes more effective for conservation organisations to stop data gathering and instead implement protection, albeit with imperfect information.

When data or data-gathering resources are scarce, surrogates are often used. There has been considerable assessment of, and debate on, the effectiveness of surrogates for the distribution of species and taxa [137–141]. Most land-use planning approaches use surrogates for mapping pressures on biodiversity and vulnerability [77], yet relatively little attention has been paid to their relative effectiveness. Surrogates for vulnerability in planning land protection included tenure and land use, environmental or spatial variables correlated with past conversion, threatened species distributions, and maps compiled from expert judgement [77]. Many assumptions are inherent in the application of these surrogates. For example, use of land tenure as a surrogate assumes vulnerability can be estimated from associated permitted land uses; surrogates based on past conversion and threatened species patterns assume that future distributions and impacts of threatening processes are indicated by those in the past [78].

Recently, attention in land-use planning research has shifted from techniques to static reserve blueprints, such as those produced by optimisation, to solving the challenge of conservation planning in the context of dynamic threats [142, 143]. Practitioners preferred not to use static optimised maps, and required tools that helped them make quick decisions for estimating marginal benefits [104]. In fact simple decision rules may have greater practical utility than detailed optimised plans when degradation rates and uncertainty are high, and implementation is carried out over a number of years [144]. These approaches acknowledge that threats are dynamic in most conservation planning situations, that prioritisation that ignores dynamism can be ineffective, and that the need for dynamic updating of conservation priorities is based on updated vulnerability data. However, a major implication of dynamic prioritisation is that solutions may be more demanding of data, and more complex to produce, than those that assume stasis [143].

This situation brings trade-offs between data gathering and conservation effectiveness into stark relief. Clearly, as with data on biodiversity distribution, diminishing returns are inherent in the gathering and validation of accurate data on expanding threats. Land-use planning tools based on less accurate data and simpler solutions may be more effective for conservation, once the cost and flexibility limitations of incorporating more accurate data are accounted for. For example, comprehensive reserve network design may be counterproductive in situations where site availability is uncertain, reserve acquisition is protracted, and rates of biodiversity loss are high [44]. In these situations, simple decision rules, such as protecting the available site with the highest irreplaceability or with the highest species richness, may be more effective for protecting biodiversity.

6. Conclusion

Most of the world's biological diversity currently exists outside protected areas and this is likely to remain true for the foreseeable future. Maintaining the ecological integrity of this matrix is essential to support biological diversity, maintain the embedded protected areas and support changing land use needs. However, achieving both conservation and resource extraction across the landscape will require careful consideration of the different trade-offs between biodiversity protection, ecosystem function, and socio-economic well-being. Future research needs to bring a quantitative approach to the general question: How best to allocate land use (or prioritise decisions) to ensure optimal delivery of both ecosystem services and biodiversity protection – more specifically, which land-use strategy delivers the greatest return on investment.

Biologists and economists have recognized that they need to work more closely to develop a framework to estimate the marginal value of biological diversity. Many countries that appear to have annual increases in gross domestic products may have stagnant or even declining economies when it comes to the costs of development on the loss of species and the depletion of ecosystems. Ecosystems deliver multiple services and many involve trade-offs. The value of biodiversity change to society depends on the net marginal effect of the change on all ecosystem services. Future work needs to quantify the marginal benefits of biodiversity (in terms of services gained) relative to marginal costs (in terms of lost).

Even with the growing literature on ecosystem services, there are still many challenges to integrate ecosystem services into decision making. Although a basic framework showing how to integrate ecosystem services may exist, the science needed to inform the link that connects decisions and ecosystems still remains unclear [3, 145]. The challenge is how to make the ecosystem services framework operational across a range of landscapes, changing conditions and management settings.

The increasing use of conservation planning is accompanied by an increasing number of software tools [146], reflecting diverse ideas about conservation objectives. As conservation planning continues to incorporate more complex and often competing objectives into the prioritisation framework, models should continue to be flexible and adaptable to these changes. The capability of most current conservation tools is limited by using a target-based approach to modelling conservation priorities. Recent research illustrates that a priority ranking rather than "satisfying targets with minimum costs" is more applicable to multi-objective conservation prioritisation, such as ecosystem services [114]. A more comprehensive treatment of the types of social and values of ecosystem services that might affect the implementation of conservation plans is a key area for improvement.

Ecosystem services prioritisation share some of the same elements as any conservation prioritisation problem – assessing the capacity for an ecosystem to meet demand, identifying ecosystem features that could supply services, determining threats to service provision, creating the potential for the future supply of services and the costs of these actions. The conceptual framework for integrating each of these components into spatial prioritization for ecosystem services is illustrated in Figure 4. A fuller characterisation of the biophysical and social context for ecosystem services should improve future prioritisation and the identification of locations where ecosystem-service management is especially important or cost effective.

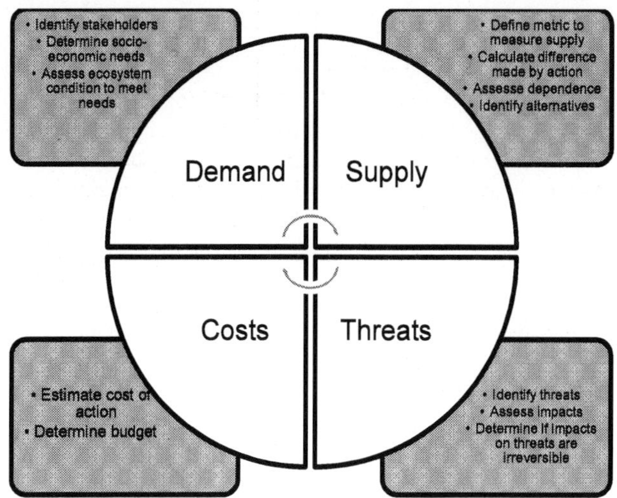

Figure 4. Key aspects to the prioritisation process considered for ecosystem services [147].

The extent to which ecosystem services can be integrated into conservation planning remains largely unknown. Although ecosystem services are increasingly acknowledged in conservation planning, the formal integration of ecosystem services into conservation assessments remains low [14, 21]. There is little research into developing methods to include ecosystem services in conservation assessments or into the extent of concordance between prioritising biodiversity features and the spatial features required for ecosystem services. As a result there is an urgent need to develop an appropriate conceptual framework, an operational model [149], and software tools for planning for ecosystem services. The first step, however, is to map the extent of ecosystem services and identify synergies and trade-offs in conserving areas of high biodiversity value and areas important for service delivery [28, 147, 148].The second step is to use this information to build a dynamic land-use model that can be integrated into a unified framework. This model should build on existing models but should also integrate habitat restoration and consider both biodiversity protection and ecosystem services. From this, further research is required to contrast the performance of habitat protection, restoration or resource management. This could be applied to divergent examples – one where the metric of performance is the persistence of threatened species, and the other where the metric of performance is an ecosystem service. Aside from finding optimal solutions to the problem the method should include rules of thumb for the general conditions and timing at which natural resource managers should shift their emphasis from management to protection or restoration.

Ideally, six key features should be incorporated into conservation planning tools in order to plan effectively for ecosystems services [110, 147, 148]. First, a new tool should incorporate different features within a single network. Second, it should include the option that targets may not be met with available resources or within the planned region. Third, an ideal tool would account, spatially and temporally, for potential impacts of management and threats on species and services. Fourth, the tool would also account for the likeliness of conflicting management practices. Fifth, a tool must consider the flow from ecosystems to user. Last, a tool should allow for "flexibility between the ends of benefit maximisation and suitability-maximizing target achievement, which will each be appropriate for individual ecosystem services in different circumstances".

In reality, coming up with a model that balances all needs is complicated by uncertain data, conflicting needs at different spatial resolutions, and the need to consider costs. There are currently two approaches to ecosystem service assessments. The first uses a broad-scale assessment of multiple services to extrapolate a few estimates of values, based on habitat types, at large spatial scales (including global assessments) [1, 150–152]. In contrast, in the second approach ecosystems services are assessed at the production level of a single service within a small area. Although this approach is more reliable than the first, it lacks both scope (number of services) and scale (geographic and temporal) to be relevant for most policy decisions. What is needed is an approach that is ecologically relevant and appropriate to the scale of land-management decisions.

Some studies have found a decline in the congruence between species richness of different higher taxa at higher spatial resolution [153, 154]. Such findings lead to the question – what is the 'ideal' spatial resolution to outweigh the benefits and costs of loss of biodiversity? It is obvious that management agreements should reflect the spatial scale of the biological processes that are important, but the challenge remains how best to fit this into the various legislative processes. Conservation planning for biodiversity has traditionally tended to adopt a vertical integration of governance to national scales [155, 156]. The emphasis on vertical integration stems from the nature of the spatial turnover in biodiversity, which is not necessarily the case in all ecosystem services (e.g. carbon storage). For ecosystems services, different relationships exist at different scales [19]. Though in general, investing in conservation that increases the value of ecosystems services is also beneficial for biodiversity [15, 64], policy should be underpinned by science that highlights the many roles biodiversity has in ecological processes at different scales

Currently, too little is known about the ecological interactions (including role of biodiversity) in major productive landscapes and about the economic value of the ecosystem services on which they rely or which they potentially provide. To address this lack of knowledge there is a need to adopt an ecological approach to current management. Crop and livestock production systems must be managed as ecosystems, with management decisions fully aware of environmental costs and benefits. Actively managing productive systems for both biodiversity protection and productivity could result in the delivery of multiple services. Many biodiversity management actions can result in multiple benefits. For example, maintaining invertebrate diversity in soils promotes fertility, plant water use efficiency, and increased carbon storage [157].Creative science should provide multiple options and a sound basis for decision.

Research must provide sufficient ecological understanding of productive systems and identify the value of important ecosystems services. Agriculture science, for example, must move beyond understanding ecological constraints to productivity and must focus on identifying the biodiversity and associated ecological processes that underpin the delivery of ecosystem services essential for maintaining a highly productive system [17, 18].

Ideally, conservation efforts in productive landscapes should contribute to national and regional conservation objectives. This means that the aim should be to go beyond enhancing "native dominance" in productive landscapes, which might simply reflect increasing the "bulk supply" of indigenous species and ecosystem types that are already common. Rather, the aim should be to increase the populations and distribution of rare and threatened species (i.e. increase "species occupancy") and the distribution of threatened ecosystems (i.e. increase "environmental representation"). The first step in achieving this is to determine which threatened species and ecosystems occur, or could potentially occur, in productive (unpro-tected) landscapes.

Ultimately an effective conservation plan is one that translates science into action, and it is increasingly recognised that systematic conservation planning must be complemented by an implementation strategy [149], or at least consider implementation issues in its design. Complex optimisation models may not only have greater information needs, but could also be more difficult to communicate to practitioners, and to embed and implement within the operational framework of an organisation. While there is a need for science to solve the problems of dynamic planning, there is also a pressing need for policy and practice to catch up with science [139, 148, 157]. Practitioners prefer not to use static optimised maps [104], and therefore require tools that help them make quick decisions for estimating marginal benefits. Simple decision rules may have greater practical utility than detailed optimised plans, particularly when land-use change is rapid, uncertainty is high, and implementation is carried out over a number of years. Fitting the solution to the practical situation is a challenge that still requires greater attention.

Acknowledgements

This work was supported by the National Land Resource Centre and Landcare Research. We particularly thank Chris Muckersie for useful comments on an earlier draft of the chapter and Anne Austin for editing.

Author details

Emily S. Weeks[1], Norm Mason[2], Anne-Gaelle E. Ausseil[2] and Alexander Herzig[2]

1 National Land Resource Centre, New Zealand

2 Landcare Research, New Zealand

References

[1] Millennium Ecosystem Assessment. *Millenium ecosystem assessment synthesis report.* Washington D.C., USA: Island Press. 2005.

[2] Hoekstra JM, Boucher TM, Ricketts TH, Roberts C. Confronting a biome crisis: global disparities of habitat loss and protection. Ecol Lett. 2005;8(1):23–9.

[3] Daily GC, Polasky S, Goldstein J, Kareiva PM, Mooney HA, Pejchar L, et al. Ecosystem services in decision making: time to deliver. Front Ecol Environ. 2009;7(1):21–8.

[4] Barnosky AD, Matzke N, Tomiya S, Wogan GO, Swartz B, Quental TB, et al. Has the Earth/'s sixth mass extinction already arrived? Nature. 2011;471(7336):51–7.

[5] Stork NE, Coddington JA, Colwell RK, Chazdon RL, Dick CW, Peres CA, et al. Vulnerability and Resilience of Tropical Forest Species to Land-Use Change. Conserv Biol. 2009;23(6):1438–47.

[6] Hooper DU, Adair EC, Cardinale BJ, Byrnes JE, Hungate BA, Matulich KL, et al. A global synthesis reveals biodiversity loss as a major driver of ecosystem change. Nature. 2012;486(7401):105–8.

[7] Haines-Young R. Land use and biodiversity relationships. Land Use Policy. 2009;26:S178–S186.

[8] Cardinale BJ, Duffy JE, Gonzalez A, Hooper DU, Perrings C, Venail P, et al. Biodiversity loss and its impact on humanity. Nature. 2012;486(7401):59–67.

[9] Rockström J, Steffen W, Noone K, Persson \AAsa, Chapin FS, Lambin EF, et al. A safe operating space for humanity. Nature. 2009;461(7263):472–5.

[10] Foley JA, DeFries R, Asner GP, Barford C, Bonan G, Carpenter SR, et al. Global consequences of land use. science. 2005;309(5734):570–4.

[11] Reidsma P, Tekelenburg T, Van den Berg M, Alkemade R. Impacts of land-use change on biodiversity: An assessment of agricultural biodiversity in the European Union. Agric Ecosyst Environ. 2006;114(1):86–102.

[12] Sala OE, Chapin FS, Armesto JJ, Berlow E, Bloomfield J, Dirzo R, et al. Global biodiversity scenarios for the year 2100. science. 2000;287(5459):1770–4.

[13] Imhoff ML, Bounoua L, Ricketts T, Loucks C, Harriss R, Lawrence WT. Global patterns in human consumption of net primary production. Nature. 2004;429(6994):870–3.

[14] Egoh B, Rouget M, Reyers B, Knight AT, Cowling RM, van Jaarsveld AS, et al. Integrating ecosystem services into conservation assessments: a review. Ecol Econ. 2007;63(4):714–21.

[15] Maestre FT, Quero JL, Gotelli NJ, Escudero A, Ochoa V, Delgado-Baquerizo M, et al. Plant species richness and ecosystem multifunctionality in global drylands. Science. 2012;335(6065):214–8.

[16] Dutton A, Edwards-Jones G, Macdonald DW. Estimating the Value of Non-Use Benefits from Small Changes in the Provision of Ecosystem Services. Conserv Biol. 2010;24(6):1479–87.

[17] Tilman D, Polasky S, Lehman C. Diversity, productivity and temporal stability in the economies of humans and nature. J Environ Econ Manag. 2005;49(3):405–26.

[18] Tilman D, Balzer C, Hill J, Befort BL. Global food demand and the sustainable intensification of agriculture. Proc Natl Acad Sci. 2011;108(50):20260–4.

[19] Schwartz MW, Brigham CA, Hoeksema JD, Lyons KG, Mills MH, Van Mantgem PJ. Linking biodiversity to ecosystem function: implications for conservation ecology. Oecologia. 2000;122(3):297–305.

[20] Chapin FS, Sala OE, Burke IC, Grime JP, Hooper DU, Lauenroth WK, et al. Ecosystem consequences of changing biodiversity. Bioscience. 1998;48(1):45–52.

[21] Egoh B, Reyers B, Rouget M, Bode M, Richardson DM. Spatial congruence between biodiversity and ecosystem services in South Africa. Biol Conserv. 2009;142(3):553–62.

[22] Fisher B, Turner RK, Morling P. Defining and classifying ecosystem services for decision making. Ecol Econ. 2009;68(3):643–53.

[23] Tallis H, Kareiva P, Marvier M, Chang A. An ecosystem services framework to support both practical conservation and economic development. Proc Natl Acad Sci. 2008;105(28):9457–64.

[24] Nelson E, Mendoza G, Regetz J, Polasky S, Tallis H, Cameron Dr, et al. Modeling multiple ecosystem services, biodiversity conservation, commodity production, and tradeoffs at landscape scales. Front Ecol Environ. 2009;7(1):4–11.

[25] Chown SL, van Rensburg BJ, Gaston KJ, Rodrigues AS, van Jaarsveld AS. Energy, species richness, and human population size: conservation implications at a national scale. Ecol Appl. 2003;13(5):1233–41.

[26] Litt AR, Gordon DR. Predictors of species richness in northwest Florida longleaf pine sandhills. Conserv Biol. 2003;17(6):1660–71.

[27] Van Rensburg BJ, Chown SL, Gaston KJ. Species richness, environmental correlates, and spatial scale: a test using South African birds. Am Nat. 2002;159(5):566–77.

[28] Naidoo R, Balmford A, Costanza R, Fisher B, Green RE, Lehner B, et al. Global mapping of ecosystem services and conservation priorities. Proc Natl Acad Sci. 2008;105(28):9495–500.

[29] Kremen C, Williams NM, Bugg RL, Fay JP, Thorp RW. The area requirements of an ecosystem service: crop pollination by native bee communities in California. Ecol Lett. 2004;7(11):1109–19.

[30] Anderson BJ, Armsworth PR, Eigenbrod F, Thomas CD, Gillings S, Heinemeyer A, et al. Spatial covariance between biodiversity and other ecosystem service priorities. J Appl Ecol. 2009;46(4):888–96.

[31] Broussard W, Turner RE. A century of changing land-use and water-quality relationships in the continental US. Front Ecol Environ. 2009;7(6):302–7.

[32] Tscharntke T, Klein AM, Kruess A, Steffan-Dewenter I, Thies C. Landscape perspectives on agricultural intensification and biodiversity–ecosystem service management. Ecol Lett. 2005;8(8):857–74.

[33] Turner IM, T Corlett R. The conservation value of small, isolated fragments of lowland tropical rain forest. Trends Ecol Evol. 1996;11(8):330–3.

[34] Chiarello AG. Density and population size of mammals in remnants of Brazilian Atlantic forest. Conserv Biol. 2000;14(6):1649–57.

[35] Jobin B, Bélanger L, Boutin C, Maisonneuve C. Conservation value of agricultural riparian strips in the Boyer River watershed, Quebec (Canada). Agric Ecosyst Environ. 2004;103(3):413–23.

[36] Hector A, Bagchi R. Biodiversity and ecosystem multifunctionality. Nature. 2007;448(7150):188–90.

[37] Bengtsson J, Angelstam P, Elmqvist T, Emanuelsson U, Folke C, Ihse M, et al. Reserves, resilience and dynamic landscapes. AMBIO J Hum Environ. 2003;32(6):389–96.

[38] Elmqvist T, Folke C, Nyström M, Peterson G, Bengtsson J, Walker B, et al. Response diversity, ecosystem change, and resilience. Front Ecol Environ. 2003;1(9):488–94.

[39] Perfecto I, Vandermeer J. Biodiversity conservation in tropical agroecosystems. Ann N Y Acad Sci. 2008;1134(1):173–200.

[40] Schroth G, Harvey CA. Biodiversity conservation in cocoa production landscapes: an overview. Biodivers Conserv. 2007;16(8):2237–44.

[41] Schroth G. Agroforestry and biodiversity conservation in tropical landscapes [Internet]. Island Press; 2004 [cited 2013 Oct 23]. Available from: http://books.google.com/books?hl=en&lr=&id=etuh8kXYMDQC&oi=fnd&pg=PR11&dq=schroth+et+al+2004+biodiversity&ots=9-5MWJFL4G&sig=8bf5xHA-2K8luhM80bsT5Uwpe58

[42] Hoctor TS, Allen WL, Carr MH. Land corridors in the southeast: connectivity to protect biodiversity and ecosystem services. J Conserv Plan. 2007;4:90–122.

[43] Dale VH, Polasky S. Measures of the effects of agricultural practices on ecosystem services. Ecol Econ. 2007;64(2):286–96.

[44] Cardinale BJ, Harvey CT, Gross K, Ives AR. Biodiversity and biocontrol: emergent impacts of a multi-enemy assemblage on pest suppression and crop yield in an agro-ecosystem. Ecol Lett. 2003;6(9):857–65.

[45] Wilby A, Thomas MB. Natural enemy diversity and pest control: patterns of pest emergence with agricultural intensification. Ecol Lett. 2002;5(3):353–60.

[46] Finke DL, Denno RF. Predator diversity and the functioning of ecosystems: the role of intraguild predation in dampening trophic cascades. Ecol Lett. 2005;8(12):1299–306.

[47] Schmidt MH, Lauer A, Purtauf T, Thies C, Schaefer M, Tscharntke T. Relative importance of predators and parasitoids for cereal aphid control. Proc R Soc Lond B Biol Sci. 2003;270(1527):1905–9.

[48] Snyder WE, Ives AR. Interactions between specialist and generalist natural enemies: parasitoids, predators, and pea aphid biocontrol. Ecology. 2003;84(1):91–107.

[49] Bianchi F, Booij CJH, Tscharntke T. Sustainable pest regulation in agricultural landscapes: a review on landscape composition, biodiversity and natural pest control. Proc R Soc B Biol Sci. 2006;273(1595):1715–27.

[50] Kleijn D, Rundlöf M, Scheper J, Smith HG, Tscharntke T. Does conservation on farmland contribute to halting the biodiversity decline? Trends Ecol Evol. 2011;26(9):474–81.

[51] Pressey RL, Humphries CJ, Margules CR, Vane-Wright RI, Williams PH. Beyond opportunism: key principles for systematic reserve selection. Trends Ecol Evol. 1993;8(4):124–8.

[52] Overton JM, Stephens RT, Cook S, Wright E. Unpublished Landcare Research Contract Research Report. Wellington, New Zealand: Landcare Research; 2010. Report No.: LC0910/064.

[53] Awimbo JA, Norton DA, Overmars FB. An evaluation of representativeness for nature conservation, Hokitika Ecological District, New Zealand. Biol Conserv. 1996;75(2):177–86.

[54] Walker S, Price R, Rutledge D, Stephens RT, Lee WG. Recent loss of indigenous cover in New Zealand. New Zealand J Ecol. 2006;30(2):169–77.

[55] Crossman ND, Bryan BA. Systematic landscape restoration using integer programming. Biol Conserv. 2006;128(3):369–83.

[56] Wilson KA, Lulow M, Burger J, Fang Y-C, Andersen C, Olson D, et al. Optimal restoration: accounting for space, time and uncertainty. J Appl Ecol. 2011;48(3):715–25.

[57] Duarte CM, Dennison WC, Orth RJ, Carruthers TJ. The charisma of coastal ecosystems: addressing the imbalance. Estuaries Coasts. 2008;31(2):233–8.

[58] Niemelä J, Baur B. Threatened species in a vanishing habitat: plants and invertebrates in calcareous grasslands in the Swiss Jura mountains. Biodivers Conserv. 1998;7(11):1407–16.

[59] Ausseil A-GE, LIindsay Chadderton W, Gerbeaux P, Theo Stephens RT, Leathwick JR. Applying systematic conservation planning principles to palustrine and inland saline wetlands of New Zealand. Freshw Biol. 2011;56(1):142–61.

[60] Mason NW, Ausseil A-GE, Dymond JR, Overton JM, Price R, Carswell FE. Will use of non-biodiversity objectives to select areas for ecological restoration always compromise biodiversity gains? Biol Conserv. 2012;155:157–68.

[61] Seppelt R, Voinov A. Optimization methodology for land use patterns using spatially explicit landscape models. Ecol Model. 2002;151(2):125–42.

[62] Groot JC, Rossing WA, Jellema A, Stobbelaar DJ, Renting H, Van Ittersum MK. Exploring multi-scale trade-offs between nature conservation, agricultural profits and landscape quality—a methodology to support discussions on land-use perspectives. Agric Ecosyst Environ. 2007;120(1):58–69.

[63] Meyer BC, Lescot J-M, Laplana R. Comparison of two spatial optimization techniques: a framework to solve multiobjective land use distribution problems. Environ Manage. 2009;43(2):264–81.

[64] Polasky S, Johnson K, Keeler B, Kovacs K, Nelson E, Pennington D, et al. Are investments to promote biodiversity conservation and ecosystem services aligned? Oxf Rev Econ Policy. 2012;28(1):139–63.

[65] Ausseil A-GE, Herzig A, Dymond J.R. Optimising land use for multiple ecosystem services objectives: A case study in the Waitaki catchment, New Zealand. In Seppelt R, Voinov AA, Lange S, Bankamp D eds, International Congress on Environmental Modelling and Software, Leipzig, Germany, July 2012. Available from: http://www.iemss.org/sites/iemss2012//proceedings/F5_0473_Ausseil_et_al.pdf

[66] Loehle C. Optimal control of spatially distributed process models. Ecol Model. 2000;131(2):79–95.

[67] Wierzbicki A, Makowski M, Wessels J. Model-based decision support methodology with environmental applications [Internet]. Kluwer Academic Publishers Dordrecht; 2000 [cited 2013 Oct 23]. Available from: http://user.iiasa.ac.at/~marek/pubs/book_s.pdf

[68] Liu X, Li X, Shi X, Huang K, Liu Y. A multi-type ant colony optimization (MACO) method for optimal land use allocation in large areas. Int J Geogr Inf Sci. 2012;26(7):1325–43.

[69] Bevers M, Hof J, Uresk DW, Schenbeck GL. Spatial optimization of prairie dog colonies for black-footed ferret recovery. Oper Res. 1997;45(4):495–507.

[70] Randhir TO, Lee JG, Engel B. Multiple criteria dynamic spatial optimization to manage water quality on a watershed scale. Trans ASAE. 2000;43(2):291–300.

[71] Herzig A. A GIS-based Module for the Multi-Objective Optimization of Areal Resource Allocation. 11th AGILE Int Conf Geogr Inf Sci [Internet]. 2008 [cited 2013 Oct 24]. Available from: http://agile.gis.geo.tu-dresden.de/web/Conference_Paper/CDs/AGILE%202008/PDF/105_DOC.pdf

[72] Herzig, A., Ausseil, A.G., & Dymond, J.R. (2014). Spatial optimisation of ecosystem services. In J.R. Dymond (Ed.), Ecosystem services in New Zealand - conditions and trends. Lincoln, New Zealand: Manaaki Whenua Press

[73] Herzig A, Ausseil A-GE, Dymond JR 2013. Sensitivity of land-use pattern optimisation to variation in input data and constraints. In: In Piantadosi, J., Anderssen, R.S. and Boland J. (eds) MODSIM2013, 20th International Congress on Modelling and Simulation. Modelling and Simulation Society of Australia and New Zealand, December 2013, pp. 1840–1846.

[74] Weeks ES, Walker SF, Dymond JR, Shepherd JD, Clarkson BD. Patterns of past and recent conversion of indigenous grasslands in the South Island, New Zealand. New Zeal J Ecol. 2013;37(1): 127–138

[75] Weeks ES, Overton JM, Walker S. Estimating patterns of vulnerability in a changing landscape: a case study of New Zealand's indigenous grasslands. Environ Conserv. 2013;1(1):1–12.

[76] Margules CR, Pressey RL. Systematic conservation planning. Nature. 2000;405(6783): 243–53.

[77] Wilson K, Pressey RL, Newton A, Burgman M, Possingham H, Weston C. Measuring and incorporating vulnerability into conservation planning. Environ Manage. 2005;35(5):527–43.

[78] Wilson KA, Westphal MI, Possingham HP, Elith J. Sensitivity of conservation planning to different approaches to using predicted species distribution data. Biol Conserv. 2005;122(1):99–112.

[79] Pontius RG, Cornell JD, Hall CA. Modeling the spatial pattern of land-use change with GEOMOD2: application and validation for Costa Rica. Agric Ecosyst Environ. 2001;85(1):191–203.

[80] Weeks ES, Walker S, Overton JM, Clarkson B. The Value of Validated Vulnerability Data for Conservation Planning in Rapidly Changing Landscapes. Environ Manage. 2013;1–12.

[81] Veldkamp A, Lambin EF. Predicting land-use change. Agric Ecosyst Environ. 2001;85(1):1–6.

[82] Theobald DM, Hobbs NT, Bearly T, Zack JA, Shenk T, Riebsame WE. Incorporating biological information in local land-use decision making: designing a system for conservation planning. Landsc Ecol. 2000;15(1):35–45.

[83] Muller MR, Middleton J. A Markov model of land-use change dynamics in the Niagara Region, Ontario, Canada. Landsc Ecol. 1994;9(2):151–7.

[84] Brown S, Hall M, Andrasko K, Ruiz F, Marzoli W, Guerrero G, et al. Baselines for land-use change in the tropics: application to avoided deforestation projects. Mitig Adapt Strat Glob Change. 2007;12(6):1001–26.

[85] Irwin EG, Geoghegan J. Theory, data, methods: developing spatially explicit economic models of land use change. Agric Ecosyst Environ. 2001;85(1):7–24.

[86] Barredo JI, Kasanko M, McCormick N, Lavalle C. Modelling dynamic spatial processes: simulation of urban future scenarios through cellular automata. Landsc Urban Plan. 2003;64(3):145–60.

[87] Pontius Jr RG, Huffaker D, Denman K. Useful techniques of validation for spatially explicit land-change models. Ecol Model. 2004;179(4):445–61.

[88] Echeverria C, Coomes DA, Hall M, Newton AC. Spatially explicit models to analyze forest loss and fragmentation between 1976 and 2020 in southern Chile. Ecol Model. 2008;212(3):439–49.

[89] Liebhold AM, Elkinton JS, Zhou C, Hohn ME, Rossi RE, Boettner GH, et al. Regional correlation of gypsy moth (Lepidoptera: Lymantriidae) defoliation with counts of egg masses, pupae, and male moths. Environ Entomol. 1995;24(2):193–203.

[90] Weng Q. Land use change analysis in the Zhujiang Delta of China using satellite remote sensing, GIS and stochastic modelling. J Environ Manage. 2002;64(3):273–84.

[91] Hall CAS, Tian H, Qi Y, Pontius G, Cornell J. Modelling spatial and temporal patterns of tropical land use change. J Biogeogr. 1995;753–7.

[92] Odum HT. Systems Ecology; an introduction. 1983 [cited 2013 Oct 24]; Available from: http://www.osti.gov/energycitations/product.biblio.jsp?osti_id=5545893

[93] Houghton RA. The worldwide extent of land-use change. BioScience. 1994;44(5):305–13.

[94] Parks PJ. Explaining" irrational" land use: risk aversion and marginal agricultural land. J Environ Econ Manag. 1995;28(1):34–47.

[95] Singh RB. Environmental consequences of agricultural development: a case study from the Green Revolution state of Haryana, India. Agric Ecosyst Environ. 2000;82(1):97–103.

[96] Verburg PH, Schot PP, Dijst MJ, Veldkamp A. Land use change modelling: current practice and research priorities. GeoJournal. 2004;61(4):309–24.

[97] Lambin EF, Rounsevell MDA, Geist HJ. Are agricultural land-use models able to predict changes in land-use intensity? Agric Ecosyst Environ. 2000;82(1):321–31.

[98] Mottet A, Ladet S, Coqué N, Gibon A. Agricultural land-use change and its drivers in mountain landscapes: A case study in the Pyrenees. Agric Ecosyst Environ. 2006;114(2):296–310.

[99] Wood EC, Tappan GG, Hadj A. Understanding the drivers of agricultural land use change in south-central Senegal. J Arid Environ. 2004;59(3):565–82.

[100] Veldkamp A, Fresco LO. CLUE: a conceptual model to study the conversion of land use and its effects. Ecol Model. 1996;85(2):253–70.

[101] Rounsevell MDA, Reginster I, Araújo MB, Carter TR, Dendoncker N, Ewert F, et al. A coherent set of future land use change scenarios for Europe. Agric Ecosyst Environ. 2006;114(1):57–68.

[102] Pressey RL, Hager TC, Ryan KM, Schwarz J, Wall S, Ferrier S, et al. Using abiotic data for conservation assessments over extensive regions: quantitative methods applied across New South Wales, Australia. Biol Conserv. 2000;96(1):55–82.

[103] Grantham HS, Moilanen A, Wilson KA, Pressey RL, Rebelo TG, Possingham HP. Diminishing return on investment for biodiversity data in conservation planning. Conserv Lett. 2008;1(4):190–8.

[104] Ferrier S, Drielsma M. Synthesis of pattern and process in biodiversity conservation assessment: a flexible whole-landscape modelling framework. Divers Distrib. 2010;16(3):386–402.

[105] Wilson KA, McBride MF, Bode M, Possingham HP. Prioritizing global conservation efforts. Nature. 2006;440(7082):337–40.

[106] Underwood EC, Shaw MR, Wilson KA, Kareiva P, Klausmeyer KR, McBride MF, et al. Protecting biodiversity when money matters: maximizing return on investment. PLoS One. 2008;3(1):e1515.

[107] Watts ME, Ball IR, Stewart RS, Klein CJ, Wilson K, Steinback C, et al. Marxan with Zones: software for optimal conservation based land-and sea-use zoning. Environ Model Softw. 2009;24(12):1513–21.

[108] Pressey RL, Watts ME, Barrett TW, Ridges MJ. The C-Plan conservation planning system: origins, applications, and possible futures. 2009 [cited 2013 Oct 24]; Available from: http://eprints.jcu.edu.au/10708/

[109] Kelley C, Garson J, Aggarwal A, Sarkar S. Place prioritization for biodiversity reserve network design: a comparison of the SITES and ResNet software packages for coverage and efficiency. Divers Distrib. 2002;8(5):297–306.

[110] Chan KM, Shaw MR, Cameron DR, Underwood EC, Daily GC. Conservation planning for ecosystem services. PLoS Biol. 2006;4(11):e379.

[111] Chan K, Klinkenberg B. Ecosystem services in conservation planning: Less costly as costs and side-benefits. J Ecosyst Manag [Internet]. 2011 [cited 2013 Oct 24];12(1). Available from: http://jem.forrex.org/index.php/jem/article/viewArticle/80

[112] Moilanen A, Franco AM, Early RI, Fox R, Wintle B, Thomas CD. Prioritizing multiple-use landscapes for conservation: methods for large multi-species planning problems. Proc R Soc B Biol Sci. 2005;272(1575):1885–91.

[113] Moilanen A, Kujala H, Leathwick J. The Zonation framework and software for conservation prioritization. Spat Conserv Prioritization. 2009;196–210.

[114] Moilanen A, Anderson BJ, Eigenbrod F, Heinemeyer A, Roy DB, Gillings S, et al. Balancing alternative land uses in conservation prioritization. Ecol Appl. 2011;21(5): 1419–26.

[115] Moilanen A, Wilson KA, Possingham HP. Spatial conservation prioritization: quantitative methods and computational tools [Internet]. Oxford University Press Oxford; 2009 [cited 2013 Oct 23]. Available from: http://library.wur.nl/WebQuery/clc/1909723

[116] Brock W, Xepapadeas A. Optimal ecosystem management when species compete for limiting resources. J Environ Econ Manag. 2002;44(2):189–220.

[117] Sodhi NS, Lee TM, Sekercioglu CH, Webb EL, Prawiradilaga DM, Lohman DJ, et al. Local people value environmental services provided by forested parks. Biodivers Conserv. 2010;19(4):1175–88.

[118] Wenny DG, Devault TL, Johnson MD, Kelly D, Sekercioglu CH, Tomback DF, et al. The need to quantify ecosystem services provided by birds. The Auk. 2011;128(1):1–14.

[119] Goldman RL, Tallis H, Kareiva P, Daily GC. Field evidence that ecosystem service projects support biodiversity and diversify options. Proc Natl Acad Sci. 2008;105(27): 9445–8.

[120] Tallis H, Goldman R, Uhl M, Brosi B. Integrating conservation and development in the field: implementing ecosystem service projects. Front Ecol Environ. 2009;7(1):12–20.

[121] Polasky S, Camm JD, Garber-Yonts B. Selecting biological reserves cost-effectively: an application to terrestrial vertebrate conservation in Oregon. Land Econ. 2001;77(1):68–78.

[122] Naidoo R, Ricketts TH. Mapping the economic costs and benefits of conservation. PLoS Biol. 2006;4(11):e360.

[123] Gutman P. Putting a price tag on conservation: cost benefit analysis of Venezuela's national parks. J Lat Am Stud. 2002;34(1):43–70.

[124] Burns BR, Floyd CG, Smale MC, Arnold GC. Effects of forest fragment management on vegetation condition and maintenance of canopy composition in a New Zealand pastoral landscape. Austral Ecol. 2011;36(2):153–66.

[125] Inness J, Overton JM, Gillies C. Predation and other factors currently limiting New Zealand forest birds. New Zealand J Ecol. 2010;34(1):86–114.

[126] Shipley B, Vile D, Garnier É. From plant traits to plant communities: a statistical mechanistic approach to biodiversity. science. 2006;314(5800):812–4.

[127] Laughlin DC, Joshi C, Bodegom PM, Bastow ZA, Fulé PZ. A predictive model of community assembly that incorporates intraspecific trait variation. Ecol Lett. 2012;15(11):1291–9.

[128] Pollock LJ, Morris WK, Vesk PA. The role of functional traits in species distributions revealed through a hierarchical model. Ecography. 2012;35(8):716–25.

[129] Lavorel S, Garnier E. Predicting changes in community composition and ecosystem functioning from plant traits: revisiting the Holy Grail. Funct Ecol. 2002;16(5):545–56.

[130] Cornelissen JHC, Lavorel S, Garnier E, Diaz S, Buchmann N, Gurvich DE, et al. A handbook of protocols for standardised and easy measurement of plant functional traits worldwide. Aust J Bot. 2003;51(4):335–80.

[131] Cleary DF, Boyle TJ, Setyawati T, Anggraeni CD, Loon EEV, Menken SB. Bird species and traits associated with logged and unlogged forest in Borneo. Ecol Appl. 2007;17(4):1184–97.

[132] Diaz S, Lavorel S, McIntyre SUE, Falczuk V, Casanoves F, Milchunas DG, et al. Plant trait responses to grazing–a global synthesis. Glob Change Biol. 2007;13(2):313–41.

[133] Barbaro L, Van Halder I. Linking bird, carabid beetle and butterfly life-history traits to habitat fragmentation in mosaic landscapes. Ecography. 2009;32(2):321–33.

[134] Brunet J, De Frenne P, Holmström E, Mayr ML. Life-history traits explain rapid colonization of young post-agricultural forests by understory herbs. For Ecol Manag. 2012;278:55–62.

[135] Bottrill MC, Joseph LN, Carwardine J, Bode M, Cook C, Game ET, et al. Is conservation triage just smart decision making? Trends Ecol Evol. 2008;23(12):649–54.

[136] McDonald-Madden E, Baxter PW, Possingham HP. Making robust decisions for conservation with restricted money and knowledge. J Appl Ecol. 2008;45(6):1630–8.

[137] Rodrigues AS, Brooks TM. Shortcuts for biodiversity conservation planning: the effectiveness of surrogates. Annu Rev Ecol Evol Syst. 2007;38:713–37.

[138] Rodrigues AS, Gaston KJ. Optimisation in reserve selection procedures—why not? Biol Conserv. 2002;107(1):123–9.

[139] Brooks T, Fonseca DA, Rodrigues AS. Species, data, and conservation planning. Conserv Biol. 2004;18(6):1682–8.

[140] Lombard AT, Cowling RM, Pressey RL, Rebelo AG. Effectiveness of land classes as surrogates for species in conservation planning for the Cape Floristic Region. Biol Conserv. 2003;112(1):45–62.

[141] Pressey RL, Watts ME, Barrett TW. Is maximizing protection the same as minimizing loss? Efficiency and retention as alternative measures of the effectiveness of proposed reserves. Ecol Lett. 2004;7(11):1035–46.

[142] Higgins SI, Richardson DM, Cowling RM. Using a dynamic landscape model for planning the management of alien plant invasions. Ecol Appl. 2000;10(6):1833–48.

[143] Pressey RL, Cabeza M, Watts ME, Cowling RM, Wilson KA. Conservation planning in a changing world. Trends Ecol Evol. 2007;22(11):583–92.

[144] Meir E, Andelman S, Possingham HP. Does conservation planning matter in a dynamic and uncertain world? Ecol Lett. 2004;7(8):615–22.

[145] De Groot RS, Alkemade R, Braat L, Hein L, Willemen L. Challenges in integrating the concept of ecosystem services and values in landscape planning, management and decision making. Ecol Complex. 2010;7(3):260–72.

[146] Sarkar S, Pressey RL, Faith DP, Margules CR, Fuller T, Stoms DM, et al. Biodiversity conservation planning tools: present status and challenges for the future. Annu Rev Env Resour. 2006;31:123–59.

[147] Luck GW, Chan KM, Klien CJ. Identifying spatial priorities for protecting ecosystem services. F1000Research [Internet]. 2012 [cited 2013 Oct 25];1. Available from: http:// f1000research.com/articles/1-17

[148] Chan KM, Hoshizaki L, Klinkenberg B. Ecosystem services in conservation planning: targeted benefits vs. co-benefits or costs? PloS One. 2011;6(9):e24378.

[149] Knight AT, Driver A, Cowling RM, Maze K, Desmet PG, Lombard AT, et al. Designing systematic conservation assessments that promote effective implementation: best practice from South Africa. Conserv Biol. 2006;20(3):739–50.

[150] Costanza R, d' Arge R, De Groot R, Farber S, Grasso M, Hannon B, et al. The value of the world's ecosystem services and natural capital. nature. 1997;387(6630):253–60.

[151] Troy A, Wilson MA. Mapping ecosystem services: practical challenges and opportunities in linking GIS and value transfer. Ecol Econ. 2006;60(2):435–49.

[152] Turner WR, Brandon K, Brooks TM, Costanza R, Da Fonseca GA, Portela R. Global conservation of biodiversity and ecosystem services. BioScience. 2007;57(10):868–73.

[153] Pearson DL, Carroll SS. Global patterns of species richness: spatial models for conservation planning using bioindicator and precipitation data. Conserv Biol. 1998;12(4): 809–21.

[154] Grenyer R, Orme CDL, Jackson SF, Thomas GH, Davies RG, Davies TJ, et al. Global distribution and conservation of rare and threatened vertebrates. Nature. 2006;444(7115):93–6.

[155] Erasmus BF, Freitag S, Gaston KJ, Erasmus BH, van Jaarsveld AS. Scale and conservation planning in the real world. Proc R Soc Lond B Biol Sci. 1999;266(1417):315–9.

[156] Strange N, Thorsen BJ, Bladt J. Optimal reserve selection in a dynamic world. Biol Conserv. 2006;131(1):33–41.

[157] Lal R. Soil carbon sequestration to mitigate climate change. Geoderma. 2004;123(1):1– 22.

Chapter 2

Biodiversity and Ecosystem Functioning: a Conceptual Model of Leaf Litter Decomposition

Dalana C. Muscardi, José H. Schoereder and
Carlos F. Sperber

Additional information is available at the end of the chapter

http://dx.doi.org/10.5772/57396

1. Introduction

In this chapter we present a brief history of studies on the relationship between biodiversity and ecosystem functioning (BEF), describing the main models used to explain this relationship, as well as the biodiversity metrics most commonly used. Furthermore, we use litter decomposition as a "process model", presenting a flowchart of mechanisms that may affect the decomposition. The flowchart represents the linking between the diversity of leaves that compose the litter, which is usually called the litter mixture, to its decomposition rates. Finally, we present a simplified flowchart of the edaphic trophic web, relating it to litter decomposition, and some perspectives for future studies in this area.

2. A brief history of studies on biodiversity and ecosystem functioning — BEF

Knowledge about biodiversity has passed through various stages in recent years, resulting from an accelerated scientific production. This scientific output, in turn, is a result of concerns arising from anthropogenic disturbances, which occur on spatial scales ranging from local to global [1].

To illustrate the changes observed in the study of biodiversity, Kevin Gaston in 1996 published a book entitled "Biodiversity: a biology of numbers and difference" [2]. In the first chapter of this book, Gaston emphasizes the relative infancy of the biodiversity issue, stating that a science can be seen by passing through three stages, as it matures [3]. The first stage of biodiversity

studies is called the stage "What?", in which scientists seek only to know what are the species that occur in a particular location. In the second stage, the stage called "How?", is characterized by the attempts to search for patterns of biodiversity, and the third stage (the stage "Why?"), seeks to explain the factors that lead to the patterns observed before. Also according to [1], when the book was written, biodiversity studies remained "emphatically in the second stage of development, with more discussions on the measures and standards, than with issues related to the mechanisms."

In just over a decade after the publication of Gaston, several mechanisms that determine and influence biodiversity were discussed [1], and today the statement made by the author would certainly be very wrong. Many studies have been carried out in different spatial and temporal scales [1], using several different biological systems and in different regions of the world. Much remains to be done, but it is not risky to affirm that little knowledge is still to be generated regarding the mechanisms responsible for determining biodiversity, especially with respect to the definition of biodiversity as the number of species in a given area.

Around the same time of the publication of the book by Gaston, there was a shift in the view of scientists on biodiversity: the thought that biodiversity was an expression of abiotic environmental conditions gave way to the recognition that the properties of the environment were also affected by the biota [4]. This recognition spurred the search for the elucidation of the effect of the loss of biodiversity on ecosystem functioning, generating more than 50 different hypotheses to explain how this relationship would be [5]. Ecosystem functioning can be understood as a set of biogeochemical processes and ecosystem functions [6], responsible for the flow of matter and energy, and it is directly related to the dynamics of resources and the stability of the ecosystem [7].

The hypotheses that explain the effect of biodiversity on ecosystem functioning may be classified into three major classes [8]. Within the first class fall the cases which species are assumed as redundant, and the loss of some species may be compensated by the presence of others, which perform the same function. Thus, to some extent, there would be no reduction of ecosystem functioning due to the lost of species and, on the other hand, there was no increase in the ecosystem functioning when species are added. In the second category of hypotheses are the cases when species are singular or unique, such as, for example, key species. According to this hypothesis, species lack redundancy and the loss or addition of some species would cause drastic changes to ecosystem functioning. Finally, there are hypotheses in which the ecosystem functioning effects of the loss or gain of species does not depend neither on the number nor the identity of the species present, but on the conditions under which this loss occurs, so that the effects of species on the functioning become idiosyncratic.

These model-hypotheses formed the necessary structure to the experimental tests of the relationship between biodiversity and ecosystem functioning [5]. The initial tests involved theoretical approaches, as well as the use of simplified micro and mesocosm laboratory experiments. Posterior studies incorporated actual environmental variation through observational and manipulative experiments, allowing higher applicability in public policy management and biodiversity conservation [9].

3. Biodiversity in studies of BEF

Biodiversity may be estimated through several different metrics, and the most traditional approach is to access it via taxonomic diversity [10]. Later in the studies of BEF, functional diversity started to be used as an additional metric [11, 12] and, more recently, phylogenetic diversity has been included as a proposal of metric to biodiversity [13-15].

Taxonomic diversity, in turn, may be translated by species abundance, richness and composition [16, 17] parameters that are easily accessed and that may give basic information to the generation of diversity indices (such as Simpson or Shannon). These diversity indices are useful to synthesize and compare the biodiversity in different environments or sites [18, 19]. However, the indiscriminate use of diversity indices, and as a goal in itself, rather than using them as a useful metric of diversity, prompted several authors to avoid its use in favor of species richness as a metric. Species may be identified from morphological or genetic traits and, when using taxonomic diversity as a metrics, it is assumed that the differences among species are determined by these aspects. Nevertheless, some authors suggest that taxonomic diversity may not be the most adequate metric [12] to evaluate the effect of organisms on ecosystem functioning. To better evaluate such effects, it would be necessary to use the functional diversity, even though it is expected that, with an increase in species richness, there would be an increase in functional diversity [20].

Functional diversity may be understood as the group of characteristics, of species or organisms, responsible by altering one or more aspects of ecosystem functioning [21]. Such characteristics may be related to the abilities that organisms have to engage or to alter ecosystem processes such as seed dispersal [22], pest biological control [23], pollination [24], nutrient cycling [25], decomposition [26], productivity [27, 28], amongst others. However, when inferring functional diversity from species richness, it is assumed that the relationship between species number and niche occupation is linear, which usually does not occur in nature [7].

Therefore, the use of species richness as an estimate of functional diversity has been criticized, and the estimate of functional diversity has been achieved by species classification by their trophic level, guild, as well as physiological and phenotypic characteristics. Functional diversity is usually used to estimate the biodiversity of plant communities, classifying plant species according to their physiognomy, phenology or photosynthetic pathways. Animals are frequently grouped in guilds based on their consumption, but commonly with a low level of resolution, due to the weak knowledge of their biology. Such characteristics are frequently considered to determine functional diversity because they supposedly relate to aspects of the niches occupied and, consequently, they may express the effects of the organisms in ecosystem processes [21, 29]. From the analysis of these characteristics it is possible to determine and to include species in functional groups, which assemble organisms that fulfill similar functions and, consequently, have similar effects on the ecosystem [30, 22]. This approach is the most usual when using functional diversity in BEF studies.

Functional diversity is considered an estimate that may express more powerfully the effects of biodiversity on ecosystem functioning, because it refers to those biodiversity components that directly affect in how ecosystem operates [22].

There are two questions regarding the use functional diversity as an estimate of biodiversity. The first relates to its teleological use, which involve purpose, such as some earlier discussion of the hypothesis of redundancy [31]. These discussions were full of anthropocentric analogies, describing redundant species as passengers in an automobile [32] or words in a phrase [31]. The replacement of purpose with process, as suggested by [33] and [31], "retains the intuition that if something is functional, it must do something". Therefore, to be functional, species must interact with ecological processes, a relationship that brings us to the second possible question regarding the use of functional diversity as an estimate of biodiversity: is it a tautological discussion?

Because functional diversity is an estimate derived from the relationship between species and ecological processes, it stays clear to us that an observed relationship between ecosystem functioning and functional biodiversity must be positive and highly significant. There are even suggestions that ecological processes and ecological function might be treated as synonymous [31], increasing the possibility of a positive relationship in the studies relating ecosystem functioning and functional diversity.

The above issues must be considered when using functional diversity as an estimate of biodiversity in studies relating ecosystem function and biodiversity. The definition of functional diversity is somewhat established since the proposal by [31], but this metric would only correlate two variables that are a priori related. The metric could be attractive to achieve a definition of the functional groups that are more prevalent in the effects on ecosystem processes, but the metrics alone lack an explanatory power. It is still necessary to question why biodiversity, if it estimated independently of functionality, interferes increasing or decreasing ecosystem functioning.

There has been a suggestion that animal traits that are predicted to influence ecosystem processes must be defined a priori [13]. The actual effect of each functional group on ecosystem processes is then tested and the predictions may be refined accordingly. The above authors also propose a hypothetical relationship between functional diversity and the number of taxa or groups. They suggest a loose relationship between functional diversity and species richness, and adding more families within a trophic level would add more functional trait variation than adding more species within a genus or family. Consequently, the increase in taxa variation would represent an increase of the expected functional diversity, and a priori classification of functional groups could be achieved by a metric that could capture phylogenetic diversity.

Recent studies observed the existence of a relationship between functional and phylogenetic diversities, mainly when considering functional groups as an estimate of functional diversity [16, 30, 34]. According to [30], assemblages with a higher phylogenetic diversity present higher functional diversity, possibly due to the complementarity of ecosystem functions among clades, contrasting with the expected redundancy among species that possess common ancestors.

Species functional characteristics, as well as all other traits, appear along their evolutionary history, affecting the way species distribute and relate to each other. Therefore, these traits tend to be shared among the species that have a common ancestor [30]. The latter authors propose that, since these characteristics may influence ecosystem processes, the phylogenetic history may mirror more accurately the effects of biodiversity on ecosystem functioning. This may occur because closely related species tend to occupy similar niches, and thus may play a similar role in the ecosystem processes. Consequently, communities composed by species that encompass a higher phylogenetic diversity (or more distantly related species), would also encompass a wider range of niches. These communities would be more efficient communities in maintaining ecosystem functioning (Figure. 1), due to the higher complementarity of species effect on ecosystem processes [33].

There are two main reasons to use phylogenetic diversity as a relevant biodiversity estimate in comprehension of ecosystem functioning, instead of functional groups: (1) the removal or addition of a functionally redundant species may have effects of ecosystem processes, highlighting important functional among species, non-captured by functional groupings; and (2) in the case of prediction of change in ecosystem processes like productivity, for example, functional groups may explain as much as categories of randomly chosen groups [34].

Phylogenetic diversity may be defined, in general, as the sum of phylogenetic branches that ling species [34], although several methods of estimating phylogenetic diversity may be recognized [35]. These metrics may be classified into two basic types. The type I metrics begin by calculating a distinctness score for all species of a regional phylogeny and, after this calculation, the distinctness scores of the focal subset of species are summed (or any other function to be used), to produce the phylogenetic diversity metric. Contrarily, the type II metrics use a local phylogeny, from which the distinctness scores are calculated to the focal species [35].

Important considerations regarding the choice of these estimates are presented [41], based on the available data, listing the metrics that are typically used in community ecology studies. These are: (1) Phylogenetic diversity (PD): sum of all branches lengths in the portion of a phylogenetic tree connecting the focal set of species; (2) Mean phylogenetic distance (MPD): mean phylogenetic distance between each pair of species in the focal set of species; (3) Sum of phylogenetic distances (SPD): sum of phylogenetic distances between each pair of species (MPD) multiplied by the number of species pairs; (4) Mean nearest neighbor distance (MNND): mean phylogenetic distance from each species to its closest relative in the focal species set. These metrics are primarily used on conservation biology studies that focus on species conservation, supporting conservation decisions. Nevertheless, in relation to biodiversity and ecosystem functioning studies we do not have a standard metric for usage and this lack comprises an important issue in BEF studies that uses phylogenetic diversity as biodiversity estimate. We recommend to researchers who want to use PD as biodiversity estimate, to search for a powerful background in choosing the metric. We expect that with the increase of studies using PD as metric in BEF researches will result in an improvement of using this metric.

However, even though the estimates of phylogenetic diversity may be promising in BEF studies, there are still no solid evidences about its relationship with ecosystem functioning [15], except those reported by [14]. These latter authors propose a model that partitions the effects of biodiversity into phylogenetic effects and other community properties.

There are three aspects in the studies of BEF relationship that have to be distinguished. The first aspect is the evaluation of which would be the appropriate metrics for biodiversity and for ecosystem functioning. This evaluation involves both methodological and mechanistic considerations, but certainly may affect the results of BEF relationship evaluation, for non-biological reasons. Therefore, one has to be especially careful to distinguish actual mechanistic relationships from tautological correlations, disconnected from biology itself. The second aspect of BEF studies is the evaluation of BEF relationship hypotheses, which necessarily involve regression models that must be tested against null hypothesis of no relationship, or spurious relationship due exclusively to methodological or mathematical issues.

These two first aspects of BEF studies have been often confused, and may lead to unwarranted conclusions. For example, studies on the relationship of phylogenetic diversity in relation to ecosystem functioning [12], have shown that phylogenetic diversity is more relevant to predict ecosystem functioning, at least when evaluated by primary productivity, than crude species richness or diversity. The authors argue for evolutionary reasons for this pattern, and further suggest that thus one should favor phylogenetic diversity rather than species diversity, as a metric to evaluate BEF. This should not, however, be interpreted as a test of the BEF relationship, but rather a methodological refinement, prior to an actual hypotheses testing.

The third aspect of BEF studies, which is the one that we proposed to illustrate in this chapter, regards the explanations for the actual relationship between biodiversity and ecosystem functioning. Disregarding the metrics used to depict biodiversity and the shape of the relationship between biodiversity and ecosystem function, the question why does ecosystem function vary with biodiversity remains. The mechanistic processes by which these two community parameters relate are then described below, using decomposition as the ecosystem function and the variety of species that compose the litter, the so-called litter mixture, as the biodiversity metrics.

4. The process model — Litter decomposition

Litter decomposition is a so-called support process of ecosystems, necessary to the mainte-nance of several ecosystem processes [36]. Due to the implications and to the importance of litter decomposition process in local scale, such as in the productivity of agricultural areas [37], and also in global scale, such as in the global carbon cycling [38], studies involving biodiversity and litter decomposition are increasingly frequent [39-43]. The central goal of these studies is to understand how the factors that drive litter decomposition interact and what are the consequences of biodiversity decrease on this process.

Litter decomposition is the process by which organic matter is progressively broken in smaller parts, since all organic molecules have been mineralized into their primary constituents: water, carbon dioxide and mineral elements [44]. Litter decomposition process is regulated by three main factors: the physico-chemical environment, the quality of the material to be decomposed and the edaphic biota [44-46].

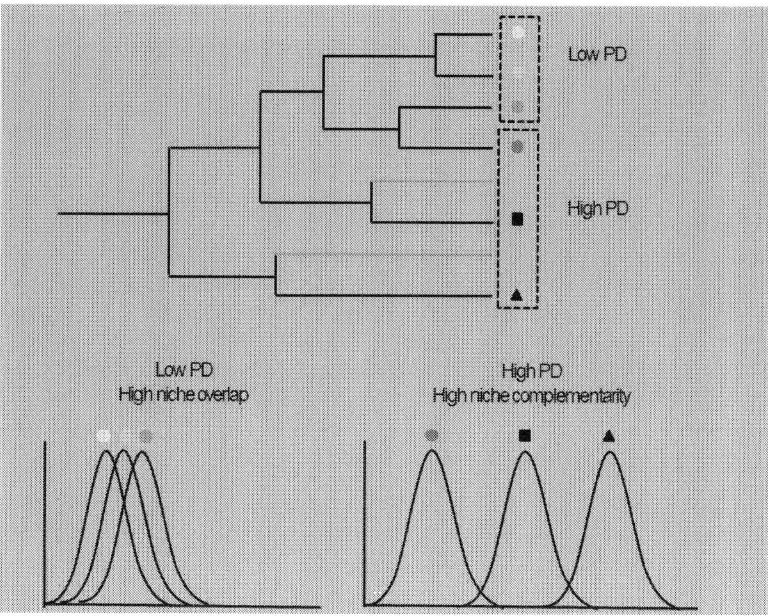

Figure 1. Relationship between functional and phylogenetic diversities in a community, considering the existence of a phylogenetic signal to the functional characteristics in question. PD: Phylogenetic diversity. Closely related species are more similar (symbols differ only in color) than distantly related species (symbols differ in shape). Adapted from [15].

The physico-chemical environment is related to climate, especially humidity and temperature [45, 47]. The climate may indirectly affect litter decomposition, changing litter characteristics, or directly, controlling the activity of decomposing organisms. Plant chemical composition may result from soil formation and from nutrient cycling, and both are regulated by the climate [47]. Therefore, nutrient mineralization may be more accelerated in hot and humid climates, resulting in higher nutrient concentration in litter, increasing its degradability and decomposition [48]. In addition, the direct effect of physico-chemical environment on decomposition occurs from favoring the activity of decomposer organisms by the higher temperature and humidity, increasing the rates of litter decomposition and nutrient release [48, 49].

The second main factor regulating litter decomposition is the quality of organic matter from litter, which is frequently associated to leaf degradability [46]. There is not a unanimity regarding a valid index of litter degradability, although certain nutrient concentrations have been usually associated to higher quality [50]. High nitrogen (N) and phosphorus (P) concentrations, as well as high proportions of easily degradable carbon compounds, such as sugars, have been associated to better litter quality [50], which translates into higher decomposition rates. Contrarily, less degradable carbon compounds (such as lignin), decrease degradability and litter quality, because these compounds require higher energy from decomposers to break the organic matter [46].

The edaphic biota is the third factor regulating decomposition and comprises a plethora of organisms, ranging from bacteria to insects. These organisms remove, mix, break and digest the organic matter, metabolizing litter constituents, mineralizing and making nutrients available to plants [51]. Even though all components of the edaphic biota may perform important roles on litter decomposition, the main decomposing agents are fungi and bacteria, being responsible by nutrient mineralization. Fungi may colonize recently fallen leaves, building up a net of hyphae that allows them the transference of carbon and nitrogen from litter to soil. Bacteria, on other hand, are the main responsible by nitrogen mineralization and availability, which make them extremely significant for the cycling of this nutrient through the soil and to the plants [52]. Several bacteria not only degrade the organic matter, but are central in nitrogen transformations, in a complex of chemical reactions of oxidation and reduction, fixing nitrogen from the air, transforming nitrites into nitrates and back to nitrites, ammonia and returning it to the air. The activity of decomposers, both fungi and bacteria, is affected by the action of detritivore arthropods, which break the litter by its ingestion and digestion, increasing the litter area available to decomposers and facilitating litter decomposition by microorganisms [53].

The diversity of plants whose leaves compose the litter, the litter mixture, may affect the decomposition process through different pathways (Figure 2). According the general findings of the relationship between biodiversity and ecosystem functioning, it is expected a positive relationship between litter mixture diversity (biodiversity) and litter decomposition (ecosystem process). Nevertheless, there are possible pathways that may conduct to the absence of such relationship (the null hypothesis), or even to a negative relationship. The importance of the flowchart depicted in Figure 2 is to generate hypotheses to explain this diversity of possible outcomes in the BEF relationship, allowing the posterior designing of experiments to test these hypotheses.

In general, the main mechanism whereby litter mixture diversity affects litter decomposition it is via resource heterogeneity [54]. In Figure 2, litter taxonomic diversity may be understood in three ways: different species richness and/or compositions, varied functional groups or phylogenetic diversity. We discussed these metrics previously in this chapter, highlighting the advantages and disadvantages of each of them, but disregarding the metrics used, we considered that an increase of each of them would result in higher litter heterogeneity.

A more diverse litter mixture would present a more varied resource supply, allowing the occurrence of a higher abundance and richness of detritivore and decomposer organisms [54]. A higher abundance and/or species richness of decomposers and detritivores would increase litter decomposition.

The heterogeneity promoted by the increase of species composing the litter mixture may occur both due to the physical and chemical plant characteristics. Leaves with certain physical characteristics, such as lower hardness and lignin content, are correlated to higher decomposition rates [55]. Chemical aspects, such as higher carbon and nitrogen concentrations, are frequently correlated to higher decomposition rates [26, 50, 55, 56], even though this pattern is not fully established [51]. According to our flowchart, there would be a positive relationship between environmental heterogeneity and the possibility of the litter mixture being explored by different decomposer species, which possess distinct nutritional needs, or different abilities to exploit resources originated from a varied plant species. The relationship between an increasing environmental heterogeneity affecting positively the species richness of the communities is a well-established pattern [57], and may be based on the amount or variety of resources that may allow the coexistence of potentially competitive species.

Theoretically, it would be possible to predict litter decomposition rates from the proportional sum of the decomposition rates observed in each plant species composing the litter mixture. However, this expected outcome does not necessarily occurs [58, 59], due to interactions between the species composing the litter mixture. When the expected prediction, decomposition rates of the litter mixture corresponds to the sum of decomposition rates of each species composing it, we say that an additive effect is occurring. However, when leaves from two or more plant species are mixed, decomposition rates of the litter mixture may not correspond to that estimated from the decomposition rates of each plant species alone, due to synergistic and antagonistic effects among species composing the litter mixture [59]. These effects may occur because nutrients may be transferred from one plant species to another one [60], altering the expected effects of diversity of litter mixture on leaf litter decomposition [26]. There are evidences that a nutrient transference from a nutrient-rich species (with a lower carbon/ nitrogen ratio) to a nutrient-poor species may increase litter mixture decomposition when compared to the sum of decomposition rates of each species considered alone [56], resulting in a synergistic effect. On the other hand, the presence of some secondary plant metabolites, such as polyphenols, may decrease, revert, or even compensate synergistic mechanisms that would be occurring simultaneously. This effect would decrease the decomposition rate of litter mixture, when compared to the sum of decomposition rates observed in each of the species composing the mixture, causing the antagonistic effect [51, 59]. Thus, plant diversity that composes the litter mixture would not always have the expected positive effect on decomposition. This is so mostly because the number of species in litter [40], the environment in which the litter is decomposing [61], the origin of leaves [62], amongst other aspects, may alter the response of decomposition to plant species diversity of the litter mixture.

Moreover, the effect of litter mixture biodiversity on decomposition rates is deeply related to the edaphic biota, its activity, abundance and composition [56]. The edaphic biota may modulate the decomposition process, mainly in tropical ecosystems, where arthropods are extremely abundant and their effect on decomposition is more consistent [49, 63].

5. Effect of the edaphic biota on decomposition — Top-down/bottom-up control

The edaphic organisms organize in an intricate trophic web, and the diversity of the litter mixture would alter the interactions among these organisms and therefore the effects of this interaction on litter decomposition. The knowledge regarding trophic webs related to the decomposition process is very weak, especially in tropical biomes. This lack of knowledge is caused primarily by the weakness of taxonomic and biological knowledge [64, 65], both

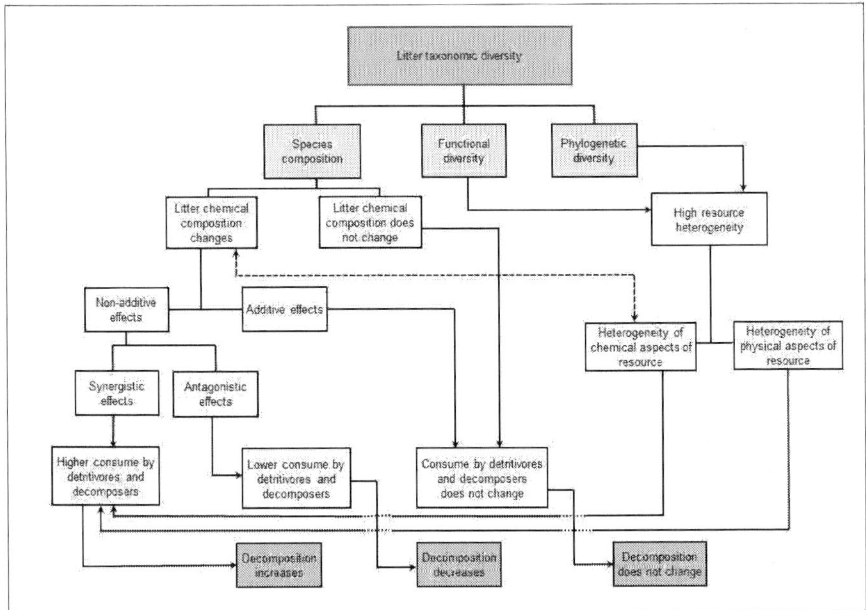

Figure 2. Factors and processes that associate plant diversity in litter mixture to litter decomposition.

regarding the decomposers themselves and their consumers, from different trophic levels, associated to the trophic web. The study of these trophic webs necessarily involves two components that may influence the litter decomposition process: (1) the diversity of litter mixture composing the litter; and (2) the diversity and the trophic links among decomposers and detritivores. Once the trophic web related to litter decomposition is, at least partially, donor-controlled – in the case, the leaf litter – it is expected these webs to be controlled predominantly by bottom-up effects, since consumers do not control resource abundance (and, in our example, nor the diversity).

In Figure 3 we represent a simplified trophic web of detritivores and decomposers, recognizing three different control pathways of the represented trophic levels. The more direct pathway, symbolized by the dotted line, represents a direct effect of the diversity of litter mixture accounting for higher resource heterogeneity to decomposers (but see Figure 2 and discussion above). This pathway may promote the increase of decomposer diversity and activity, directly augmenting the decomposition process, without the interference of organisms from other trophic levels. However, such effects would only be possible if the effects promoted by the increase of litter diversity on the edaphic biota are positive (see Figure 2 and discussion above).

The activities of the animals that inhabit litter may interfere in the above described patterns in different ways. Following the dashed line in Figure 3, the activity of litter breaking by arthropods may increase the resource availability to the decomposer community, since the passage of litter along the digestive tract of detritivores facilitates the degradation by the enzymes of decomposer organisms [66, 67]. Thus, the action of animals may increase the abundance and/or species richness of microorganisms, with a consequent increase in decomposition rates.

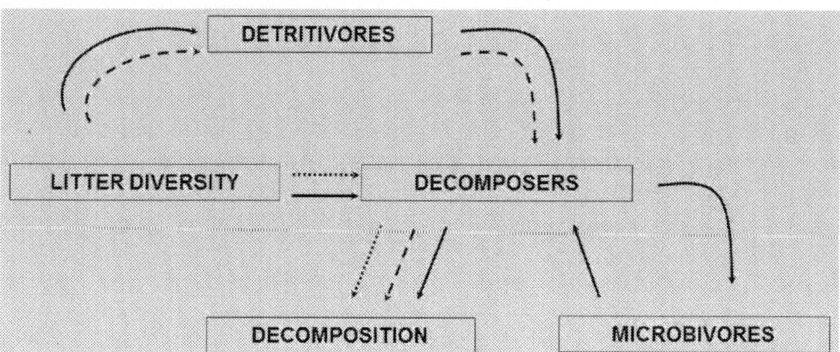

Figure 3. Simplified flowchart of the edaphic trophic web, depicting possible pathways of the effect of diversity of litter mixture on litter decomposition. See text for further information.

On the other hand, the higher abundance and/or diversity of decomposers may lead to an increase of abundance and/or species richness of fungivores and bacteriovores (called here microbivores), due to an increase of resources available to these animals. An increase of microbivores may augment a predation pressure on decomposers, decreasing their abundance and activity. As a consequence, there would be a decrease of litter decomposition (continuous line in Figure 3). This control pathway of the decomposers by the microbivores could represent a top-down control in the decomposer trophic level.

Moreover, predators, not depicted in the trophic web of Figure 3, may control top-down both detritivores and microbivores, making this trophic web more complex and difficult to interpret. The study of edaphic trophic webs, its influence on litter decomposition process and the effects of litter mixture diversity on this web must thus be studied through well designed experiments. Correlational studies may give weak evidences regarding the relative role of each group of organisms within this trophic web, since several of the involved factors in these webs tend to covariate and confound data interpretation [65]. The exclusion of some groups or trophic levels may represent, hence, an option to study these trophic webs, provided that the techniques involved in the exclusion process do not interfere in the decomposition process or in other organisms. A critique of the use of fumigation as an arthropod exclusion method is given by [65], due to the nutrient addition to litter caused by the technique.

6. Conclusions and perspectives

According to the above discussions, it is possible to notice that a single response relating the effects of litter diversity on decomposition does not exist. The interactions among the factors that determine litter decomposition is complex, and several pathways may occur. The effects of litter diversity on decomposition rates are dependent on several features, and the experimental control of these factors is absolutely necessary to clarify their relative roles and interactions.

Therefore, studies controlling, simultaneously, litter and edaphic fauna diversities, may give evidence to their joined effects on decomposition rates, allowing a better understanding of the relationship between these two factors. Furthermore, since several elements external to litter diversity (such as soil use, habitat fragmentation and others) may affect the effect of the discussed factors on litter decomposition, the incorporation of such elements would add information to the process. To do this it is necessary to design studies with manipulative control, which would render more informative studies than those that limit to compare decomposition rates in different litter mixtures.

There are some alternative for these studies, and each alternative may achieve an answer that would clarify one or more of the above discussed points. One proposal it is the use of several plant species, among which different number of species are drawn, giving an idea of the relationship between the number of species in litter mixture and decomposition. In the array of litter mixtures other parameters may also be tested, such as arthropod species richness and abundance, as well as microbial biomass and activity. This approach tends to create situations more close to the actual environments studied, since usually the plant species used are the same as those occurring in the study site. On the other hand, this approach may increase very much the variability among repetitions, creating a noise that may impair data interpretation. The alternative approach would be the use of the same plant species in all repetitions, decreasing variability among them and giving more interpretable data, although decreasing their realism towards the studied environments. These studies are particularly common in habitats involving crops, associated or not to other plants, using decomposition litterbags containing leaves of the cultivated plant, alone and together with other plant species.

In order to study the relative role of the trophic web components, manipulative studies are the most usual approach. In these studies, detritivore species are added or removed, and their effect on other species and/or on decomposition rates are observed. This approach suffers from the same problem of the manipulative studies involving litter mixtures: the distancing from biological reality. Furthermore, these studies are virtually impossible to carry out in hyper diversity habitats (as it is usual in most tropical biomes), and when the knowledge regarding taxonomic identity and/or biology of the involved species is weak. In such cases, experiments excluding litter fauna may be carried out through methods as fumigation, or by the use or other biocides, such as naphthalene. These substances, however, may cause impacts in the decomposer community, modifying decomposition process through more than one pathway of the activity of organisms, impairing once more data interpretation. Another option to exclude fauna is decomposition litterbags with different sized mesh, which may exclude selectively the fauna and modify their species composition. Nevertheless, most options to exclude fauna deal with species richness of the edaphic trophic web, and not on its species composition, weakening the interpretation of the effects of the functional and phylogenetic diversity on litter decomposition process.

Therefore, the studies involving the relationship between the plant diversity composing litter mixtures and its effects on the decomposition process, including the possible effects on the edaphic trophic webs, still need much more study and explanation. The knowledge acquisition in these subjects must necessarily be based on well-designed experiments, to elucidate the relative role of each factor and of the interactions between these factors. We hope to have contributed for future studies by the production of the flowchart of factors and processes, facilitating the isolated tests of each of the connections shown in it.

Acknowledgements

We thank Sabrina Almeida, Júlio Louzada, Ricardo Ildefonso Campos and Elizabeth Nichols for contributions in manuscript idealization. Financial support of Fundação de Amparo à Pesquisa do Estado de Minas Gerais – FAPEMIG and Conselho Nacional de Desenvolvimento Científico e Tecnológico – CNPq are gratefully acknowledged.

Author details

Dalana C. Muscardi[1*], José H. Schoereder[2] and Carlos F. Sperber[2]

*Address all correspondence to:

1 Postgraduate Program in Entomology, Department of Entomology, Federal University of ViçosaViçosa, Minas Gerais, Brazil

2 Department of General Biology, Federal University of Viçosa, Viçosa, Minas Gerais, Brazil

References

[1] Ricklefs R E. A comprehensive framework for global patterns in biodiversity. Ecology Letters 2004; 7 1–15.

[2] Gaston K J., editor. Biodiversity: a biology of numbers and difference. Oxford: Blackwell Science Ltd; 1996.

[3] Wiegert R G. Holism and reductionism in ecology: hypotheses, scale and systems. Oikos 1988; 53 267-269.

[4] Chapin F S, Sala O E, Burke I C, Grime P J, Hooper D U, Lauenroth W K, Lombard A, Mooney H A, Mosier A R, Naeem S, Pacala S W, Roy J, Steffen W L, Tilman D. Ecosystem consequences of changing biodiversity. BioScience 1998; 48 45-52.

[5] Schläpfer F, Schmid B. Ecosystem effects of biodiversity: a classification of hypotheses and exploration of empirical results. Ecological Applications 1999; 9 893–912.

[6] Naeem S. Biodiversity, ecosystem functioning, and ecosystem services. In: Levin S A, Carpenter S R, Godfray H C J, Kinzig A P, Loreau M, Losos J B, Walker B, Wilcove D S. (eds.) The Princeton guide to ecology. Princeton: Princeton University Press; 2009. p367-375.

[7] Díaz S, Cabido M. Vive la différence: plant functional diversity matters to ecosystem processes. Trends in Ecology & Evolution 2001; 16 646–655.

[8] Naeem S, Loreau M, Inchausti P. Biodiversity and ecosystem functioning: the emergence of a synthetic ecological framework. In: Loreau M, Naeem S, Inchausti P. (eds) Biodiversity and Ecosystem Functioning: synthesis and perspectives. Oxford: Oxford University Press; 2002. p3-11.

[9] Solan M, Godbold J A, Symstad A, Flynn D F B, Bunker D. E. Biodiversity-ecosystem function research and biodiversity futures: early bird catches the worm or a day late and a dollar short? In: Naeem S, Bunker D E, Hector A, Loreau M, Perrings C (eds) Biodiversity, Ecosystem Functioning, and Human Wellbeing: An Ecological and Economic Perspective. Oxford: Oxford University Press; 2009. p30-45.

[10] Hou P C L, Zou X, Huang C Y, Chien H J. Plant litter decomposition influenced by soil animals and disturbance in a subtropical rainforest of Taiwan. Pedobiologia 2005; 49 539–547.

[11] Scherer-Lorenzen M, Functional diversity affects decomposition processes in experimental grasslands. Functional Ecology 2008; 22 547–555.

[12] Naeem S, Wright J P. Disentangling biodiversity effects on ecosystem functioning: deriving solutions to a seemingly insurmountable problem. Ecology Letters 2003; 6 567–579.

[13] Hooper D U, Solan M, Symstad A, Diaz S, Gessner M O, Buchmann N, Degrange V, Grime P, Hulot F, Mermilod-Blondin F, Roy J, Spehn E, van Peer L. Species diversity, functional diversity, and ecosystem functioning. In: Loreau M, Naeem S, Inchausti P. Biodiversity and Ecosystem Functioning: synthesis and perspectives. Oxford: Oxford University Press; 2002. p195-208.

[14] Connolly J C, Cadotte M W, Brophy C, Dooley A, Finn J, Kirwan L, Roscher C, Weigelt A. Phylogenetically diverse grasslands are associated with pairwise interspecific processes that increase biomass. Ecology 2011; 92 1385–1392.

[15] Srivastava D S, Cadotte M W, MacDonald A A M, Marushia R G, Mirotchnick N. Phylogenetic diversity and the functioning of ecosystems. Ecology Letters 2012; 15 637–648.

[16] Mikola J, Salonen V, Setälä H. Studying the effects of plant species richness on ecosystem functioning: does the choice of experimental design matter? Oecologia 2002; 133 594–598.

[17] Pacini A, Mazzoleni S, Battisti C, Ricotta C. More rich means more diverse: Extending the "environmental heterogeneity hypothesis" to taxonomic diversity. Ecological Indicators 2009; 9 1271–1274.

[18] Pielou E C. Shannon's formula as a measure of specific diversity: its use and misuse. The American Naturalist 1966; 100 463–465.

[19] Simpson, E H. Measurement of diversity. Nature 1949; 163 688.

[20] Petchey O L, Gaston K J. Functional diversity (FD), species richness and community composition. Ecology Letters 2002; 5 402–411.

[21] Tilman D. Functional Diversity. In: Levin S A. (ed.) Enciclopedia of Biodiversity. Vol. 3. San Diego, Academic Press; 2001. p109–121.

[22] Gorb S N, Gorb E V, Punttila P. Effects of redispersal of seeds by ants on the vegetation pattern in a deciduous forest: A case study. Acta Oecologica 2000; 21 293–301.

[23] Gordon C E, McGill B, Ibarra-Núñez G, Greenberg R, Perfecto I. Simplification of a coffee foliage-dwelling beetle community under low-shade management. Basic and Applied Ecology 2009; 10 246–254.

[24] Vergara C, Badano E. Pollinator diversity increases fruit production in Mexican coffee plantations: The importance of rustic management systems. Agriculture, Ecosystems & Environment 2009; 129 117–123.

[25] Nichols E, Spector S, Louzada J, Larsen T, Amezquitad S, Favila M E, The Scarabaeinae Research Network. Ecological functions and ecosystem services provided by Scarabaeinae dung beetles. Biological Conservation 2008; 141 1461–1474.

[26] Wang S, Ruan H, Wang B. Effects of soil microarthropods on plant litter decomposition across an elevation gradient in the Wuyi Mountains. Soil Biology and Biochemistry 2009; 41 891–897.

[27] Balvanera P, Pfisterer A B, Buchmann N, He J-S, Nakashizuka T, Raffaelli D, Schmid B. Quantifying the evidence for biodiversity effects on ecosystem functioning and services. Ecology Letters 2006; 9 1146–56.

[28] Poisot T, Mouquet N, Gravel D. Trophic complementarity drives the biodiversity-ecosystem functioning relationship in food webs. Ecology Letters 2013; 16 853–61.

[29] Cadotte M W, Cardinale B J, Oakley T H. Evolutionary history and the effect of biodiversity on plant productivity. Proceedings of the National Academy of Sciences of the United States of America 2008; 105 17012–17017.

[30] Martinez N D. Defining and measuring functional aspects of biodiversity. In: Gaston K J. (ed.) Biodiversity: a biology of numbers and difference. Oxford, Blackwell Science. 1996. p114-148.

[31] Walker B H. Biodiversity and ecological redundancy. Conservation Biology 1992; 6 18-23.

[32] Lawton J H, Brown V K. Functional redundancy. In: Schulze E D, Mooney H A. (eds.) Biodiversity and ecosystem function. Berlin: Springer-Verlag; 1993. p255-270.

[33] Flynn D, Mirotchnick N, Jain M, Palmer M, Naeem S. Functional and phylogenetic diversity as predictors of biodiversity-ecosystem-function relationships. Ecology 2011; 92 1573–1581.

[34] Cadotte M W, Cavender-Bares J, Tilman D, Oakley T H. Using phylogenetic, functional and trait diversity to understand patterns of plant community productivity. Plos One 2009; 4 e5695.

[35] Vellend M, Cornwell W K, Magnuson-Ford K, Mooers A. Measuring phylogenetic biodiversity. In: Magurran A E, McGill B J (eds). Biological Diversity: Frontiers in Measurement and Assessment. Oxford: Oxford University Press; 2011. p194-207.

[36] Corvalan C, Hales S, McMichael A. Ecosystems and human well-being: health synthesis: a report of the Millennium Ecosystem Assessment. Switzerland: World Health Organization Press; 2005.

[37] Seneviratne G. Litter quality and nitrogen release in tropical agriculture: a synthesis. Biology and Fertility of Soils 2000; 31 60–64.

[38] Fearnside P M. Global warming and tropical land-use change: greenhouse gas emissions from biomass burning, decomposition and soils in forest conversion, shifting cultivation and secondary vegetation. Climatic Change 2000; 46 115–158.

[39] Cragg R G, Bardgett R D. How changes in soil faunal diversity and composition within a trophic group influence decomposition processes. Soil Biology and Biochemistry 2001; 33 2073–2081.

[40] Harguindeguy N P, Blundo C M, Gurvich D E, Díaz S, Cuevas E. More than the sum of its parts? Assessing litter heterogeneity effects on the decomposition of litter mixtures through leaf chemistry. Plant and Soil 2007; 303 151–159.

[41] Pettit N E, Davies T, Fellman J B, Grierson P F, Warfe D M, Davies P M. Leaf litter chemistry, decomposition and assimilation by macroinvertebrates in two tropical streams. Hydrobiologia 2011; 680 63–77.

[42] Szanser M, Ilieva-Makulec K, Kajak A, Górska E, Kusińska A, Kisiel M, Olejniczak I, Russel S, Sieminiak D, Wojewoda D. Impact of litter species diversity on decomposition processes and communities of soil organisms. Soil Biology and Biochemistry 2011; 43 9–19.

[43] Yang X, Chen J. Plant litter quality influences the contribution of soil fauna to litter decomposition in humid tropical forests, southwestern China. Soil Biology and Biochemistry 2009; 41 910–918.

[44] Swift M J, Aheal O W, Anderson J M. Decomposition in terrestrial ecosystems. Berkeley: University of California Press; 1979.

[45] Cotrufo M F, Del Gado I, Piermatteo D. Litter decomposition: concepts, methods and future perspectives. In: Werner L K, Bahn M, Heinemeyer A. (Eds.) Soil Carbon Dinamics. An integrated methodology. New York: Cambridge University Press; 2009. p76-90.

[46] Melillo J M, Aber J D, Muratore J F. Nitrogen and lignin control of hardwood leaf litter decomposition dynamics. Ecology 1982; 63 621–626.

[47] Aerts R. Climate, leaf litter chemistry and leaf litter decomposition in terrestrial ecosystems: a triangular relationship. Oikos 1997; 439–449.

[48] Hättenschwiler S, Coq S, Barantal S, Handa I T. Leaf traits and decomposition in tropical rainforests: revisiting some commonly held views and towards a new hypothesis. New Phytologist 2011; 189 950–965.

[49] González G, Seastedt T R. Soil fauna and plant litter decomposition in tropical and subalpine forests. Ecology 2001; 82 955–964.

[50] Li L J, Zeng D H, Yu Z Y, Fan Z P, Yang D, Liu Y X. Impact of litter quality and soil nutrient availability on leaf decomposition rate in a semi-arid grassland of Northeast China. Journal of Arid Environments 2011; 75 787–792.

[51] Hättenschwiler S, Gasser P. Soil animals alter plant litter diversity effects on decomposition. PNAS 2005; 102 1519–1524.

[52] Andrade A G, Tavares S R L, Costa C H L. Contribuição da serrapilheira para recuperação de áreas degradadas e para manutenção da sustentabilidade de sistemas agroecológicos. Agroecologia 2003; 24 55–63.

[53] Pramanik R, Sarkar K, Joy V C. Efficiency of detritivore soil arthropods in mobilizing nutrients from leaf litter. Tropical Ecology 2001; 42 51–58.

[54] Kaneko N, Salamanca E F. Mixed leaf litter effects on decomposition rates and soil microarthropod communities in an oak – pine stand in Japan. Ecological Research 1999; 14 131–138.

[55] Harguindeguy N P, Díaz S, Cornelissen J H C, Vendramini F, Cabido M, Castellanos A. Chemistry and toughness predict leaf litter decomposition rates over a wide spectrum of functional types and taxa in central Argentina. Plant and Soil 2000; 218 21–30.

[56] Song F, Fan X, Song R. Review of mixed forest litter decomposition researches. Acta Ecologica Sinica 2010; 30 221–225.

[57] Ribas C R, Schoereder J H, Pic M, Soares S M. Tree heterogeneity, resource availability, and larger scale processes regulating arboreal ant species richness. Austral Ecology 2003; 28 305–314.

[58] Gartner T B, Cardon Z G. Decomposition dynamics in mixed-species leaf litter. Oikos 2004; 2 230–246

[59] Hättenschwiler S, Tiunov A V, Scheu S. Biodiversity and litter decomposition in terrestrial ecosystems. Annual Review of Ecology, Evolution, and Systematics 2005; 36 191–218.

[60] Seastedt T. The role of microarthropods in decomposition and mineralization processes. Annual Review of Entomology 1984; 29 25–46.

[61] Louzada J N C, Schoereder J H, De Marco P. Litter decomposition in semideciduous forest and Eucalyptus spp. crop in Brazil: a comparison. Forest Ecology and Management 1997; 94 31–36.

[62] Gholz H L, Wedin D A, Smitherman S M, Harmon M E, Parton W J. Long-term dynamics of pine and hardwood litter in contrasting environments: toward a global model of decomposition. Global Change Biology 2000; 6 751–765.

[63] Smith V C, Bradford M A. Do non-additive effects on decomposition in litter-mix experiments result from differences in resource quality between litters? Oikos 2003; 2 235–242.

[64] De Ruiter P C, Griffiths B, Moore J C. Biodiversity and stability in soil ecosystems: patterns, processes and the effects of disturbance. In: Loreau M, Naeem S, Inchausti P. (Eds.) Biodiversity and ecosystem functioning: Synthesis and perspectives. Oxford: Oxford University Press; 2002. p102-113.

[65] Mikola J, Bardgett R D, Hedlund K. Biodiversity, ecosystem functioning and soil decomposer food webs. In: Loreau M, Naeem S, Inchausti P. (Eds.) Biodiversity and ecosystem functioning: Synthesis and perspectives. Oxford: Oxford University Press; 2002. p169-180.

[66] Hopkin S P. Biology of springtails. Oxford: Oxford University Press; 1997.

[67] Krantz G W, Walter D E. A manual of acarology. Texas: Texas Tech University Press; 2009.

Chapter 3

Modern Taxonomy for Microbial Diversity

Indira P. Sarethy, Sharadwata Pan and
Michael K. Danquah

Additional information is available at the end of the chapter

http://dx.doi.org/10.5772/57407

1. Introduction

Microorganisms are actually composed of very different and taxonomically diverse groups of communities: archaea, bacteria, fungi and viruses. The members of these groups or taxa are distinct in terms of their morphology, physiology and phylogeny and fall into both prokaryotic and eukaryotic domains. They constitute a broad group of life system inhabiting the known ecosystems on earth: terrestrial and marine; including geographical locations considered to be extreme or inimical to life. The latter comprise of such areas as habitats with high salinity, alkalinity, acidity, high and low temperatures, high pressure, and high radiation. Considering the adaptability of microorganisms to grow and survive under varied physico-chemical conditions and their contribution in maintaining the balance in ecosystems, it is pertinent to catalogue their diversity as it exists. The inability to visualize them with the naked eye precludes effective classification. As such, using the available tools, microorganisms are broadly classified into prokaryotes and eukaryotes and subsequently into various taxonomical units depending on the resources available and required.

The sustenance of life on earth depends on maintaining the diversity of microorganisms. Human intervention is resulting in depletion of biodiversity and many hotspots are also fast losing their endemic biodiversity. While specific data is hard to come by, it is likely that loss of macro life forms also results in loss of the associated microbial species: symbionts as well as the rhizosphere-colonizing microbes. The significant contribution made by microorganisms in ecosystem sustainability as well as the industrially important biomolecules obtained from them: antibiotics, anti-cancer drugs, enzymes, biofuel and various other compounds, implies that cataloguing them is imperative. However, a simple and effective microbial identification system is still far off. The available tools for classification and identification of microorganisms rely on a number of different technologies. This chapter provides an overview of taxonomy

tools for understanding prokaryotic and eukaryotic microbial diversity. Taxonomy (or biosystematics) consists of three main parts: *classification* (arrangement of organisms based on similarity), *nomenclature* (naming of the organisms) and *identification* (determining whether an organism belongs to the group under which it is classified and named). Modern biosystematics also includes *phylogeny* as an integral part of the classification process [1].

2. Biogeography of microbial diversity

The diversity of microbial communities varies within habitats as much as between habitats [2]. This variation can even occur within a few millimetres, suggesting that microbial diversity encompasses more than the documented evidence available. Hence, biogeography is gaining importance as a field of study from microbial diversity point of interest. Many reasons have been postulated to explain this phenomenon. Due to the innately small size of the microorganisms, environmental complexity plays a major role in determining diversity. Spatial heterogeneity is likely to lead to the formation of many niches within a habitat [3]. Recent tools like metagenomics aid in biogeography studies by providing information on nucleic acid sequence data, thereby directly identifying microorganisms (see Section 9). Therefore the phylogenetic information can be used to compare microbial diversity profile across habitats [2].

Generally, diversity within a particular location and in a community is called alpha diversity. Beta diversity measures the community composition between two or more locations while gamma diversity applies to a region, across continents and biomes and is larger in size than that used for measuring alpha diversity [4].

3. Microbial evolution

The evolutionary relationship of microorganisms is called phylogeny. Understanding phylogenetic profiles of microbes becomes a daunting task because of their small size and the lack of particular indicators that could serve as markers. Some proteins and genes are considered as evolutionary chronometers which measure the evolutionary change [5]. Currently, the 16S rDNA sequence is considered to be most reliable for measuring evolutionary relationships in bacteria and archaea (detailed in Section 7.2.1) and the 18S sequence for fungi (see Section 8). However, it is necessary to choose the correct protein or gene for such studies. Such a gene or protein should have certain features which make it most appropriate for deriving evolutionary relationship. The most important criterion is that it should be present in all members of the target group and be functionally homologous in the organisms. The molecule must contain regions of conserved sequences for comparison purposes. The changes in sequence data must be at a slow enough rate to permit measurement so that it may also reflect evolutionary change for the entire group [5].

In the current system of classification, based on the 16S rDNA sequence, evolutionary relationships form the basis for division and three major domains have been recognized, out of which two comprise of bacteria and archaea (prokaryotes) and the third domain is of eukaryotes [6]. It is important to understand evolution in the context of biodiversity. Evolution leading to new ecotypes/species is achieved in many ways. Some species with quick generation times also undergo mutation frequently leading to novel species or strains [3]. Horizontal gene transfer (HGT), via transformation, transduction or conjugation, also accounts for introduction of genes into distantly related organisms, thereby introducing new traits and also impacting on interaction between species and thereby ecosystem processes [7]. It has also been hypothesized that large population sizes of microbes and their low extinction rates may also play a role in maintaining biodiversity, though measurement of such extinction rates is difficult ([8] and references therein).

4. Microbial phylogeny

The phylogenetic tree representing all living organisms shows that, evolution of current forms of life occurred from a common ancestor (the universal ancestor), depicted by the root (see Figure 1). Two domains are of prokaryotic systems of life: the archaea and Eubacteria; in contrast to previous systems of classification, wherein, the prokaryotes were confined to a single kingdom. However, it is intriguing to note that, genomic studies have shown the archaea to contain unique gene sequences which are not present in bacteria or eukaryotes. Certain

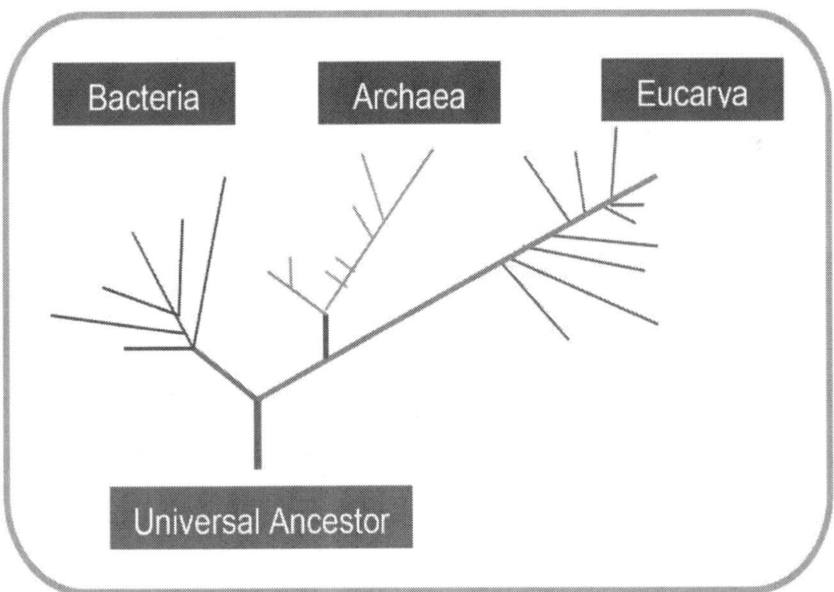

Figure 1. The universal phylogenetic tree based on comparative ribosomal RNA sequences (adapted from [6]).

genes are also shared between all the three domains. The genes required for core cellular functions are the ones which are necessary for survival of a cell and could have arisen from the common ancestor.

The divergence of the organisms represents the differences in genetic sequences which could have become fixed in each group as they evolved. It is also postulated that earlier, HGT played a key role in transfer of genes between organisms early in the evolutionary history [6]. It occurs as a response to any change in the environment and provides for better adaptation ([9] and references therein). Subsequently, reproductive isolation could have prevented extensive exchange of genes, though it continues to occur amongst prokaryotes.

5. The prokaryotic microbes — Bacteria and archaea

The bacteria and archaea have evolved along different lines though both are essentially prokaryotic. The archaea are considered to be the most primitive and are common inhabitants of the so-called extreme habitats (hot springs, deep sea hydrothermal vents, alkaline and acidic habitats). Though the bacteria and archaea share certain common features, the archaea also share similarities with eukaryotes which are further exemplified by the 16S rDNA-based phylogenetic analyses.

6. The species concept

An array of diverse definitions have been proposed to describe microbial species. Currently, a polyphasic approach is used to define a microbial species using phenotypic and genotypic properties [1, 10]. Whenever a new taxon is proposed, it is essential that the organism be isolated in pure culture and its characteristic features be tested under standard conditions [11]. Whether an organism constitutes a member of a common species is primarily based on whether its DNA-DNA re-association values are more than 70% and melting temperature (ΔT_m) is less than 5°C, the experiments being performed under standard conditions [12]. All the strains within a species must show similar phenotypes. A designated type strain of a species constitutes the reference specimen for that species [13]. A species description must preferably be based on the characteristics of more than one type strain. To be assigned a different species name, members must show at least one and is governed by the [12]. If the 16S rDNA sequences of organisms are ≤ 98.7% or ≤ 97% identical, they are members of different species. This is considered even in the absence of DNA-DNA hybridization experiments since this level of divergence in 16S rDNA sequences constitutes less than 70% DNA-DNA similarity [14]. Uncultured microbes cannot be assigned to a definite species since their phenotype is not known; however, they can be assigned a 'Candidatus' designation provided their 16S rRNA sequence subscribes to the principles of identity with known species [15]. A concept applying to a taxon lower than that of the strain is the ecotype – those microorganisms that occupy an ecological niche and are adapted to the conditions of that niche [16]. It is important to remember here that the nomenclature of a taxon is very important as it serves to maintain effective communication across microbiological disciplines and it is governed by the Bacteriological Code [17, 18].

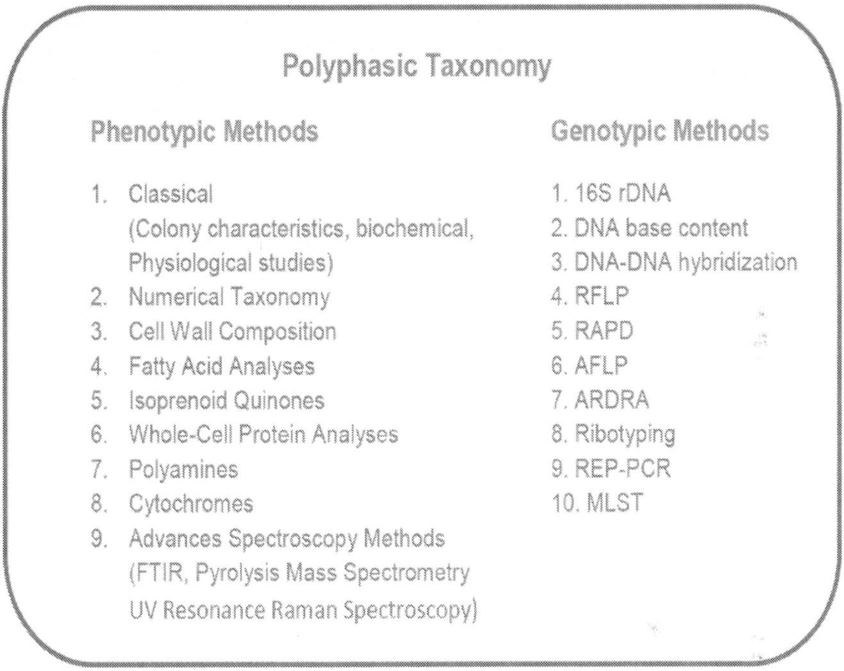

Polyphasic Taxonomy

Phenotypic Methods

1. Classical
   (Colony characteristics, biochemical, Physiological studies)
2. Numerical Taxonomy
3. Cell Wall Composition
4. Fatty Acid Analyses
5. Isoprenoid Quinones
6. Whole-Cell Protein Analyses
7. Polyamines
8. Cytochromes
9. Advances Spectroscopy Methods
   (FTIR, Pyrolysis Mass Spectrometry UV Resonance Raman Spectroscopy)

Genotypic Methods

1. 16S rDNA
2. DNA base content
3. DNA-DNA hybridization
4. RFLP
5. RAPD
6. AFLP
7. ARDRA
8. Ribotyping
9. REP-PCR
10. MLST

Figure 2. Various techniques used in polyphasic taxonomy for characterization of prokaryotes.

7. Polyphasic taxonomy

While the species is accepted as the basic unit of taxonomy (see Section 6), sub-species, strains and ecotypes occupy lower distinctive taxonomic levels for certain groups of organisms and are not mandatory for all. When classifying a new taxon, it is essential to describe phenotypic, genotypic and phylogenetic information as accurately as possible. This constitutes the polyphasic approach of taxonomy [1] and is shown in Figure 2. The phenotypic information comes from the colony characteristics, cell type, cell wall-type, pigmentation patterns, proteins and other chemotaxonomic markers while genotypic features are derived from the nucleic acids (DNA / RNA). Phylogenetic information is obtained from studying sequence similarities of the 16S rRNA or 23 S rRNA genes in case of bacteria and 18S rRNA in case of fungi. Many types of molecules are used for delineating and describing a taxon; some are mandatory (16S rRNA genes, phenotypes, chemotaxonomy) while others are optional (amino-acid sequencing of certain protein products, DNA-DNA hybridization), unless required for appropriate description.

7.1. Phenotypic techniques

The phenotypic methods are all those that do not include the DNA/RNA sequencing or their typing methods. Study of morphological characteristics and chemotaxonomic profiles is broadly associated with phenotypic characterization.

7.1.1. Classical: Colony characteristics, biochemical and physiological analyses

The phenotypic features are the foundation for description of taxa. The morphological, biochemical and physiological characteristics provide in-depth information on a taxon. The morphology can include the colony characteristics (colour, shape, pigmentation, production of slime etc.). Further, the features of the cell are described as to shape, size, Gram reaction, extracellular material like capsule, presence of endospores, flagella presence and location, motility and inclusion bodies. Light microscopy is generally used to describe the broad cell features; however electron microscopy is recommended for high resolution images [18]. The biochemical and physiological features describe growth of the organism at different ranges of temperature, pH, salinity and atmospheric conditions, growth in presence of anti-microbial agents, production of various enzymes and growth in presence of different sole carbon and nitrogen sources [1]. These tests have to be carried out using standardized procedures to obtain results that are reproducible within and between laboratories [19].

7.1.2. Numerical taxonomy

Analyses of huge volumes of phenotypic data to derive meaningful relationships amongst a large number of microorganisms can be carried out using computer programs [20]. This system of analyses is called numerical taxonomy. Giving numerical weightage to each trait is followed by analyses of the data by the computer programs generating data matrices between each pair of isolates according to the degree of similarity. Based on the similarity data, cluster analyses are carried out (based on different algorithms) and dendrograms ('trees') are generated showing the overall pattern of similarity/dissimilarity amongst the various organisms being studied. While 16S rDNA sequences have garnered attention in recent times as sole means of bringing out the uniqueness of a species; numerical taxonomy (based on phenotypic traits of a large number of species) compares favourably with that of genotypic data and, indeed, is in alignment with the latter [1].

7.1.3. Cell wall composition

The peptidoglycan component of cell walls of bacteria does not provide much information except for classifying into Gram-positive, Gram-negative and acid-fast bacterial types. However, those in Gram-positive cells contain different types of peptidoglycan depending on the genus or species [21]. The peptidoglycan structure can be analysed by determining its type

(A or B), mode of cross-linking (whether it is directly linked or via interpeptide bridge and with amino acids in the bridge), and the composition of amino acids (especially the diaminoacid) of the side chain [18]. While the mode of cross-linkage can vary within a species and also between strains, the amino acid composition is common to all species within a genus. In archaea, pseudomurein is present where N-acetyl muramic acid is replaced by N-acetyl talosuronic acid [22].

7.1.4. Fatty acid analyses

Different types of lipids are present in bacterial cells. Polar lipids are present in the lipid bilayer of the cytoplasmic membrane. The diversity of polar lipids is known to be large and many are yet to be structurally elucidated. While in archaea, polar lipids are of types phospholipids, aminophospholipids, glycolipids and phosphoglycolipids, in bacteria, apart from the ones seen in archaea, there are also lipids derived from amino acids, capnines, sphingolipids (glycol or phosphosphingolipids) and hopanoids [18]. In Gram-negative bacteria, lipopolysaccharides are present in the outer membranes. The type of sugar present and the fatty acid type, the linkage of the fatty acid to the sugar (amide or ester linkage) provide information on characteristic of the cell [23–25]. However, determination of lipopolysaccharides is not routinely used in recent times. The total cellular fatty acids are extracted, esterified and the methyl ester content is analyzed by gas chromatography. Under standard conditions, this provides a reliable estimate of taxonomy up to genus and sometimes species level. The technique has been automated and the Sherlock MIS system (MIDI Inc.) has developed an extensive database. Though incomplete, it still is the most widely used system in recent times [18].

7.1.5. Isoprenoid quinones

The respiratory isoprenoid quinones are components that occur in cytoplasmic membranes of prokaryotes (archaea as well as bacteria). The naphthoquinones (with sub-types phylloquinone and menaquinone) and benzoquinones (ubiquinones, rhodoquinones and plastoquinones) are the major types of quinones. The variability they depict in their side chains in terms of length (5-15 isoprenoid units known till date), degree and position of saturation are of taxonomic significance and help in characterization to various levels of genus and species [26]. These features generally also mirror the 16S rDNA groupings. Isoprenoid ether-linked side chains are present in members of the archaea. They are of different types such as diethers, hydroxylated diethers, macrocyclic diethers, tetraethers, and polyol derivatives of the tetraether [18]. Non-isoprenoid based-ether-linked lipids are present in bacteria and can be straight chain or with simple branched side chains or with mono-unsaturated derivatives [18].

7.1.6. Other diagnostic methods

Other than the principal diagnostic methods described above, other techniques used to lesser levels or for comparison between species or strains comprise of the following:

a. Whole cell protein analyses, wherein protein is extracted from the cells and analysed by sodium dodecyl sulfate-polyacrylamide gel electrophoresis (SDS-PAGE). This can help in comparison between related strains ([1] and references therein) and also shows congruency with DNA-DNA hybridization results [27]. However, since the identity of the protein bands is not known, this technique suffers from a drawback that is not associated with fatty acid analysis.

b.   Polyamines are a group of compounds in the cytoplasm that provide stability to the DNA and maintain osmolarity in the cell. They are useful to distinguish above genus level and at species level too [28].

c.   Cytochromes are associated with the cytoplasmic membrane and are involved in respiratory and photosynthetic electron transfer. They are 'heme' proteins with a 'heme' prosthetic group attached to a protein. These are not used alone in identification since there are limited types of distinct cytochromes [29].

7.1.7. Advanced spectroscopy and spectrometric methods

Many advanced analytical techniques such as Pyrolysis Mass Spectrometry, Fourier Transform Infrared spectroscopy and UV Resonance Raman Spectroscopy have been employed to determine chemical composition of bacterial cells, primarily the bioactive metabolites from drug discovery point of view and relate it to characteristics of microbes from which the metabolites are obtained [30, 31]. Pyrolysis Mass Spectrometry is a high-resolution technique, wherein microbial colonies are carefully picked and placed onto a iron-nickel foil, vacuum desiccated, heated rapidly and the pyrolysate bombarded with low-energy electrons. The ionized fragments are separated on the basis of their mass to charge ratio (m/z), detected and amplified by an electron multiplier [30]. Metabolites with m/z ratio of 51-200 constitute degradation products useful for taxonomical discrimination, since those with m/z less than 50 are produced by most biological materials and those with m/z ratio greater than 200 are not useful in discriminating taxons. The multivariate data is further analysed by Principal Components Analysis (PCA) to understand the variance while reducing the dimensionality of the data.

Fourier Transform Infrared spectroscopy (FTIR) is a simple and cost-efficient method and has been applied to identify discriminative features of strains. Most cellular components (fatty acids, proteins, carbohydrates and nucleic acids) can be analysed by this method to reveal strain-specific features [32]. Five IR spectral regions or 'windows': W1 (3000–2800 cm^{-1}) for fatty acids, W2 (1700–1500 cm^{-1}) for amide I and II bands of proteins and peptides, W3 (1500–1200 cm^{-1}) for a mixed region of fatty acid bending vibrations, proteins, and phosphate-carrying compounds, W4 (1200–900 cm^{-1}) for carbohydrates of cell walls and W5 (900–700 cm^{-1}) which is the 'fingerprint region' with unique absorbances specific for different taxa [33]. The differences in spectra are resolved using multivariate tools such as cluster analysis, discriminant analysis etc. [34].

UV Resonance Raman Spectroscopy (UVRR) uses a frequency of Raman spectra when biological materials are not subjected to fluorescent background while using IR or visible excitation [35]. Where an IR spectroscopy needs hundreds of cells, Raman spectroscopy does not require this [33, 36]. It can also be used with single cells, as demonstrated by Rösch et al.

[37], in combination with a data classification approach [38] and can also provide information about the Gram-type of a bacterium [39] as well as relate to moles G + C content [40].

7.2. Genotypic techniques

Modern taxonomy has been influenced by genetic methods and indeed, much of the classification and identification is predicated on specific gene sequences. All the techniques involving DNA or RNA fall under genotypic methods.

7.2.1. 16S rDNA-based analyses

The technique, which is very nearly a gold standard for taxonomic purposes today, is sequencing of the 16S rRNA gene of bacteria. The 23S rRNA gene sequence is also considered in many studies but lack of comprehensive databases for comparison is a drawback. Since the 16S rRNA is present in all bacteria, is functionally constant and is composed of conserved and variable regions, it has consistently served as a good taxonomic marker for deriving taxonomic relationships [1]. While it has proved to be the foundation for modern taxonomy, there are certain caveats and it has to be considered along with other techniques for formal identification purposes, especially at the species level. Generally universal sets of primers are used to amplify the 16S rRNA gene product gene (~1.5 kB). Subsequently, the amplified product is sequenced and quality of the sequence checked. The sequence is then aligned against high quality sequence data from curated databases [ARB (www.arb-home.de), RDP (http://rdp.cme.msu.edu/), SILVA (www.arb-silva.de) and LTP (www.arb-silva.de/projects/living-tree/)]. Multiple alignment programs such as CLUSTAL_X, CLUSTAL W, CLUSTAL X2, CLUSTAL W2, MEGA, T-COFFEE, MUSCLE) can also be used with manual editing [18].

As mentioned earlier, it has been shown that < 97% similarity of two 16S rDNA sequences implies a different species ([18] and references therein). This cut-off value is generally considered for ecological studies [41]. For actual taxonomic studies, a 98.5% cut-off value is considered [42]. However, the values should be based on high-quality and almost full-length sequences. When the similarity values are ~95%, other methods must also be included to justify the creation of a new genus. The descriptions must also provide information on the differences between the potentially new and existing genera. Subsequent to alignment, phylogenetic trees or dendrograms have to be constructed to reveal the taxonomic position of an organism. Different treeing algorithms such as maximum-parsimony, maximum-likelihood methods are preferred for evaluation of taxonomic position. Inclusion of type strain or type species is essential.

7.2.2. DNA base content

Determination of moles percent guanosine and cytosine constitutes a classical method of establishment of genomic content. This is now being used along with other genotyping methods to establish taxonomic position of an organism [1]. Within species, the G+C content ranges within 3% and within genera 10% [42]. Overall, the G+C content ranges from 24-76% in bacteria.

7.2.3. DNA-DNA hybridization

This method is an indirect measurement of sequence similarity between genomes. A cut-off value of 70% similarity is considered for establishment of species [1]. However, the method has to be reproducible between laboratories and performed under standardized conditions, which is often a drawback. Hence it is applied only where 16S rRNA gene sequences show similarity values above 98%. There have been reports where 16S rRNA gene sequence has shown 99% similarity and yet DNA-DNA hybridization values have been 60% or less. Hence, this method has to be used with caution and performed under highly standardized conditions.

7.2.4. Other genotyping methods

Earlier, sub-typing was done on the basis of biochemical profile (biotyping), serological profile (serotyping), phase susceptibility (phage typing) or antibiotic susceptibility. But currently DNA-typing methods are preferred due to their reproducibility, ease of performance and high level of discrimination between strains [1]. Genotyping methods such as Restriction Fragment Length Polymorphism (RFLP), Randomly Amplified Polymorphic DNA (RAPD), Amplified Fragment Length Polymorphism (AFLP), Amplified Ribosomal DNA Restriction Analysis (ARDRA), Repetitive Element-Polymerase Chain Reaction (REP-PCR), Ribotyping and Multi Locus Sequence Analyses (MLSA) are some of the newer methods to characterize a taxon.

RFLP was one of the earliest methods to be used and consisted of extraction of whole-genome DNA, restriction digestion using specific restriction enzymes and visualization of the DNA bands using gel electrophoresis. However, complex patterns can be generated, making comparison difficult. AFLP makes use of restriction analyses followed by PCR amplification to select for particular DNA fragments. Two restriction enzymes are used and fragments are amplified which can be genus and species-specific. Thus the method,though tedious, can be used for identification and typing purposes [1]. A PCR-based methodology makes use of random oligonucleotide primers 10 bases in length (RAPD), followed by amplification under specified conditions and analyses of the bands for similarity in size after gel electrophoresis. Ribotyping is another technique in which total genomic DNA is extracted, followed by restriction digestion and separation by electrophoresis [43]. Subsequently, the fragments are hybridized with a radiolabeled ribosomal operon probe (genes for 16S, 23S, tRNA and 5S rRNA) and analysed for presence of bands by autoradiography [44]. A simpler technique is amplifying the 16S rDNA (with or without spacer regions) using universal primers and restriction analysis (ARDRA). It can be used to discriminate at species level ([45]; [1] and references therein). Ribotyping and ARDRA have been shown to produce reproducible and congruent results in *Lactobacillus sp.* [46]. Consensus sequences complementary to repetitive sequences in the genomes of bacteria are used as primers, PCR-amplified and visualized as distinctive bands (REP-PCR). This method is discriminatory and rapid up to species or intraspecific level but is based on a library-based approach [47]. MLSA focuses on sequencing of housekeeping (6-8 protein-coding) genes and phylogenetic analyses of the same. It is a new method for characterizing a species [48]; however, databases are limited for realising the full-extent of utility of this method. Average Nucleotide Identity (ANI) of all orthologous genes in complete genome sequences has been proposed as a method to define species. Limited studies show that 95% ANI corresponds roughly to 70% DNA-DNA similarity values [49]. However, more datasets are required to implement this method.

8. The eukaryotic microbes — Fungi

Fungi are important from industrial point of view as well as the increasing numbers of pathogens that are arising. Primarily, fungi are classified on the basis of appearance, the structure appearing above ground. The below-ground vegetative structures are difficult to identify [50]. The focus is on the asexual stage. The concept of species is somewhat difficult to interpret in the case of fungi since sexual mating does not occur in all fungi; meiosis occurs only in sexual fungi and even where mating occurs, the product of fusion could either be sterile or fertile. The biological concept of species can therefore be applied to sexual fungi. In the case of asexual fungi, similarities in characteristics provide a system for classification. Modern developments such as analyses of DNA by sequencing have brought about a paradigm shift in fungal taxonomy. The phylogenetic species concept is being favoured over the other definitions, especially where asexual fungi are considered. The evolutionary relationships amongst fungi have not been well-delineated. Traditionally, fungal phylogeny has been classified using morphology (involving primarily the fruiting body), cell wall composition [51], cell ultra-structure [52], cytology [53] and metabolism [54] and even based on the study of fossils [55]. Nomenclature also creates confusion since fungi can exist in sexual and asexual stages and can develop at different times in different substrates and relationship between the two states has to be established first.

Currently, modern developments underscoring the importance of phylogeny in bacterial classification has also been used for fungal taxonomy. Molecular genotyping methods such as restriction analysis of internal transcribed spacer (ITS) region, 18S rDNA and RFLP are being used to classify fungi [50]. However, the datasets are not extensive enough to permit effective identification [56]. Hibbett et al. [57] proposed a phylogenetic-based classification for fungi with 195 taxa. The 5S, 5.8S, 18S and 25S rDNAs are considered for phylogenetic studies. Of these, the 18S rDNA has been more extensively used for filamentous fungi [58]. The D1 and D2 variable regions of 25S rDNA are used for yeasts. Limited datasets are available where both 18S and 25 S rDNAs have been sequenced leading to difficulties in comparison [59]. The 25S rDNA allows for comparison to species level [60] while 18S rDNA has to be nearly completely sequenced to obtain pertinent information. The ITS region is suitable to reveal close relationships [61]. The 5S sequence provides information suitable for order level [62].

9. Uncultured microorganisms

Metagenomics has emerged as a promising field of interest, where identification of uncultured microorganisms is attempted. Since 99% of the microbial population is considered to be uncultivable, metagenomics assumes importance [63]. Next generation Sequencing (NGS) has fuelled interest in this field. Classical metagenomics analyses samples by extracting environmental DNA followed by *de novo* sequencing, or amplification of 16S/18S rDNA using specific primers while functional metagenomics focuses on amplification of genes of interest (generally antibiotics, enzymes etc.) and their cloning into select target microorganisms to produce the metabolite of interest. Roche 454 pyrosequencing and Illumina are the most widely used NGS technologies [4]. DNA bar-coding approach is gaining popularity for assessing microbial diversity [64]. Though only limited datasets (especially for eukaryotic microbes) are currently available, the scenario is improving due to faster and cheaper sequencing methods. Inherent differences in microbial evolution rates, chimeric DNA sequences, artefacts generated during PCR or sequencing and non-universality of primers preclude derivation of a common threshold for taxonomic units [65]. However, bioinformatics handles some of these issues. Sequence quality is checked for series of Ns (nucleotides that are unresolved), errors in primer sequences are checked and verified, and sequences where length varies from the expected length [66] are assessed. Programs have been developed to remove pyrosequencing as well as PCR errors ([4] and references therein). After error-checking and trimming, the sequences are aligned, distance matrices calculated and used for clustering the Operational Taxonomic Units (OTUs) using programs such as MOTHUR [67]. OTUs represented by single sequences (singletons) are also documented and can overestimate diversity. Removal of singletons has not been shown to affect alpha diversity much [4] though more studies are required in this regard. Beta diversity remains conserved without singletons but diversity patterns may change in their presence [68].

10. Taxonomy of viruses

The definition of a virus 'species' is: "A virus species is a polythetic class of viruses that constitutes a replicating lineage and occupies a particular ecological niche" [69]. A virus isolate can refer to any virus as long as the virus has existed for some time. Viruses are not considered to be either prokaryotes or eukaryotes but have implication from health point of view; hence characterization of viruses has increased considerably. Where earlier, only electron microscopy was used, today sequencing of viral genomes constitutes advancements and the database is increasing. According to International Committee on Taxonomy of Viruses (ICTV), proposals are afoot to accept online descriptions of viral taxa based on taxonomical details such as : dsDNA, ssDNA, rtDNA, rtRNA, dsRNA, ssNRNA, ssPRNA, SAT (Satellites), VIR (Viroids), UN (unassigned).

11. Conclusions & perspectives

Zinger et al. state (see Pg. 2 of Ref. [4]): "In its broadest meaning, measuring biodiversity consists of characterizing the number, composition and variation in taxonomic or functional units over a wide range of biological organizations (from genes to communities)". The taxonomical classification of microorganisms has been difficult due to their small size, short generation times and confounded by genetic exchange between unrelated organisms. These limitations have been largely overcome by modern developments of sequencing technologies and the recognition of rDNA sequences as a cornerstone for identification purposes. Overall, it is important to recognize that microbial diversity is intricately linked to its environment and this correlation has to be established by description of environmental parameters whenever sampling is carried out. It is also important to study the phenotypic characteristics and link them to the observations obtained from genotyping techniques. The link between habitat and diversity then becomes easier to understand for future studies.

Author details

Indira P. Sarethy[1], Sharadwata Pan[2] and Michael K. Danquah[3*]

*Address all correspondence to: mkdanquah@curtin.edu.my

1 Department of Biotechnology, Jaypee Institute of Information Technology, Noida (U.P.), India

2 Department of Chemical Engineering, Indian Institute of Technology Bombay, Mumbai, India

3 Department of Chemical and Petroleum Engineering, Curtin University of Technology, Sarawak, Malaysia

References

[1] Vandamme P, Pot B, Gillis M, De Vos P, Kersters K, Swings J. Polyphasic taxonomy, a consensus approach to bacterial systematics, Microbiological Reviews 1996; 60 (2): 407-438.

[2] Fierer N. Microbial biogeography: patterns in microbial diversity across space and time. In: Zengler K. (ed.), Accessing Uncultivated Microorganisms: from the Environment to Organisms and Genomes and Back. Washington DC: ASM Press; 2008. p95-115.

[3] Kassen R, Rainey P. The ecology and genetics of microbial diversity, Annual Review of Microbiology 2004; 58: 207 – 231.

[4] Zinger L, Gobet A, Pommier T. Two decades of describing the unseen majority of aquatic microbial diversity, Molecular Ecology 2012; 21(8): 1878–1896.

[5] Madigan MT, Martinko JM, Stahl DA, Clark DP. Brock biology of microorganisms 13th Edition; Ch. 16, pp. 446-474; Benjamin Cummings; 2010.

[6] Woese CR. (2000). Interpreting the universal phylogenetic tree, Proceedings of National Academy of Sciences U.S.A. 2000; 97(15): 8392-8396.

[7] Boucher Y, Douady CJ, Papke RT, Walsh DA, Boudreau MER, Nesbo CL, Case RJ, Doolittle WF. Lateral gene transfer and the origins of prokaryotic groups, Annual Review of Genetics 2003; 37: 283 – 328.

[8] Fierer N, Lennon JT. The Generation and Maintenance of Diversity in Microbial Communities, American Journal of Botany 2011; 98(3): 439–448.

[9] Jain R, Rivera MC, Moore JE, Lake JA. Horizontal Gene Transfer accelerates genome innovation and evolution, Molecular Biology and Evolution 2003; 20(10): 1598–1602.

[10] Achtman M, Wagner M. Microbial diversity and the genetic nature of microbial species, Nature Reviews 2008; 6: 431-440.

[11] Stackebrandt E, Frederiksen W, Garrity GM, Grimont PA, Kämpfer P, Maiden MC et al. Report of the ad hoc committee for the re-evaluation of the species definition in bacteriology, International Journal of Systematic and Evolutionary Microbiology 2002; 52(Pt 3): 1043–1047.

[12] Wayne LG, Brenner DJ, Colwell RR, Grimont PAD, Kandler O, Krichevsky MI et al. Report of the ad hoc committee on reconciliation of approaches to bacterial systematic, International Journal of Systematic Bacteriology 1987; 37(4): 463–464.

[13] Staley JT, Krieg NJ. Classification of prokaryotic organisms: an overview. In: Krieg NR, Holt JG. (eds.), Bergey's Manual of Systematic Bacteriology Vol. 1; Baltimore: The Williams & Wilkins Co. 1984. p1–3.

[14] Stackebrandt E, Ebers J. Taxonomic parameters revisited: tarnished gold standards, Microbiology Today 2006; 33: 152–155.

[15] Murray RG, Stackebrandt E. Taxonomic note: implementation of the provisional status Candidatus for incompletely described prokaryotes, International Journal of Systematic Bacteriology 1995; 45(1): 186–187.

[16] Koeppel A, Perry EB, Sikorski J, Krizanc D, Warner A, Ward DM. et al. Identifying the fundamental units of bacterial diversity: a paradigm shift to incorporate ecology into bacterial systematics, Proceedings of National Academy of Sciences USA 2008; 105(7): 2504–2509.

[17] Lapage SP, Sneath PHA, Lessel EF, Skerman VBD, Seeliger HPR, Clark WA. (eds.) International Code of Nomenclature of Bacteria (1990 Revision). Bacteriological Code. Washington, DC: American Society for Microbiology, 1992.

[18] Tindall BJ, Rosselló-Móra R, Busse HJ, Ludwig W, Kämpfer P. Notes on the characterization of prokaryote strains for taxonomic purposes, International Journal of Systematic and Evolutionary Microbiology 2010; 60(Pt 1): 249–266.

[19] On SL, Holmes B. Reproducibility of tolerance tests that are useful in the identification of campylobacteria, Journal of Clinical Microbiology 1991; 29(9): 1785–1788.

[20] Sneath P. Numerical taxonomy. In: Krieg NR, Holt JG. (eds.), Bergey's Manual of Systematic Bacteriology Vol. 1.; Baltimore: The Williams & Wilkins Co. 1984. p111-118.

[21] Schleifer KH, Ludwig W. Phylogenetic relationships of bacteria. In: Fernholm B, Bremer K, Jornvall H. (eds.), The hierarchy of life. Amsterdam: Elsevier Science Publishers B. V. 1989. p103-117.

[22] König H, Kralik R, Kandler O. Structure and modifications of the pseudomurein in Methanobacteriales, Zentralblatt fur Bakteriologie.Allgemeine Angewandte und Okologische Microbiologie Abt.1 Orig.C Hyg. 1982; 3(2): 179–191.

[23] Hase S, Rietschel ET. Isolation and analysis of the lipid A backbone. Lipid A structure of lipopolysaccharides from various bacterial groups. European Journal of Biochemistry 1976; 63(1): 101–107.

[24] Weckesser J, Mayer H. Different lipid A types in lipopolysaccharides of phototrophic and related non-phototrophic bacteria, FEMS Microbiology Letters 1988; 54(2): 143–153.

[25] Mayer H, Masoud H, Urbanik-Sypniewska T, Weckesser J. Lipid A composition and phylogeny of Gram-negative bacteria, Bull Japan Federation Culture Collections 1989; 5: 19–25.

[26] Collins MD, Jones D. Distribution of isoprenoid quinine structural types in bacteria and their taxonomic implications, Microbiological Reviews 1981; 45(2): 316–354.

[27] Costas M. (1992). Classification, identification, and typing of bacteria by the analysis of their one-dimensional polyacrylamide gel electrophoretic protein patterns. In: Charmbach A, Dunn MJ, Radola BJ. (eds.), VCH, Weinheim, Germany: Advances in Electrophoresis Vol. 5. 1992. p351–408.

[28] Busse J, Auling G. Polyamine pattern as a chemotaxonomic marker within the Proteobacteria, Systematic and Applied Microbiology 1988; 11(1): 1–8.

[29] Jones D, Krieg NR. In: Krieg NR. (ed.), Bergey's Manual of Systematic Bacteriology Vol. 1, Baltimore: Williams and Wilkins 1984. p15-18.

[30] Goodacre R, Kell DB. Pyrolysis mass spectrometry and its applications in biotechnology, Current Opinion in Biotechnology 1996; 7(1): 20-28.

[31] Magee J. Whole-organism fingerprinting. In: Goodfellow M, O'Donnell AG. (eds.), Handbook of new bacterial systematic. London: Academic Press Ltd. 1993. p.383-427.

[32] Dziuba B, Nalepa B. Propionic and LAB Identification by FTIR and ANN, Food Technology and Biotechnology 2012; 50(4): 399–405.

[33] Naumann D, Helm D, Labischinski H, Giesbrecht P. The characterization of microorganisms by Fourier-transform infrared spectroscopy (FT-IR). In: Nelson WH. (ed.), Modern techniques for rapid microbiological analysis. New York: VCH Publishers 1991. p.43-96.

[34] Samelis J, Bleicher A, Delbès-Paus C, Kakouri A, Neuhaus K, Montel MC. FTIR-based polyphasic identification of lactic acid bacteria isolated from traditional Greek Graviera cheese, Food Microbiology 2011; 28(1): 76–83.

[35] Spiro TG. Resonance Raman spectroscopy: a new structural probe for biological chromophores. Accounts of Chemical Research 1974; 7(10): 339-344.

[36] Naumann D, Keller S, Helm D, Schultz C, Schrader B. FT-IR spectroscopy and FT-Raman spectroscopy are powerful analytical tools for the noninvasive characterization of intact microbial cells, Journal of Molecular Structure 1995; 347: 399–405.

[37] Rösch P, Harz M, Schmitt M, Peschke KD, Ronneberger O, Burkhardt H. et al. Chemotaxonomic identification of single bacteria by micro-Raman spectroscopy: application to clean-room-relevant biological contaminations, Applied Environmental Microbiology 2005; 71(3): 1626-1637.

[38] López-Díez EC, Goodacre R. Characterization of microorganisms using UV resonance Raman spectroscopy and chemometrics, Analytical Chemistry 2004; 76(3): 585-591.

[39] Manoharan R, Ghiamati E, Dalterio RA, Britton KA, Nelson W H, Sperry JF. UV resonance Raman spectra of bacteria, bacterial spores, protoplasts, and calcium dipicolinate, Journal of Microbiological Methods 1990; 11(1): 1-15.

[40] Nelson WH, Manoharan R, Sperry JF. UV Resonance Raman Studies of Bacteria, Applied Spectroscopic Reviews 1992; 27(1): 67-124.

[41] Pedrós-Alió C. Marine microbial diversity: can it be determined? Trends in Microbiology 2006; 14(6): 257–263.

[42] Stackebrandt E, Liesack W. Nucleic acids and classification. In: Goodfellow M, O'Donnell AG. (eds.), Handbook of new bacterial systematics. London: Academic Press Ltd. 1993. p.151-194.

[43] Grimont F, Grimont PAD. Ribosomal ribonucleic acid gene restriction as potential taxonomic tools, Annales de l'Institut Pasteur/Microbiologie 1986; 137B(2): 165-175.

[44] Bouchet V, Huot H, Goldstein R. Molecular genetic basis of ribotyping, Clinical Microbiology Reviews 2008; 21 (2): 262-273.

[45] Kostman JR, Edlind TD, LiPuma JJ, Stull TL. Molecular epidemiology of Pseudomonas cepacia determined by polymerase chain reaction ribotyping, Journal of Clinical Microbiology 1992; 30(8): 2084-2087.

[46] Miteva V, Boudakov I, Ivanova-Stoyancheva G, Marinova B, Mitev V, Mengaud J. Differentiation of Lactobacillus delbrueckii subspecies by ribotyping and amplified

ribosomal DNA restriction analysis (ARDRA), Journal of Applied Microbiology 2001; 90(6): 909-918.

[47] Kon T, Weir SC, Howell ET., Lee H, Trevors JT. Repetitive element (REP)-polymerase chain reaction (PCR) analysis of Escherichia coli isolates from recreational waters of southeastern Lake Huron, Canadian Journal of Microbiology 2009; 55(3): 269-276.

[48] Hanage WP, Fraser C, Spratt BG. Sequences, sequence clusters and bacterial species, Philosophical Transactions of the Royal Society of London B 2006; 361P: 1917–1927.

[49] Goris J, Konstantinidis KT, Klappenbach JA, Coenye T, Vandamme P, Tiedje JM. DNA–DNA hybridization values and their relationship to whole-genome sequence similarities, International Journal of Systematic and Evolutionary Microbiology 2007; 57(Pt 1): 81–91.

[50] Horton TR. Molecular approaches to ectomycorrhizal diversity studies: variation in ITS at a local scale. Plant and Soil 2002; 244(1-2), 29– 39.

[51] Bartnicki-Garcia S. Cell wall composition and other biochemical markers in fungal phylogeny. In: Harbone JB. (ed.), Phytochemical phylogeny. London: Academic Press, Ltd. 1970. p.81-103.

[52] Heath IB. Nuclear division: a marker for protist phylogeny. Progress in Protistology 1986; 1: 115–162.

[53] Taylor FJR. Problems in the development of an explicit hypothetical phylogeny of the lower eukaryotes, BioSystems 1978; 10(1-2): 67–89.

[54] Vogel HJ. Distribution of lysine pathways among fungi: evolutionary implications, The American Naturalist 1964; 98(903): 435–446.

[55] Hawksworth DL, Kirk PM, Sutton BC, Pegler DN. Ainsworth and Bisby's dictionary of the fungi, 8th ed. International Mycological Institute, Egham, U.K. 1995.

[56] Kirk JL, Beaudette LA, Hart M, Moutoglis P, Klironomos JN, Lee H. et al. Methods of studying soil microbial diversity, Journal of Microbiological Methods 2004; 58 (2): 169-188.

[57] Hibbett DS, Binder M, Bischoff JF, Blackwell M, Cannon PF, Eriksson OE. et al. A higher-level phylogenetic classification of the Fungi, Mycological Research 2007; 111 (5): 509-547.

[58] Woese CR, Kandler O, Wheelis ML. Towards a natural system of organisms: proposal for the domains Archaea, Bacteria, and Eucarya, Proceedings of National Academy of Sciences USA 1990; 87(12): 4576–4579.

[59] Guarro J, Gené J, Stchigel AM. Developments in fungal taxonomy, Clinical Microbiology Reviews 1999; 12 (3): 454-500.

[60] Guého E, Improvisi L, Christen R, de Hoog GS. Phylogenetic relationships of Cryptococcus neoformans and some related basidiomycetous yeasts determined from parti-

al large subunit rRNA sequences, Antonie van Leeuwenhoek, International Journal of General and Molecular Microbiology 1993; 63(2):175–189.

[61] de Hoog GS, Gerrits van den Ende AH. Molecular diagnostics of clinical strains of filamentous Basidiomycetes, Mycoses 1998; 41(5-6): 183–189.

[62] Walker WF, Doolittle WF. Redividing the basidiomycetes on the basis of 5S rRNA sequences, Nature 1982; 299: 723–724.

[63] Brady SF. Construction of soil environmental DNA cosmid libraries and screening for clones that produce biologically active small molecules. Nature Protocols 2007; 2: 1297-1305.

[64] Valentini A, Pompanon F, Taberlet P. DNA barcoding for ecologists, Trends in Ecology and Evolution 2009; 24(2): 110–117.

[65] Thornhill DJ, Lajeunesse TC, Santos SR. Measuring rDNA diversity in eukaryotic microbial systems: how intragenomic variation, pseudogenes, and PCR artifacts confound biodiversity estimates, Molecular Ecology 227; 16(24): 5326–5340.

[66] Huse SM, Huber JA, Morrison HG, Sogin ML, Mark Welch D. Accuracy and quality of massively parallel DNA pyrosequencing, Genome Biology 2007; 8: R143.

[67] Schloss PD, Westcott SL, Ryabin T, Hall JR, Hartmann M, Hollister EB. et al. Introducing mothur: open-source, platform-independent, community supported software for describing and comparing microbial communities, Applied and Environmental Microbiology 2009; 75(23): 7537–7541.

[68] Agogué H, Lamy D, Neal PR, Sogin ML, Herndl GJ. Water mass-specificity of bacterial communities in the North Atlantic revealed by massively parallel sequencing, Molecular Ecology 2011; 20(2): 258–274.

[69] Fauquet CM, Fargette D. International Committee on Taxonomy of Viruses and the 3,142 unassigned species, Virology Journal 2005; 2: 64.

Chapter 4

Marine Biodiversity and Chemodiversity — The Treasure Troves of the Future

Stéphane La Barre

Additional information is available at the end of the chapter

http://dx.doi.org/10.5772/57394

1. Introduction

Chemodiversity usually refers to small molecules that have a signaling (offensive or defensive) function, sometimes protective. This leaves aside larger molecules that are purely structural and those that participate in essential metabolic functions, and make up the bulk of the organic body mass of living organisms.

From cyanobacteria and bacteria to the largest metazoans, chemistry is the preferred mode of aquatic communication, thanks to the extraordinary solvation properties of water. Bacteria create biofilms inside which they communicate using their own chemical repertoire before colonizing new media, substrates or organisms. Microalgae form blooms which are maintained by releasing semiochemicals for cell-cell recognition. Fish rely on their extraordinary sense of smell to hunt or to migrate to some specific breeding spot. The extraordinary biodiversity of coral reefs is maintained by a highly complex chemical network of toxins and pheromones, some soluble, some dispersed with a mucus carrier or surface-coated. But not only: the amazing colors used for warning or for camouflage, the bioluminescence used in the dark correspond to very sophisticated assemblages of pigments, small metabolites or proteins, each organism having its own strategy to be visually recognized or to blend into the background.

Humans have only recently been aware of the extraordinary potential marine molecules for the design of new drugs, cosmetics and nutraceutics. Well over 20000 natural molecules have been studied so far, and several have responded to the need for novel anticancer, antibiotic, anti-inflammatory or anti-pain agents etc. The necessity to preserve this exceptional resource, however, has only manifested itself in the delineation of protected areas and in the implementation of codes of good practices regarding non-destructive boating and durable sampling protocols.

Over the last three decades, warning messages have been sent to the community about the destructive consequences worldwide economic development will have on biodiversity, both terrestrial and marine, during the 21st century. Direct impacts are caused by overexploitation and mismanagement of natural resources and improper recycling and disposal of waste products. Indirect impacts are caused by the accelerating volatilization of greenhouse molecules and their accumulation in the atmosphere where they may undergo undesirable speciation. Restitution of sulfur emissions to land may cause acidic rains and transfer of carbon-containing emissions to seawater increases its acidity, both leading to biodiversity destructive scenarios. Not to mention the release of man-made (synthetic) molecules, some of which like CFCs destroy the anti-UV ozone shield, others like PCBs accumulating along food chains and eventually killing top consumers. Synthetic molecules may respond to specific needs and criteria, but they will never replace natural molecules, in the same way as genetically transformed organisms will never replace wildlife diversity. Moreover, freak biological or chemical species should be eliminated safely once the purpose for which they were created has been fulfilled.

To-date, very little is said or written on the fate of natural chemodiversity within the context of local or general biodiversity collapse, both terrestrial and marine. After a brief historical account of the intricate connections between chemodiversity and biodiversity since life appeared on our planet, this chapter attempts to demonstrate that natural molecular diversity is a treasure to preserve for future generations, using a series of marine examples.

2. Early chemodiversity was non-biotic

2.1. From single elements to simple organic molecules in a mineral world

Chemistry is as ancient as the observable universe, ca. 13.7 billion years old according to present estimates. The simplest elements of the periodic table (namely, hydrogen, helium and traces of lithium) were formed as an immediate consequence of the Big Bang.

The *nuclear chemistry* of the galaxies and stars which developed during the following few billion years is the chemistry of very high energies, yielding a small and finite number of chemical species, some very short-lived. Carbon, oxygen and nitrogen came to existence. The data on star-forming elements is of spectroscopic nature. When decomposed and analyzed, the light emitted by these objects reveals the presence of spectral lines at set wavelengths, and their association forms fingerprints typical of individual elements when heated to incandescence. The solar system was formed about 4.6 billion years ago by the accretion of *simple molecular substances* from a giant cloud at the center of which most the concentrating matter formed the sun, the rest forming a disc from which primary planets, then satellites planets emerged under various scenarios.

Under very hot temperatures and no atmosphere, the molecular chemistry associated with proto-planets is dominated by inorganic entities, with very limited diversity in carbon chemistry. Carbon dioxide and water probably existed as soon as oxygen, a stellar product, had been available to combine with hydrogen, yet early planet Earth was most likely very hot and dry, precluding life and any of the complex chemistries it produces. This characterized the Hadean, the first geologic eon (4.6 to 4.0 billion years until conflicting evidences such as the presence of high $d^{18}O$ (oxygen isotopic ratios) in zircons tentatively dated as early as 4.4 billion years led to a reassessment of the transition between the "molten" planet (and formation of the moon) and the "solid" planet with a solid crust, a low temperature, and an atmosphere of sorts, and the possible presence of liquid water. As well as this "cool early earth" hypothesis [1], a partial explanation about the formation of oceans is that asteroids and perhaps comets carrying huge amounts of ice collided with our planet, adding to the putative *de novo* condensation into water during the degassing of rocks of the cooling planet. In the primitive atmosphere, heavier carbon dioxide became partly quenched by dissolving in oceanic waters, and partly trapped under plate tectonic movements.

2.2. The chemical origin of life on earth — Chemodiversity goes organic

Any life form necessitates the capacity to harvest energy for its own benefit, to manage chemical reactions within molecular boundaries that define it as an entity distinct from the environment, i.e. *self* vs. *non-self*, and to have the potential to replicate itself.

Irrespective of when the right conditions were first met for life to emerge, a small set of small organic molecules (reactants) is a prerequisite for the abiotic generation of life-essential molecules, in the presence of water and of some catalytic trigger, and later of thermodynamically favorable conditions for polymers to be built.

Different approaches have attempted to address the question of the chemical origin of life, leading to apparently conflicting conclusions: the "prebiotic soup theory" and the "pioneer organism theory", fueling heated debate among specialists, and also reflecting the uncertainties that still remain on the greatest mystery in science. The following paragraph attempts to follow the most generally accepted views and the most plausible scenario.

2.2.1. The Miller-Urey laboratory experiments

The prebiotic soup theory stems from the early experiments of Stanly Miller and Harold Urey [2] which led to the condensation within days of a suite of five amino acids by exposing a sterile mixture of methane (CH_4), ammonia (NH_3), carbon monoxide (CO) and hydrogen (H_2) - i.e. the "primitive gases", to a constant flux of hot water (H_2O), while discharging electric sparks (to simulate thunder) into the gas mixture. New species such as hydrogen cyanide (HCN), formaldehyde (HCHO) and reactive intermediates (acetylene, cyanoacetylene etc.) are formed in one-step, while amino acids and other biomolecules are formed from these reactants under reducing conditions (Fig. 1), i.e. the Strecker synthesis.

Some 50 years later, Bada and collaborators (in [3]) reanalyzed archived samples of Stanly Miller's experiments in which the gaseous mixture included hydrogen sulphide (H_2S), methane (CH_4), ammonia (NH_3), and carbon dioxide (CO_2). In the case of this experiment (unreported by Miller at the time), no less than 27 compounds, including the 20 "regular" amino acids and 3 new ones were found. Specifically, the presence of hydrogen sulphide and carbon dioxide as reactants - originally proposed to simulate the influence of volcanic emissions - had enriched the diversity of the reaction products with seven several sulphur-containing compounds, including methionine and cysteine), and four amines as well.

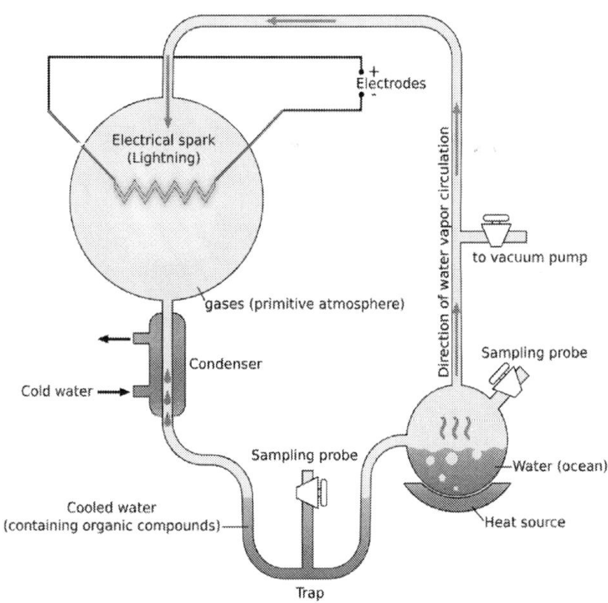

Figure 1. The apparatus used by Miller and Urey in their 1953 simulated early-earth experiment on the chemical origins of life. In the original experiment, the gases of the primitive atmosphere were: NH_3, CH_4, H_2 and water vapor. Other experiments were carried out under non-reducing conditions (using CO_2, N_2 and water) and UV exposure led to poor yields. In experiments simulating volcanic environments, Miller and Urey used H_2S, CH_4, NH_3 and CO_2 (see text). (Graphics by Yassine Mrabet in http://en.wikipedia.org/wiki/Miller-urey_experiment)

2.2.2. Other "prebiotic" laboratory experiments

Oró and collaborators [4] using similar experimental setups, established that adenine and other nucleic bases, as well as several amino acids, could be formed from HCN and NH_3 in water, under reducing conditions.

In addition, simple sugars can be formed under putative prebiotic conditions [5], an essential feature for the emergence of nucleosides [6] and RNA (qv. the RNA world). The Butlerow reaction, i.e. the synthesis of a complex mixture of sugars (including ribose and arabinose) from formaldehyde by the action of catalysts such as calcium hydroxide, has been known since the 19th century under laboratory conditions.

2.2.3. Transposing to "real" prebiotic conditions

In order to transpose from lab experiments to real life situation, several criteria must be met in order to initiate amino acid synthesis: the reactants must be concentrated enough under an aqueous environment under favorable temperature and pH conditions.

The primitive soup theory relies on a stochastic occurrence of optimal conditions for the emergence of organic life. Polymerization necessitates suitable absorbing substrates and catalysts such as metal cations and imidazole derivatives and several others, to proceed in laboratory conditions. Charged submicronic montmorillonite clay particles would have helped activated monomers to selectively concentrate (electrostatic bonds) and induce covalent polymerization. Shallow lagoons in tidal zones may theoretically provide basic chemistry. Evaporation induces concentration and promotes the formation of eutectic complexes.

Contenders of the "primitive soup" scenario advocate that life may have arisen through chemoautotrophic processes occurring in oceanic depths in the vicinity of hydrothermal settings which would provide all necessary starting conditions [7]. In this "pioneer metabolism" scenario, the generation of homochiral metalloenzymes of extant organisms from inorganic transition metal precipitates (by chelation of alpha-hydroxyl and alpha-amino acids ligands) follows a stepwise evolution by autocatalytic feedback. This "hot volcanic" prebiotic chemistry is often opposed to the "cool oceanic" chemistry that typifies the primitive soup scenario.

It appears that the composition of the primitive atmosphere is not what it was thought to be at the time of Miller's experiments, certainly not as reducing, and the conditions afforded in the hydrothermal vent environments appear more amenable to prebiotic chemistry. Oligomers could have then been formed at liquid-solid interfaces [8].

However, prebiotic chemistry under early atmospheric conditions has opened the issue towards exobiology. Scientists of repute now speculate on an extraterrestrial origin of life, or at least in the alien seeding of life essential molecular building blocks on planet earth. But this is another story.

2.2.4. Before and during the RNA world

The recent finding that ribonucleic acids (RNAs) can perform a variety of hitherto unsuspected structural and metabolic functions in cells has given credit to Walter Gilbert's prediction [9] that an all-RNA world had preceded the nucleic acid-protein world as we know it today. The fact that RNA takes evolutionary precedence on DNA (itself regarded by some as a modified RNA better suited for the conservation of genetic information), or the fact that RNAs as a catalysts (ribozymes) take precedence on enzyme catalysts, stimulate intense experimental interest.

Nucleotide monomers can theoretically be surface-assembled as oligomers [10]. The longest strands serving as templates, direct synthesis of a complementary strand starting from monomers or short oligomers, and double-stranded RNA molecules can accumulate. Dissociation of strands, one of which endowed with RNA polymerase activity, would lead to successive replication processes. Finally, the RNA world would have emerged from a mixture of activated nucleotides. However, the precise molecular mechanisms which initiated RNA oligomers in the first place, remain obscure. The monomers (nucleotides) must be activated and homochiral, before they can be assembled as strands on an absorbing mineral surface (e.g. montmorillonite clay) acting as a catalyst.

Interestingly, other nucleotide-like structures that may be formed under prebiotic conditions can reasonably qualify as RNA functional analogues and have been termed "alternative genetic systems" [11].

The existence in prebiotic conditions of AMP-derived cofactors that have acquired amino acid-like properties through post-translational modifications is possible, since these cofactors participate at all levels in the metabolism of extant life forms. As Maurel and Haenni [12] put it," Coenzymes would be vestiges of catalytic nucleic enzymes that preceded ribosomal protein synthesis, and tRNAs can be viewed as large coenzymes participating in the transfer of amino acids."

Ligand-substrate recognition must have followed from the selection of *aptamers* (molecules that bind specifically to one target molecule) from a pool of oligonucleic acids or peptides. Artificial selection of adenine-dependent ribozyme aptamers [13] indicates that the use of small exogenous cofactors (adenine) by ribozymes could have actively contributed to the expansion of the catalytic and metabolic repertoires of RNA species in the prebiotic RNA world. Modern catalytic equivalents of ribozymes are metalloenzymes, and adenine fills the same cofactor function as histidine.

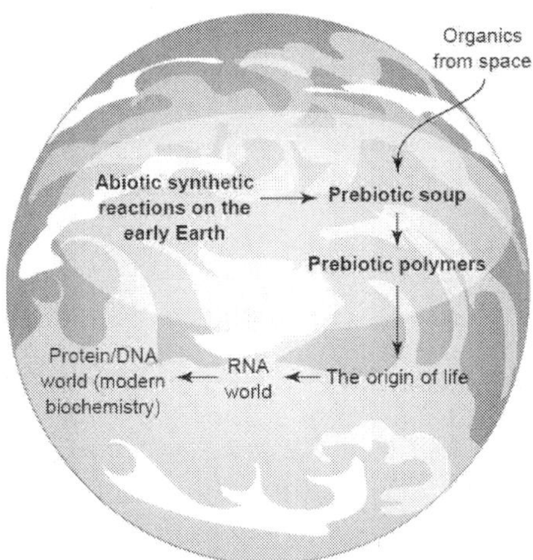

Figure 2. A depiction of the step-by step scenario involved in the origin of life on earth, attempting to integrate the "metabolist" scenario into the "primitive soup" scheme (from: Bada and Lazcano, [14])

Indeed, the RNA world was probably much more chemodiverse than sometimes described (a uniform soup), notwithstanding the fact that, as said before, tidal lagoons with the presence of catalytic elements such as clay submicronic particles, or the vicinity of hydrothermal vents would greatly favor polymerization and functional gains of prebiotic organic molecules.

Extra-terrestrial chemistry going organic

Recent investigations [15] have led to the identification of many (> 100) prebiotic molecules (HCN, HCHO, glycoaldehyde etc.) in the composition of interstellar clouds and in the gas-phase chemical evolution in the atmospheres of various planets (i.e. corresponding to the first step of the Miller-Urey experiment). If the radicalar processes leading to the formation of these intermediate species are simple compared to the sophisticated biochemistry involved in terrestrial chemodiversity, they remain largely unknown and should help us better understand prebiotic chemistry. Different types of molecular processes are thought to be involved, including radiative association and recombination, surface-induced processes, photon or particle induced ionization, ion-molecule reactions, photon or particle induced dissociation and radical-molecule interactions.

Moreover, amino acids are known to occur extra-terrestrially since the 1970 discovery in the Murchison meteorite, of over 70 common and exogenous species under non-racemic proportions, some under relative abundances similar to those found with the original 1953 Miller-Urey experiment.

This was made possible through the activation of a multiple-component system [16].

This raises the problem of chirality, since on our planet, natural amino acids are all left-handed (L-amino acids). It is speculated that the prevalence of the L- form may have been influenced by polarized radiations from outer space [17]. In the laboratory, crystallization experiments in a racemic mixture always follow the same enantiomeric form as that of the initiating nucleus, further amplifying the "preference" for the L-form. Homochirality is an essential feature of biopolymers, for which correct folding must be required for proper function [18].

2.3. From proto-cells to living entities

2.3.1. Self-assembling vesicles

Self-organization of amphiphilic molecules (i.e. long chain fatty acids) into bilayers or into vesicles can be readily observed in laboratory conditions. In live cells, membranes provide the most elementary delineation of the self from the non self, i.e. it defines the basic identity of an individual from its environment. In non-living systems, vesicles can selectively separate solutes according to their affinities (polar or apolar) and even to their chiral properties [19].

2.3.2. The amphiphilic double layer

In particular, the double layer which separates the vesicular contents from the outside can selectively extract lipophilic substances from the environment and arrange them around the inner (hydrophobic) tails. Experimentally, this feature promotes concentration of e.g. apolar amino acids and their non-enzymatic condensation into dipeptides or small oligopeptides within the lipid bilayer. In addition, vectorial properties in phospholipidic vesicles are thought to influence the inner pH (making it lower than outer pH) in small vesicles, due to differences in the behavior of water molecules.

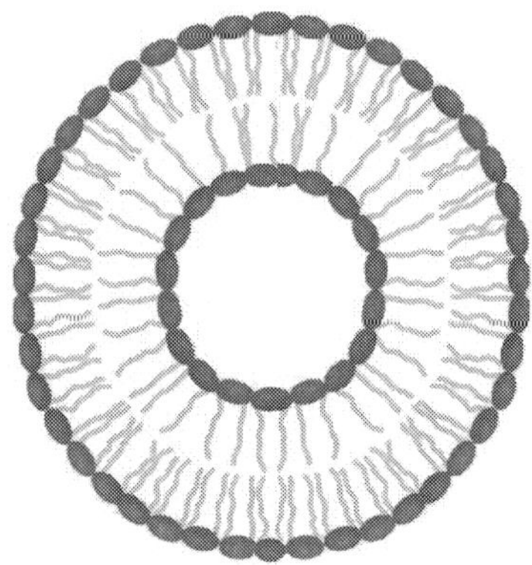

Figure 3. A phospholipid vesicle, with a double-layered membrane. Each layer includes a hydrophilic head (dark knobs) and a hydrophobic tail made up of long lipid chains. The aqueous environment inside the vesicle isolates entrapped metabolites, biopolymers and catalysts, the whole structure acting as a reaction vessel.

2.3.3. Vesicles as proto-cells

The total synthesis of a simple cell is possibly the ultimate challenge in synthetic biology. It is experimentally possible to entrap nucleic acids inside self - forming phospholipid vesicles, acting as reaction vessels for e.g. polymerase chain reaction, enzymatic RNA replication and even protein synthesis, as demonstrated by Oberholzer and collaborators (reviewed in [19]). The molecular pump scenario [20], in which peptides were initially synthesized through a molecular engine could also have taken place inside such structures. In nature, the inclusion of clay microparticles acting as catalysts is highly probable. Thus, phospholipid vesicles can be regarded as useful experimental models of proto-cells [21], but many questions still need addressing before considering them as early precursors of living cells.

Protobiology uses a bottom up approach when designing complex and functional protocells from simple molecules and organic catalysts.

Synthetic biology uses a top-down approach to identify the minimal components of living cells that will qualify, in other words to rediscover the basic cell before it underwent the complexities of Darwinian evolution. Thus protocells are of exciting biotechnological interest in the quest to optimize chemotherapies (cellular target recognition, kinetics of drug delivery).

Prebiotic chemistry is certainly more complex than previously thought, and chemodiversity has undergone periods of expansion, especially through stochastic combinations during the so-called RNA world, and during the development of proto-cells.

Chemodiversity as we know it today is mostly organic and is the result of complex biochemical processes within living organisms that must constantly adapt to changing environments and compete for their survival.

3. Chemodiversity explodes with the emergence of life

3.1. "Cool" carbon chemistry

Our chemodiversity is usually classified as mineral or organic. Exploitable minerals are solidified in the earth crust or deposited as sediments or conglomerates - geochemical cycles connect the molten inner core of the planet with the biosphere through the displacement of continental plates and volcanic activity. Organic chemodiversity is concentrated in the biosphere, as the result of the present and past metabolic processes of marine and terrestrial organisms. Chemically, organic molecules are carbon atoms and scaffolds of connected carbon atoms, linked to hydrogen and covalently to heteroatoms e.g. oxygen, nitrogen, atoms but also to sulfur, phosphorus and halogens or metal complexes (*coordination chemistry*) to form functional groups. Larger organic molecules often adopt three-dimensional architectures that confer them with the exquisite specificity required for recognition and binding to a defined molecular target. Organic natural products are by far the largest contributors to marine and terrestrial chemodiversity, and also the most labile in the face of anthropic and natural influences that affect the stability of our ecosystems.

3.2. "Cool" oxygen biochemistry

The chemistry of our oceans has evolved considerably [22], especially with the gradual production of oxygen, which initially oxidized mantle rocks and dissolved reduced metals such as iron then tended to precipitate, as evidenced by red strata in ancient sediment beds. Free oxygen then started to rise in atmospheric concentrations, some of which was split by cosmic radiations to form the protective ozone layer. Respiration became possible and some life forms moved out of the oceans, and both aquatic and terrestrial species diversification started, and with it, organic chemodiversity. It should be emphasized that combustion occurs at oxygen levels only above 15% and organic matter would spontaneously burn at 25% concentrations - which explains that life is possible only if the present level (21%) is maintained around 20% and if genotoxic radiations can be efficiently filtered out. Oxygen is also toxic at the cellular level, if the production of (or exposure to) reactive oxygen species (ROS) or oxygen radicals cannot be counterbalanced by antioxidants or if the resulting damages cannot be repaired. When involved in regular intracellular signaling, ROS can be involved in the activation of defense metabolic pathways, and contribute to the enrichment of the metabolic repertoire of planktonic and benthic organisms.

3.3. Biochemistry operates under very specific conditions

With the exception of very few extremophiles that can live under elevated temperatures (above 80°C) and hyper acidic environments pH < 2), most microorganisms are not tolerant to conditions found outside the biosphere. The biosphere represents a very narrow region less than one thousand times thinner than the globe's diameter, and comparable conditions have not been found on any other planet so far. This does not mean that dormant bacterial spores could not withstand intersidereal journeys - lichens have successfully resisted prolonged exposures in outer space during space craft experimentations. High biodiversity environments are usually found in warm, stable environments where nutrient cycling is efficient (without accumulation of wastes) - this is at least true for tropical rainforests and shallow water coral reefs.

Biodiversity explosions and mass extinctions (from [26])

"Any lasting change in the biogeochemistry of any of the three components (atmosphere, seawater and land) will disrupt the interfacial equilibrium that supports the many thousands of life forms that interact constantly within an ecosystem. This has happened several times in the history of our planet since it became life-supporting. Changes in soil mineral strata indicate the occurrence of biodiversity-modifying events such as occupation by seas or the occurrence of an ice-age. Discrete organic layers may indicate the presence of a tropical rainforest or of a dry land savannah. Datable fossil evidence within these strata, together with paleontological reconstructions, point out the floristic and faunistic peculiarities of the times. Core drills in ice provide datable evidence of biogeoclimatic episodes within the last few millennia, while core drills in massive scleractinian corals give accurate calendar-like records of recurrent or of accidental climatic events affecting their biotope.

Speciation usually goes along with occupation of new territories and new habitats, the first colonizers having acquired the necessary adaptations to cope with evolving external demands – the Cambrian explosion (545 million years ago) being the most dramatic example of such adaptive diversification at all scales.

Along with this, evidences of accidental episodes of massive species extinctions are witnessed by the sudden "disappearance" of terrestrial and of marine life, that are attributable to tectonic, telluric or meteoritic impacts and to their profound and lasting climatologic and geochemical consequences. The most significant mass extinction is undoubtedly the Permian-Triassic Great Dying [23] where a 96% loss of all non-microbial marine life occurred within ten million years. The precise causes of mass extinction events may be in connection with continuous tectonic movements with their telluric and volcanic outbreaks and their climatic consequences, to collisions with meteoritic bodies, and to a lesser extent to the appearance of dominant predators, parasites or microbial diseases, or to combinations thereof. Common with many extinction events, however, is the massive release of greenhouse and of toxic gases (carbon dioxide, methane, hydrogen sulfide etc.). The water solubility of CO_2 being nearly 30 times that of oxygen, water acidification occurs that impacts preferentially all calcifying organisms with low metabolic rates and weak respiratory systems: most coral genera died out during the Great Dying, along with calcareous sponges, calcifying algae, echinoderms, bryozoans etc. [24].

Interestingly, profound taxonomic changes in all major phyla seem to follow extinction events, resulting in a better adapted biodiversity. Nothing is known, however, on the consequence of such changes in microbial life or on the putative role microbial associates had on the *reinvention* (understand: adaptive evolution) of new species. As Falkowski [25] puts it, animals and plants are merely new incarnations of ancient metabolic processes, but the ultimate key to biodiversity may be held by bacteria ferrying the set of core genes that are necessary for life to express itself ".

4. Marine biodiversity and chemodiversity today

Water is an extraordinary medium and, in many respects, mysterious. Its outstanding solvation properties are due to the capacity of the water dipoles (H-OH) to dissociate, reorientate and accommodate salts and polar compounds in a monophasic system [27]. Salts are essential to maintain osmotic balance and membrane polarization. Hydration is essential for the catalytic properties of enzymes. Semiotic (communication) substances are usually released as pheromones (intra-specific signaling) or allomones (against defense or predation). In addition, apolar (long chains or cyclic) molecules can be associated with mucus and dispersed. Some chemical cues are surface-adsorbed on dead substratum and serve as trans-generation signaling, e.g. for the settlement of larvae near adult colonies. Fish have extraordinarily sensitive and selective receptors that allow them to detect specific chemical signatures that influence their behavior.

Water participates in all life-essential molecular processes, from the most basic (e.g. Miller's experiment) to the most complex enzymatic processes, and it has allowed virtually all types of interatomic and molecular interactions that have resulted in the highly complex and diverse chemical diversity observed in our oceans today.

4.1. Minerals and geochemical cycles

The mineral world is massively involved in global geochemical processes, and in the composition of solutes in the oceans, as salts and inorganic carbon sources. The cycling of elements is both tectonic and biogenic. Metals are essential catalysts of many organic reactions, and salts are actively involved in osmotic balance and membrane polarization. The availability of heteroatoms (P, S, halogens in particular) is (and has been originally) fundamental in the evolution of natural organic chemistry, because of their involvement in life-essential processes.

4.2. Prokaryotes and prokaryote chemistry

"Creative" chemistry is the work of living creatures, and Eubacteria and Archaea are probably the best organic chemists ever. Microbes are the drivers of global biogeochemical processes, and their genes have survived the great extinction events [28]. They have set the stage for other organisms to adapt and evolve. The capacity for pioneer organisms to modify their environment for the benefit of other life forms is termed metabiosis, and it has been largely documented in soil biology [29], but somewhat overlooked in aquatic ecosystems. They can occupy every niche of the biosphere, utilize different carbon and/or energy sources available in order to generate ATP, the energy currency of all biochemical processes.

A distinction is made between chemotrophs and phototrophs as primary producers of organic molecules.

4.2.1. Chemotrophs

Chemotrophic Archaea and Bacteria that live in oxygen-depleted environments (e.g. in the vicinity of hydrothermal vents) derive their energy from the oxidation of reduced inorganic compounds and use (i) carbon dioxide as sole carbon source or (ii) lipids, sugars or proteins to form their own organic compounds. Many scientists believe that life originated in the oceans, in the absence of light and in the vicinity of underwater tectonic/volcanic activity, under similar conditions to post-Hadean times.

Archaea are recognized among the most abundant living entities in the oceans, participating in major biocheochemical cycles, such as the Methanogens that produce most the atmospheric methane greenhouse gas. Halophilic Archaea thrive in hyper saline environments like in the Dead Sea, sometimes in hyper alkaline conditions (pH around 12). On the other hand, extremophilic Archaea can be found in hot sulfur springs under pHs as low as 2. They are either anaerobic and reduce hydrogen sulfide to sulfur, or aerobic and oxidize sulfur to sulfuric acid. Thermophilic (50-70°C) and hyperthermophilic (70 to 118°C) Archaea produce thermo-stable enzymes of major biotechnological importance, e.g. in DNA amplification using polymerase chain reaction (PCR), in food industry, etc.

Chemotrophic Eubacteria include chemoautotrophs which utilize inorganic sources and environmental CO_2, (like nitrogen-fixing soil bacteria, iron and manganese-oxidizing hydro-thermal bacteria) and chemoheterotrophs which degrade existing organic substrates. Many bacteria (Gram-positive Actinomycetes, Bacillus, Gram-negative gamma-Proteobacteria) are chemotrophs, and produce a wide array of bioactive compounds, many of which have strong antibiotic potential (examples are given in [30] vol.1, chapter 7). Interestingly, hydrothermal vents are crowded with invertebrates (mollusks, crustaceans and worms) that cultivate chemotrophic bacteria that provide them with organic nutrients.

4.2.2. Phototrophs

Phototrophs harvest solar photons as the energy source to carry out their cellular energy processes.

Most phototrophs carry out photosynthesis, a process by which carbon dioxide is converted into organic material, used structurally (generally as polymers or supramolecular complexes) or functionally for the regulation of cellular processes, or stored as reserves.

A trans-membrane electrochemical gradient is created, which is utilized by ATP synthase, to create ATP (adenosine triphosphate), the key molecule that fuels biochemical processes in the cell, from the oxidation of glucose. Oxygen is generated in the process. Archaea may fix carbon and some use light as energy source to create metabolic energy but none is capable is capable of photosynthesis with the use of complex electron- transfer chains. Halophilic archaea use bacteriorhodopsin instead - a pigment that activates a transmembrane proton pump.

Photosynthesis began as an anoxygenic process that later evolved to produce oxygen, thus allowing the evolution of aerobic metabolism. Photoautotrophs include most producers of primary biomass, from cyanobacteria to algae, plants and are involved as photosymbionts in vegetal (e.g. lichen) and animal (e.g. coral) biological photosystems. They are the base of simple or highly complex marine and terrestrial food chains.

4.2.3. Heterotrophs - the good, the bad and the ugly

Many bacteria derive their energy from the breakdown of organic substances produced by living or dead multicellular organisms. They are instrumental in the recycling of dead organic matter, and many strains live as commensals or symbionts of most eukaryotes, to help the host's digestion with nutrient assimilation or with the energy-yielding catabolism of proteins, carbohydrates and lipids into small, usually water-soluble monomers. A highly diverse microflora is generally associated with good health at the single host's level as well as whole ecosystems. Pathogens are maintained but their numbers are regulated by complex peer-induced antibiotic and bacteriostatic pressure. In corals, metabolic efforts to adapt or compensate lasting stress (temperature, salinity, hypoxia sometimes in combination) will affect the host's resistance, favor the dominance of opportunistic and aggressive strains and dramatically reduce microbial diversity, leading to a range of necrotic diseases. In humans, obesity and consequential diabetes and cardiovascular diseases strongly correlate with poor feeding habits and deficient gut microflora [31].

Some strains are downright lethal after acquiring multi-resistance to existing antibiotics, and if provided with conditions that trigger their inherent but normally unexpressed pathogenicity. Some exciting scientific advances have been made since the discovery of quorum sensing in bacteria, the detailed study of microbial biofilms, and in the understanding of lethal epidemics (e.g. caused by *Vibrio* species). However, by and large, metagenomic studies have clearly shown that eukaryotic and prokaryotic biodiversity (and hence chemodiversity) are strongly correlated.

4.3. Eukaryotes and Eukaryote-associated chemistry

Cyanobacteria are thought to have enriched the primitive atmosphere in oxygen, promoting biodiversity through repeated endosymbioses, i.e. successive fusion-acquisition processes that led to the complex metabolic machineries of higher plants (origin of plastid organelles) while ancestral bacteria became incorporated as mitochondria in most eukaryotes. Other organelles are suspected to have generated from similar mechanisms. Many eukaryotes have successfully colonized non-aquatic environments.

This increase in biodiversity has itself led to competition for existing resources: food, space, access to light etc.

In non-motile organisms (algae, halophytes and some benthic invertebrates) in which escape responses are not possible, species-specific assortments of physical protections and/or chemical defenses have evolved to discourage predators. Their remains are collectively termed biominerals. Some growth forms (encrusting, massive, cryptic) may also resist whole-organism or whole-colony predation, or restrict their access.

Algae are generally non-toxic, but may contain antifeedants such as organohalogens or polyphenols that are distasteful to fish [32].

Not all sessile, slow-moving and sedentary organisms enjoy an efficient physical protection. Fleshy invertebrate species and those that must be exposed to sunlight to perform photosynthesis may produce toxic or distasteful substances to discourage feeding by fish, crustaceans or other carnivores. Alcyonarian octocorals shelter photosymbiotic dinoflagellates, and are able to produce complex bouquets of cyclic terpenoids that are dissuasive against predators and efficient against space competitors. Sponges shelter biodiverse archaean, eubacterial and cyanobacterial communities and produce an amazing variety of metabolites (many are nitrogen containing and halogenated) as a functional consortium. Bryozoans shelter bacteria that produce highly complex and toxic molecules (e.g. bryostatins). Ascidians can produce highly bioactive molecules, such as the depsipeptidic didemnins, in association with prochloron, i.e. cyanobacteria which they cultivate in their tissues. The purpose here is not to list the thousands of amazing molecules that are produced by benthic marine invertebrates, to which we must add the recently discovered highly diverse repertoire of planktonic organisms: [31] and its updated version [33] in press) and [34] are probably the most recent and comprehensive texts on the subject.

Evolutionary reflections of chemical defenses in marine symbiotic systems have been recently proposed by [35]. Symbiotic and photosymbiotic systems that associate an invertebrate or algal host and its specific microbial consortium are common in tropical shallow water reefs, which concentrate at least one third of the total marine biodiversity and certainly the largest proportion of all known "secondary" metabolites [26]. To these must be added marine fungi and actinobacteria that live in marine sediments.

A generally overlooked component of marine chemodiversity is mucus, sometimes constantly and abundantly produced by epithelial goblet cells of cnidarians, and also found in other sessile invertebrates, and also in fish. No two muci are alike. In fish, it helps protecting scales from unwanted fouling by encrusters or parasites, as well as having intrinsic antibacterial and antiviral properties. Coral mucus has species-specific composition and plays a role equivalent to the organ-forming mesoderm of triploblastic organisms, in managing a highly biodiverse microbial fauna that recycles carbon, nitrogen and sulfur and provides an energy complement to symbiotic photosynthesis. To the coral associated bacteria mucus plays the role of a biofilm in which complex antibiotic interactions are at play to maintain microbial diversity.

Behavioral adaptations like "advertising" (aposematic) colors and body shapes make use of colored metabolites borrowed from the prey organisms and concentrated into superficial diverticula. Specialist predators like shell-less mollusks like dorid nudibranchs are beautifully colored and highly toxic. Several aeolid nudibranchs on the other hand use camouflage for similar reasons.

5. Uses of marine natural products

Marine biogenic molecules are used in a large variety of applications, industrial, medical, pharmaceutical, cosmetological, as food or associated to food, decoration…

5.1. Housing and transportation matters

These include the use of fossilized or detritic materials for construction and as chemical energy sources.

5.1.1. Construction materials

Biogenic sand and rocks (e.g. sandstone) are used to create building, roads and associated structures. Urban architecture is mostly made of concrete structures, i.e. cemented biogenic or mineral sand and crushed rocks or debris.

5.1.2. Fossil fuels and biofuels

Fossil fuels are consumed at the rate of 88 million barrels a day, in heat engines, to produce electricity, domestic heating, and in petrochemical chemistry (plastics, clothing, lubricants, polymers). Fossil fuels are non-renewable energy, since the production rate is about one-million times slower the consumption rate.

5.2. Health and well-being

This includes the use of biomineral or bioorganic materials in surgery, of cosmetics and skin care substances, and most of all of anticancer, antiviral, anti-pain, antibiotic, anti-inflamma-tory, as well as insecticidal, anti-malarial or anthelminthic molecules, not to mention other domains of growing interest, such as Alzheimer's disease and schizophrenia.

The exploitation of marine natural products in folk medicine is very limited and concerns a few organisms that did not necessitate special underwater investigations, nor entail storage and conservation problems - of paramount importance since desiccation, oxidation and fermentation occur so rapidly. Exceptions are the occasional use of specific algae as insecticides and for ringworm treatment, or the ceremonial use of toxic or venomous organisms. Fixed and fleshy organisms were generally distasteful or toxic, sparing them for being considered as remedies. Corals, sponges and a few others have no nutritional value and represent the largest biomass of high-diversity systems in tropical zones.

It is only since just after WW2 that marine organisms emerged as a complementary source to traditional plant remedies, thanks to the co-occurrence of several factors, among which (i) the recent discovery of fungal antibiotics and the use of bacterial enzymes in bioprocesses, (ii) the development of adequate screening procedures, (iii) SCUBA exploration and later robot-sampling in oceanic depths. The development of synthetic chemistry soon became necessary to provide alternative sources for rare organisms, but also to find better analogues (more active, less toxic, more soluble, better tolerated, easier and cheaper to synthesize…).

A new biotechnological turn has recently been taken with the high capacity screening of vast libraries of synthetic analogues, with the development of biomimetic chemistry and with the use of genetic tools that allow bacterial carriers like *Escherichia coli* to produce a compound of interest after insertion of the set of genes sampled from the original biological source.

Marine molecular sciences have now come of age, but the exploration of novel natural products and of new biological activities must go on, especially in relation to marine microbes and to planktonic organisms.

Many (98 to 99%) of the known marine bacteria and archaea still cannot be cultivated using the existing apparatus and growth media. Knowing that invertebrates such as sponges, ascidians, bryozoans act as biological incubators of specific bacterial strains, isolated in specialized cells called bacteriocytes or allowed to interact synergistically, there is a world of possibilities opened for us to investigate. After all, we only discovered recently that bacteria have a form of social life within biofilms, have developed a chemical language that allows their populations to be regulated, have an instrumental role in carbon and nitrogen cycling in holobiont photosystems such as corals, lichens etc. and in biogeochemical cycles at large [28].

Indeed we still have no idea of the importance bacteria have in the production or in the modulation of the genetic expression of key metabolites hitherto attributed to their host organisms, except through limited attempts to cultivate the latter axenically. Using a systems biology approach, stress transcriptomics studies will undoubtedly help us achieve better control of the production of key enzymes or key defense metabolites from host organisms, as well as evaluate the optimal conditions for the maintenance of a rich and diverse microflora using metagenomics.

The purpose here is not to list the bioactive molecules that have been discovered, elucidated, synthesized, nor to detail their bioactivities. Excellent reviews are regularly updated on the subject, presented by natural products chemists and pharmacologists. Interactive databases include standard spectral information, and sophisticated algorithms are available as aids for structure elucidation and as dereplication tools. Two recent and outstanding information sources are [31] and its updated version [33], and [34].

Well over 20,000 secondary metabolites have been screened so far, but only a handful is now legally approved for use as prescription drugs or treatments by the Food and Drugs Administration (FDA) and by the European Union since 1974 (see [36] and updates thereof in: http://marinepharmacology.midwestern.edu/clinPipeline.htm). Successful molecules and analogues have to undergo in-vitro, in-vivo (preclinical) tests and finally several rounds of clinical trials on patients, a process known as the clinical pipeline. On average, each development costs about one billion dollars over a period of ten years, and there is great pressure to discover and develop novel molecules, especially against various types of cancer, antibiotics against resistant strains, and molecules for the treatment of neurodegenerative diseases. With energy, food and housing, pharmaceuticals represent a primary resource in modern economy.

5.3. High-tech

Siliceous skeletons are transformed from dissolved silicic acid into very elaborate structures by diatoms and radiolarians. They not only serve a structural function: they allow sunlight to penetrate for photosynthesis, and there is evidence that harmful radiations are filtered out somewhat like our sunglasses, using embedded mycosporine-like amino acids as DNA-protective sunscreens [37]. Enzymes that are capable of creating optical-grade glass at room temperature are under investigation. Siliceous sponges produce glass spicules which are endowed with mechanical and optical properties unknown to manufacturers, and novel enzymatic approaches using sponge silicateins are investigated for nanotechnological and biomedical applications [38].

Shells of bivalve and gastropod mollusks, other than their interest as decorative items, often present amazing properties. Oyster nacre has osteo-inductive properties that could be used in bone tissue regeneration [39]. Abalone shells offer strong mechanical resistance to impact, thanks to its dual organic-limestone microscopically layered structure, inspiring novel bulletproof materials. Skeletons that are produced by sessile marine invertebrates are a great source of inspiration: some have tremendous flexibility and resilience to currents, like the horny skeleton of some gorgonians [40]; coral skeletons are used as temporary bone implants [41], coralline algae cements have amazing particle-aggregating properties, etc. Novel adhesives with high tensile strength are inspired from byssus filaments of mussels. Crab shells can provide a sustainable source of chitin and chitosan, which are used in a variety of medical and laboratory applications [42].

Bacteria can be modified genetically to produce pure toxins, antibiotics, polymers and enzymes with high added value. Cultivatable bacterial strains offer exciting possibilities for the transformation or the recycling of a wide range of industrial materials and agricultural waste products. Experimental biofilms can produce weak currents with potential applications in sewage treatment plants.

6. Management of biodiversity and chemodiversity

Global anthropic influence is articulated on the exploitation of natural resources and the generation of wastes. Here are a few examples of man's use of biogenic resources, mineral and organic, fossil or live.

6.1. Urban development

The use of biogenic sand and limestone concretions usually involves no direct utilization of live organisms. Indirectly, the overexploitation of sand beds from littoral zones represents a double problem: the eradication of the epibenthic wildlife and with time, erosive changes in coastal profiling. Sand particles offer extended surfaces that are essential for microorganisms to recycle decaying organic debris and avoid the accumulation of toxic nitrites and nitrates. The exploitation of sands from coral reef atolls, together with the silting of coastal bays and leeward zones due to mining activities spillage are major sources of biodiversity loss.

6.2. Energy

The use of fossil hydrocarbons extracted from the Mexican Gulf, the Northern Sea etc., as energy sources results in the volatilization of carbon in the atmosphere at rates up to one million times faster than it took for phytoplankton to fix and sequester it into sediments as oil beds [25]. Atmospheric accumulation of carbon dioxide, methane, sulphur dioxide and other combustion products, create an artificial greenhouse effect after only 150 years of exploitation (now largely marine), adding to natural volcanic activity. Global warming and ocean acidification are both regarded as major potential sources of marine biodiversity destruction [43].

Biodiversity loss seems to be the inevitable toll to pay for a global economic expansion relying heavily on the use of biogenic productions. When such exploitations occur at the vicinity of marine biodiversity hotspots (coral reefs for example), which concentrate much of the useful and yet unexplored chemical diversity [26], we run into the risk of losing a lot of useful model molecules and end up with a hostile, impoverished microbial world instead [44].

6.3. Drugs and ethics

Drugs are basically dangerous when ingested, injected or applied externally without caution, and man has always made the difference between edibles and toxic or venomous substances, or foodstuff corrupted with microbes. Medicine men and witch doctors knew empirically how to use fresh plants and how to blend them to cure fevers, wounds and other ailments, and this knowledge was passed on exclusively to younger initiates who henceforth gained access to special social status and were highly respected. Today, bioactive substances are patented and the administration of prescription drugs is tightly controlled. As mentioned earlier, the drug industry is one of the very few primary sources of revenue in any country's economy. Exploration of novel sources of natural products in biodiversity hotspots, terrestrial or marine, is a touchy subject. Intellectual property should be shared in the benefit of local populations, eventually leading to their active participation in the cultivation and the preparation and purification of extracts, on a non-destructive and sustainable basis. This is especially true for developments that were guided by ethnopharmacological surveys, i.e. the participation of local medicine men. Marine bioexploration, however, is almost exclusively based on systematic screening of marine plants, invertebrates and microbes, but it should be subjected to territorial laws like the exploitation of forests, fishing zones or mines in a participative manner. More specifically, endangered or rare species should be identified, and sampling procedures be conducted in accordance with internationally established practices [45].

6.4. Food

We have purposely ignored aspects directly related to food, i.e. the exploitation of marine resources through fisheries and the farming of selected fish and seafood species. The sustainable management of natural stocks and the influence of global warming on natural biodiversity profiles is a major concern in the 21st century. Human population has almost trebled since WWII, and *Homo sapiens* dominates terrestrial vertebrate biomass (half a billion tons) to the point that meat protein supplies will be scarcer. After being a hunter-gatherer, man has become a farmer and he is now turning towards artificially produced biomass that has food value, grown from cultured transgenic cells. What next?

The management of natural sanctuaries that are subjected to minimal interference, with sizes large enough to have a self-regenerating biodiversity (and chemodiversity), and tightly controlled environmental connectivity, should be encouraged.

7. Pollution

Man does not always recycle the wastes he generates, far from it.

7.1. Urban developments

There are two major issues to consider: one is structural and the other is biological.

Structural issues are linked to urban planning and the exploitation of natural zones, and to the fate of obsolete constructions and roads that cannot be recycled back into their original components - the net result being a shift from marine destructive exploitation (sand) to terrestrial pollution (discharges). Wherever possible, alternatives to marine sand must be sought for housing, to save the resource and perhaps use recycled wood-based products and recyclable organic polymers.

Biological issues emerge from the disposal of untreated sewage, of fertilizers, pesticides, drugs, household maintenance products and industrial by-products. Regarding sewage, local enrichment of waterways and coastal areas, especially in nitrogen and phosphorus, causes toxic phytoplankton blooms, and unwanted algal proliferations that tend to displace indigenous species. This is especially true for complex oligotrophic ecosystems in which nutrients are efficiently used and recycled. In coral reefs, bacterial charges are introduced directly or via invasive algal proliferations [46], favoring the dominance of pathogenic strains and reducing the natural microbial diversity [47], [43].

7.2. Energy

Regarding energy-related matters, losses due extraction and transport (ship degassing, black tides due to wreckages etc.) are the cause of episodic but massive biodiversity destructions in littoral zones, by asphyxiating wildlife and by destroying skin protection barrier. The exploitation of light hydrocarbons from shales is currently the subject of intense debates, especially regarding its consequences on water quality [48]. The exploitation of gas hydrates represents an attractive energy source, but its exploitation may prove difficult to control, and its release from the thawing of the permafrost due to global warming difficult to evaluate [49].

7.3. Synthetic drugs, artificial molecules, genetically modified organisms

The modern world is global and highly competitive. Being positioned on an economic niche requires achieving results and outcompeting contenders, no matter what the human and environmental consequences are in the long term. Thousands of synthetic molecules - insecticides, herbicides, pharmaceuticals, conditioning agents, food additives, fuel additives, lubricants, household products, plasticizers, flame retardants, to name a few- have been synthesized without sufficiently questioning their degradability and speciation once released in the environment or abandoned, or offering detoxication, storage/disposal or reusability solutions. Highly reactive pesticides can cause immediate non-specific destruction to useful pollinizing insects and kill birds, others may insidiously find their way up the food chains and eventually poison top predators - this is the case of highly stable and carcinogenic biphenyl organohalogens. Chronic allergies and cancers are documented through the regular exposure to more and more food additives and household products. Specific information can be obtained from associative websites.

If there are no simple replacement solutions in most cases, perhaps revisiting global commercial practices, promoting local products and values and favoring educational resources (preferably non-virtual) centered on the practical resolution of local issues, will help establish respectable ethics and restore some dignity.

8. Rearranging clouds in times of global changes?

We are in the midst of the worldwide race for economic development and global trading, with their extraordinary demands on natural resources and their inevitable environmental issues. While striving to maintain social equilibrium at the same time, why not take a pause and ask ourselves not only where we are going, but simply how we are going? How can we know how we are going?

The prestigious graduate schools, the most popular jobs in western and industrialized countries are linked with the world of finance. Get closer to the sources, react faster and make the optimal decision before anyone else does. Sophisticated algorithms are being developed by highly paid experts to better analyze tendencies and make predictions, from a phenomenal amount of cross-analyzed sources of diverse natures. Lists of potential customer's addresses are sold, exchanged, stolen. Cloud metadata and then triturated and sent back to the average human as a continuous shower of commercial ads, enticing e-mails, spurious spams, in the hope of creating habits and develop ruinous compulsions? Along the same line, potential crime scenes can now be anticipated with amazing probability rates by the Californian police, thanks to the work of metadata experts.

Global changes are on the way, noticeable since the70's with the accelerating atmospheric enrichment of greenhouse gases. Science and technology have made extraordinary progress at the same time, but mostly dedicated to the continuing exploration of the unknown "before it is too late", and too little to addressing pressing environmental issues with the huge losses in biodiversity that are envisioned by many in the next decades.Cutting edge medical research has been fuelled by the need to face new challenges, in cancerology, in microbial pathologies, in immune or rare diseases. Highly efficient molecular investigations from gene mining to the knowledge and control of their expression repertoire, whole organism approaches, continuously updated cutting-edge equipment - are being applied to exploratory science and projections into future scenarios. Huge supercomputers are piling up petabytes of data waiting for bioinformatics advances to organize and make sense of them, and in the quest of the reassuring "big picture" that will once for all settle intellectual disputes, quench the thirst for the unknown, and with the hope that, ultimately, some biogeoclimatic equilibrium will be automatically restored no matter what.

How most of us are doing depends not on fluctuations of the stock market, nor on the price of the latest hi-tech widget, nor on whether the next crime scene is going to take place in our neighborhood, but on how fast our natural environment is going to degrade. Why not redirect our priorities and efforts to better understand how our biosphere functions, how species interact, how biomes are connected, how biodiversity is maintained in marine and terrestrial environments?

We generate enormous amounts of data of genomic, chemical, taxonomic, pharmacological, biotechnological value that can be better analyzed.

Dealing with environmental issues, the first question is the preparation of environmental databases, our "clouds"- the crucial initial step. What "standard" environment will be suitable to monitor long-term changes? Climatic changes being *global* (temperature rises, ocean acidification etc.) and heavily influenced by human activities, one can only define a study zone as representative of a "standard environment" if it is not *locally* impacted by *direct* human interference (habitat destruction, introduction of alien species, chemical and microbiological pollution…). Fluctuations in parameters of interest must not be subjected to amplitudes higher than those of experimental (read: impacted) study zones, and sampling must be large and statistically significant for the "standard" database to be robust and reliable enough to evidence mild to severe changes in same parameters from experimental study zones. This applies not only to physico-chemical parameters, but also to genomics, transcriptomics or metabolomics metadata (e.g. microbial metagenomics). We need markers within variables, sentinel species within populations, signatures within metabolomes, early responses within complete tran-scriptomic repertoires, and so on. Some advocate the use of artificial biological systems, a synthetic biology approach, to standardize sampling procedures. Whole-area imaging can make use of drones that interfere very little with wildlife and biological processes. Coupled with appropriate mathematical tools, automation can be achieved and time-dependent evolution of e.g. natural processes can be predicted.

Thus, whether comparing experimental vs. control areas, or monitoring the evolution of the same locality in time, or analyzing *connectivity*, enormous amounts of data are being generated that need trimming, arranging, cross-checking to be usefully interpreted. And beyond assessing the evolution of a situation, we need to predict, for example, the degradation in the biodiversity of impacted areas if no corrective action is taken. Neural networks that are "educated" by so-called genetic algorithms are currently very successfully exploited, in economics, in business, and in forensic science, to open what is known as "windows of opportunities" (to seize, but also to take corrective measures). These were created by scientists to treat scientific questions in the first place! Beyond our biosphere's apparent complexity, what it contains are quantitatively and qualitatively finite entities, subjected to predictable dynamics. There is no reason not to apply these analytical methods to create ready-to-use tools legislation can rely upon to regulate the appropriation of natural resources: water, wood, sand, fossil fuel, minerals… and wildlife. In the same way as DNA tests are used to help solve criminal investigation by 'showing irrefutable evidence', molecular tools applied to systems biology in association with multi-scale imaging tools can be tailored by analysts to guide our choices towards a more responsible management of natural resources.

We need a generation of motivated young analysts to use their skills to extract the useful information and blow away the dust. The big pictures might then emerge at last, as the clouds dissipate, ready to inspire us with a better scenario. The following phrase by Salvatore Manga no [50] *"Genetic Algorithms are good at taking large, potentially huge search spaces and navigating them, looking for optimal combinations of things, solutions you might not otherwise find in a lifetime"* makes needle-in-the haystack problems tractable. After all, we live on a planet with finite resources, our only problem now is to make them last for the benefit of the generations to come. If machines can learn, hopefully humans can too....

Acknowledgements

I wish to thank Catherine Boyen of UMR 7139 at Station Biologique de Roscoff for financial assistance, and I wish to dedicate this work to John Saville Waid (1927-2012), an internationally known Professor and specialist of soil microbiology, whose friendliness and contagious smile will always be remembered by those who knew him.

Author details

Stéphane La Barre[1,2]

1 Université Pierre et Marie Curie-Paris 6, UMR 7139 Végétaux marins et Biomolécules, Station Biologique F-29680, Roscoff, France

2 CNRS, UMR 7139 Végétaux marins et Biomolécules, Station Biologique F-29680, Roscoff, France

References

[1] Valley JW, Peck WH, King EM, Wilde SA. A cool early earth. Geology 2002;30: 351-354.

[2] Miller SL A production of amino acids under possible primitive earth conditions. Science 1953;117: 528–529.

[3] Parker ET, Henderson JC, Dworkin JP, Glavin DP, Callahan M, Aubrey A, Lazcano A, Bada JL. Primordial synthesis of amines and amino acids in a 1958 Miller H_2S-rich spark discharge experiment. PNAS 2011;108(14): 5526-5531.

[4] Oró J, Kamat SS. Amino-acid synthesis from hydrogen cyanide under possible primitive earth conditions. Nature 1961;190 (4774): 442–3. Bibcode:1961 Natur.190.442O. doi:10.1038/190442a0

[5] Ritson D, Sutherland JD. Prebiotic synthesis of simple sugars by photoredox systems chemistry. Nature Chemistry 2012;4: 895-899.

[6] Shapiro R. Prebiotic ribose synthesis: a critical analysis. Origins of Life and Evolution of the Biosphere 1988;18: 71-85.

[7] Huber C, Wächterhäuser G. α-hydroxy and α-amino acids under possible Hadean, volcanic origin-of-life conditions. Science 2010;314: 630-632.

[8] Raulin F, Coll P, Navarro-González R. Prebiotic Chemistry: Laboratory Experiments and Planetary Observation. In: Gargaud M. et al. (eds.) Lectures in Astrobiology. Berlin, Heidelberg: Springer-Verlag; 2005;1: p449-471. ISBN: 978-3-540-22315-3

[9] Gilbert W. The RNA world. Nature 1986;319: 618.

[10] Ferris JP, Hill AR, Liu R, Orgel E. Synthesis of long prebiotic oligomers on mineral surfaces. Nature 1996;381, 59–61.

[11] Eschenmoser A. Chemical etiology of nucleic acid structure. Science 1999;284, 2118–2124.

[12] Maurel C, Haenni M-L. The RNA World: Hypotheses, Facts and Experimental Results. In: Gargaud M. et al. (eds.) Lectures in Astrobiology. Berlin, Heidelberg: Springer-Verlag; 2005;1: p571-594. ISBN: 978-3-540-22315-3.

[13] Meli M, Vergne J, Maurel M-C (2003). *In vitro* selection of adenine-dependent hairpin ribozymes. J. Biol. Chem., 2003;278(11): 9835–9842.

[14] Bada JL, Lazcano L. Some like it hot, but not the first biomolecules. Science 2002;296: 1982-1983.

[15] Balucani N. Elementary reactions and their role in gas-phase prebiotic chemistry. International Journal Of Molecular Sciences, 2009;10: 2304-2335; doi: 10.3390/ijms10052304.

[16] Commeyras A, Boiteau L, Vandenabeele-Trambouze O, Selsis F. Peptide emergence, evolution and selection on the primitive earth. I. Convergent Formation of N -Carbamoyl Amino Acids Rather than Free α -Amino Acids?. In: Gargaud M. et al. (eds.) Lectures in Astrobiology. Berlin, Heidelberg: Springer-Verlag; 2005;1: p517-545. ISBN: 978-3-540-22315-3.

[17] Chrysostomou A, Ménard F, Gledhill TM, Clark S, Hough JH, McCall A, Tamura M. Circular Polarization in Star Formation Regions: Implications for Biomolecular Homochirality. Science 1998;281(5377): 672.

[18] Cronin J, Reisse J. Chirality and the Origin of Homochirality. In: Gargaud M. et al. (eds.) Lectures in Astrobiology. Berlin, Heidelberg: Springer-Verlag; 2005;1, p473-515. ISBN: 978-3-540-22315-3

[19] Ourisson G, Nakatani Y. A Rational Approach to the Origin of Life: from Amphiphilic Molecules to Protocells. Some Plausible Solutions and Some Real Problems. In: Gargaud M. et al. (eds.) Lectures in Astrobiology. Berlin, Heidelberg: Springer-Verlag; 2005;1: p429-448. ISBN: 978-3-540-22315-3

[20] Commeyras A, Boiteau L, Vandenabeele-Trambouze O, Selsis F. Peptide Emergence, Evolution and Selection on the Primitive Earth. II - The Primary Pump Scenario. In: Gargaud M. et al. (eds.) Lectures in Astrobiology. Berlin, Heidelberg: Springer-Verlag; 2005;1: p547-569. ISBN: 978-3-540-22315-3

[21] Szostak JW, Bartel DP, Luisi PL. Synthesizing life. Nature 2001;409: 387-390.

[22] Pinti M. The Origin and Evolution of the Oceans. In: Lectures in Astrobiology, vol. 1, 83-112, M. Gargaud et al. (Ed.), Springer-Verlag Berlin Heidelberg 2005 ISBN: 978-3-540-22315-3

[23] Benton MJ. When Life Nearly Died: the Greatest Mass Extinction of all Time. London: Thames & Hudson 2005, 336 pages. ISBN-13: 978-0500285732.

[24] Knoll AH, Bambach RK, Canfield DE, Grotzinger JP. Comparative Earth history and Late Permian mass extinction. Science 1996;273(5274): 452-457.

[25] Falkowski P. Tenth Annual Roger Revelle Commemorative Lecture: The once and future ocean. Oceanography 2009;22(2): 246-251, doi:10.5670/oceang.2009.57.

[26] La Barre S. Novel tools for the evaluation of the health status of coral reefs and for the prediction of their biodiversity in the face of climatic changes. In: Zambianchi E. (ed.) Topics in Oceanography. Rijeka: InTech; 2013. p127-155.

[27] Dill KA, Truskett TM, Vlachy V, Hribar-Lee B. Modelling water, the hydrophobic effect and ion solvation. Annu. Rev. Biophys. Biomol. Struct. 2005;34: 173–99 doi: 10.1146/annurev.biophys.34.040204.144517

[28] Falkowski, PG, Fenchel; T, Delong, EF. The microbial engines that drive earth's biogeochemical cycles. Science 2008;320: 1034-1039.

[29] Waid JS. Does soil biodiversity depend upon metabiotic activity and influences? Applied Soil Ecology 1999;13: 151-158.

[30] Kornprobst J-M. Encyclopedia of Marine Natural Products. Weinheim: Wiley-Blackwell Verlag GmbH & Co. KGaA; 2010. ISBN; 978-3-527-32703-4.

[31] Cani PD, Delzenne NM, Amar J, Burcelin R. Role of gut microflora in the development of obsesity and insulin resistance following high-fat feeding. Pathologie Biologie 2008;56: 305-309, doi: 10.1016/j.patbio.2007.09.008

[32] Hay ME, Fenical W. Marine plan-herbivore interactions: the ecology of chemical defense. Annual Review of Ecology and Systematics, 1988;19: 111-145.

[33] Kornprobst J-M. Encyclopedia of Marine Natural Products. Weinheim: Wiley-Blackwell Verlag GmbH & Co. KGaA; in press; ISBN; 978-3-527-33429-2 (printed version) and ISBN; 978-3-527-33585-5 (electronic version).

[34] Fattorusso E; Gerwick WH, Tagliatela-Scafati O (editors). Handbook of Marine Natural products, 2011, 2 volumes, 691 Figures, 45 Tables, Springer Science, Print and electronic bundle under ISBN 978-90-481-3856-2, DOI: 10.1007/978-90-481-3834-0

[35] Lopanik, N.B. (2013) Chemical defensive symbioses in the marine environment. Functional Ecology, *in press*. doi: 10.1111/1365-2435.12160

[36] Mayer A, Glaser KB, Cuevas C., Jacobs, RS; Kem W, Little RD, McIntosh JM, Newman DJ, Potts BC, Shuster DE. The odyssey of marine pharmaceuticals: a current pipeline perspective. Trends in Pharmacological Sciences 2010;31: 255–265. doi: 10.1016/j.tips.2010.02.005

[37] Ingalls AE, Whitehead K, Bridoux MC. Tinted windows: the presence of the UV absorbing compounds called mycosporine-like amino acids embedded in the frustules of marine diatoms. Geochimica et Cosmochimica Acta 2010;74: 104–115. doi:10.1016/j.gca.2009.09.012.

[38] Müller EG, Wang X, Cui F-Z, Jochum KP, Tremel W, Bill J, Schröder HC, Natalio F, Schloßmacher, Wiens M. Sponge spicules as blueprints for the biofabrication of inorganic–organic composites and biomaterials. Appl Microbiol Biotechnol 2009;83: 397–413.

[39] Rousseau M. Nacre, a natural biomaterial. In: Pignatello R. (ed.) Biomaterials Applications for Nanomedicine. Rijeka: InTech, 2011, p661-664.

[40] Jeyasuria P, Lewis JC. Mechanical properties of the axial skeleton in gorgonians. Coral Reefs 1987;5, 213-219.

[41] Demers C, Hamdy CR, Corsi K, Chellat F, Tabrizian M, Yahia L. Natural coral exoskeleton as a bone graft substitute: a review. Bio-Medical Materials and Engineering, 2002;12(1): 15-35.

[42] Hirano S. Chitin biotechnology applications. Biotechnology Annual Review, 1996;2: 237-258.

[43] La Barre S. Coral reef biodiversity in the face of climatic changes. In: Grillo, O. (ed.) Biodiversity Loss in a Changing Planet. Rijeka: InTech; 2011. p77-112.

[44] Rohwer F. Coral reefs in the microbial seas. USA: Plaid Press; 2010, 201 pages, ISBN: 978-0-9827012-0-1.

[45] Crag GM, Katz F, Newman DJ, Rosenthal J. Legal and ethical issues involving marine biodiversity and development. In: Fattorusso E, Gerwick WH, Tagliatela-Scafati

O (editors). Handbook of Marine Natural Products. Springer Science 2011;2: p1315-1342. Print and electronic bundle under ISBN 978-90-481-3856-2, DOI: 10.1007/978-90-481-3834-0.

[46] Hay ME, Rasher DB. Coral reefs in crisis: reversing the biotic death spiral. F1000 Biology Reports 2010;2(71) (doi:10.3410/B2-71).

[47] Mouchka ME, Hewson I, Harvell D. Coral-associated bacterial assemblages: current knowledge and the potential for climate-driven impacts. Integrative and Comparative Biology, 2010;50(4): 662–674. ISSN: 1540-7063.

[48] Al T, Butler K, Cunjak R, MacQuarrie K. Opinion: Potential Impact of Shale Gas Exploitation on Water Resources. 2012. University of New Brunswick Report.

[49] Kvenvolden KA. Gas hydrates - geological perspective and global change. Reviews of Geophysics 1993;31(2): 173-187.

[50] Mangano S. Computer Design (Powerpoint presentation, 1995) https://www.cs.drexel.edu/~spiros/teaching/SE320/slides/ga.pdf.

Chapter 5

Protection of Riparian Habitats to Conserve Keystone Species with Reference to *Terminalia arjuna* – A Case Study from South India

B.C. Nagaraja, C. Sunil and R.K. Somashekar

Additional information is available at the end of the chapter

http://dx.doi.org/10.5772/58355

1. Introduction

Riparian forests (RF) growing along streams, rivers and lakes have special functions in the landscape as the interface between the terrestrial and the aquatic ecosystem (Malanson 1993). They are distinctly different from the surrounding lands because of unique soil and vegetation characteristics that are strongly influenced by free or unbound water in the soil. Riparian zones are usually a diverse mosaic of landforms, communities and environments within landscapes and they serve as a framework for understanding the dynamics of communities associated with fluvial ecosystems (Gregory *et al.*, 1991; Naiman *et al.*, 1993; Naiman *et al.*, 2005). Being a transition zone between aquatic and terrestrial area where structural and functional properties change with space and time, discontinuously. Typical examples of riparian zones would include flood plains, stream banks and lake shores. The interfaces in riparian zones possess physical and chemical attributes, biotic processes, material flow processes, but they are unique in their interactions with adjacent ecological systems.. Riparian zones are habitats of critical conservation concern worldwide, as they are known to filter agricultural contaminants, buffer landscapes against erosion, and provide habitat for high numbers of species (John *et al.*, 2005). The Riparian forests are habitats for a large number of forest species including many of the rare species that depend on water and as such serve as important areas for biodiversity (Gundersen et al., 2010; Darveau et al. 1995; Hylander 2006).

Riparian lands can also include intermittent streams gullies and dips which sometimes run with water. The vegetation ranges from emergent aquatic and semi-aquatic plants through to terrestrial understorey and canopy species (Parsons 1991). Further, the zone can be seen as an interface between terrestrial and aquatic systems and is described as a series of ecotones

between these systems (Risser 1990). Riparian vegetation plays an important role in the maintenance of stream and foreshore stability. Streams and rivers are essentially dynamic systems, their path and flow constantly changes with the time (Warner 1983). The presence of vegetation in riparian areas acts to reduce the rate of change and therefore maintain a level of stability.

Figure 1. Overview of Riparian forest in the banks of river Cauvery

2. Ecology and biodiversity in riparian forest

Plant communities in large river flood plains are amongst the most productive and diverse in the world and frequently support higher number of plant species arranged in vegetation associations of greater complexity than surrounding landscaping units (Menges and Waller, 1983; Tockner and Stanford, 2002). Water level patterns are critical for the successful establishment of new plants (both exotic and native species) following dispersal of seeds or other propagules by water, wind, animal vectors or other dispersal agents. Flow has been determined as primary factors for determining plant community composition and structure along the riparian zone (Blom *et al.*, 1990; Ferreira, 1997). Many plant species depended particularly on the flow for dispersal of their propagules a process referred to as 'hydrochory' (Nilsson *et al.*, 1991). Types of propagules include sexually derived seeds as well as vegetative fragments (mechanically sheared or physiologically abscised branch or root segments) that can re-sprout to result in asexual propagation. Propagative dispersal typically occurs in a downstream direction along streams but may be wind-aided along lakes or reservoirs. Thus, hydrochory may occur in multiple directions along relatively stationary water bodies. Propagative dispersal by water is an effective adaptation of native plants but also provides a major mechanism for invasion by exotic weeds, of which noxious species can have severe ecological and economic impacts (Braatne *et al.*, 2002).

Riparian vegetation changes continuously from the beginning of a river in the mountains up to the river mouth with the changing environmental parameters like altitude, humidity, soil conditions and also in the conditions of water like quantum and flow, temperature, pH, salinity. In a tropical countries, the riparian vegetation in a first order stream in the mountain may be ferns and other associated herbaceous plants in the rock crevices. When coming down, evergreen forest samples can be observed in the riparian zone as a quantum and the lateral influence of the water increases. Further going down the bed conditions of the river changes from rocky to sandy especially in the floodplains. Here the soil becomes looser, sedimentation rate will be high, and a good amount of alluvium can be found. In these areas the water influence on the vegetation may be more. Herbaceous, grass and hydrophytic plant communities will be abundant in these zones (Amitha, 2003).

Ripairan areas acts as a migratory corridor and routes for many wildlife as it has been used for regular daily movements and seasonal migration. Riparian zones offers an three critical resources for wildlife: cover, food and water in one space. The undisturbed stands of age old woody species provide habitat for nesting birds resided in the forests. Riparian zones are utilized by wildlife as a sort of "natural highway". They are important to mammals and birds as they journey up and down the river during daily movements besides seasonal migrations. Much wildlife is found to be associated on floodplains than in any other landscape unit in most regions of the world (Klement and Stanford, 2002). In the Pacific coastal ecoregion (USA), for example, approximately 29% of wildlife species found in riparian forests are riparian obligates (Kelsey & West 1998). It provides habitat for more species of breeding birds than any other vegetation association. For example, of all bird species breeding in northern Colorado, 82% occur in riparian vegetation, and about half of south-western species depend upon riparian vegetation (Knopf & Samson 1994). Riparian areas in semiarid zones are critical in providing stopover areas for *en route* migrants (acting as 'dispersal filters'), and therefore affect the breeding success of northern bird populations (Skagen *et al.* 1998). In Europe, 30% of threatened bird species are inland wetland-dependent species and 69% of the important breeding areas for birds contain wetland habitats, primarily flood plains (Tiker & Evans 1997). In Switzerland, 10% of the entire fauna is restricted in its occurrence to riverine flood plains, although flood plains only cover 0.26% of the country's surface. Among 10%, 28% of the fauna frequently uses flood plains and about 44% is occasionally found in flood plains. A high proportion of the riparian obligates (47%) is listed as endangered, compared to 28% for the entire fauna (Walter *et al.* 1998).

3. Ecosystem services of riparian vegetation

Riparian forests performs an array of functions in its buffer are which are beneficial to regional ecosystem to meet the some of their essential needs for their survival in the ecosystem. Some specific species stand unique in portraying their services in the particular ecosystem due to its morphological and phenological nature where their life cycle influences to protect stability of several flora and fauna in the ecosystem. Besides these functions, several species of riparian vegetation render services to the humans, as they provide several direct and indirect economic supports to run their livelihoods.

4. Ecological significance

The riparian plant species improves the microclimatic condition thereby allowing the other associated species to to grow in the community. The forks of old trees in the riparian zone provide wantage points to epiphytes.

Figure 2. Epiphyte *Acampe praemorsa* growing on forks of tree species *Terminalia arjuna* and Orchids laden on tree branches of *Madhuca latifolia*.

Riparian species develops typical root modifications to withstand during the flood events. Such typical modifications of plant root systems are called as buttressed root systems. The buttressed root systems provide the strength to the tree species and to facilitates a suitable site to other riparian species to grow. Rivers combined with such root systems in conjunction with other herbaceous vegetation dissipate stream energy, resulting in less erosion and a reduction in flood damage. A 5 cm deep root system resists erosion up to 20,000 times better than bare soil stream banks. A woody root mat is the "re-bat' of stream banks. The riparian canopy provides organic matter via litter fall; surfaces of submerged leaves are sites of primary and secondary production by micro algae and bacteria, which can rival that of phytoplankton and bactereophils in water column. The Logs of riparian vegetation play an important role in the dynamics of stream morphology and serve as substrates for biological activity by microbial and invertebrate organisms. On land the riparian stream ecosystem is the single most productive type of wildlife habitat. The Riparian areas act as a corridor for big game migratory animals between summer and winter range.

5. Social significance

Past civilizations came up on river banks, the followed generations used rivers as a source of water and food. The flood plains of the Indus, the Nile delta, and the fertile crescent of the Tigris and Euphrates rivers provided man with all his basic necessities. They can be considered the pillars of human civilization as they have formed the nuclei for human settlements from the very origins of mankind. Fishing is a major means of livelihood for the people who resided in and around the riparian zones. Many of the tribal's depend upon the river for fishing. The riparian vegetation decrease soil erosion and support silt thereby avoiding the pollutant input to the river. The shade, fruits and flowers offered by the riparian vegetations promotes the fish abundance in the aquatic ecosystems. The riparian vegetation provides Non Wood Forest Products for the dependent communities especially tribals who use the riparian forest to make their huts (Mainly *Bamboo* and *Ochlandra*), honey collections, timber, manure for farming and medicinal plants etc.

6. General overview of Cauvery riverine ecosystem

The Cauvery river originates at Talakaveri (12° 25' N, 75° 34' E) in the Western Ghats at an altitude of 1341m. It is the 8th largest river in the subcontinent and ranks as a medium river on a global scale. The Cauvery River basin is estimated to occupy 81155 km^2 area occuping nearly 2.5% of the total geographical area of the country. The Cauvery river basin areas have a large floristic wealth enough to constitute as a separate phyto-geographic unit. The vegetation of the entire peninsular India excluding Western Ghats is adequately represented in this tract alone (Jayaram, 2000). The known flora of the basin comprises 2037 species from 990 genera belonging to 180 families. The Cauvery river system harbors 1050 species belonging 128 families. 504 herbs (48%), 270 shrubs (25.7%), 170 trees (16.2%) other plant forms like climber, twinners etc constitutes 10%. The river basin is in human use since the beginning of the human civilization. As increase in the population growth intensified demands keep putting pressure on these riparian areas for agricultural development, recreational uses, commercial development, housing development and others.

The Cauvery river basin from headwater reaches to outlet exhibits remarkable habitat heterogeneity. The river is reserved by guilds of fish species. Headwater support more endangered fish which is confined to rock stream types having high gradients and predominantly bedrock substrates (Smakhtin *et al.*, 2006; Lakra *et al.*, 2010). The riparian zone in the sacred landscape provides habitat for wildlife such as Asian elephants (*Elephas maximus*), Otter species (*Amblonyx cinereus*) (near threatened) (Shenoy, 2005), Endangered Nilgiri languar (*Trachypithecus johnii*) (Sunderraj and Johnsingh 2001), Indian civet (*Viverricula indica*), Lion-tailed Macaque (*Macaca silenus*) and so on. The forest landscapes here act as corridors for wildlife, as they are in contiguous with large protected areas such as Nagarahole National Park, Talacauvery, Brahmagiri and Pushpagiri Wildlife Sanctuaries.

The river bordering the Cauvery Wildlife Sanctuary in lower reaches of the river has a population of otters, crocodiles and many varieties of fishes along with the famous Masheer. This area is the breeding ground for a number of reptilian species like crocodiles, turtles, python, cobra, russell's viper, banded krait and masheer fish besides wild boar, barking deer, four-horned antelope, green-billed malkoha, white-browed bulbul, pigmy woodpecker. Around 1000 elephants (*Elepha maximus*) graze through these riparian areas, as it also provides connectivity to Biligiri Rangan Hills Temple (BRT) wildlife Sanctuary and Mudumalai Tiger Reserve, which are in conjunction with Mysore – Nilgiri corridor (largest population of Asian elephants is found here).

7. *Terminalia arjuna* as a keystone species in Cauvery riverine ecosystem

Distributed throughout moist deciduous places of southern India, frequenting the banks of the water courses. Identified by thick grey smooth bark, exfoliating in large thin irregular sheets and buttressed trunk. It thrives best on loose moist, fertile alluvial loams and light deep sandy soils, often overlying more or less imprevious rock. The soil should have ample water supplies but should normally be well-drained. The soil under this tree becomes rich in calcium as the leaves are rich in this element. *Terminalia arjuna* species is deciduous, dominant canopy species and a representative riparian elements in riparian forests in lower reaches of Cauvery river. It can live grows to approximately 30m-45 in height, with a diameter at breast height (DBH) ranging from 300 cm – 600 cm. The *Terminalia arjuna* species is well adapted in the riparian zone by developing the buttressed type of root system to withstand the flood events.

Terminalia arjuna scattered along the lower of stretch of riparian forest is identified as a Keystone species. These scattered trees will acts as keystone structures as it supports wide array of species groups (e.g. arthropods, birds or mammals) for food resource and as shelter or nesting site (Munzbergova and Ward 2002; Plieninger *et al.*, 2003; Tews *et al.*, 2004).

8. Ecological significance of *Terminalia arjuna* in Cauvery river

8.1. Ecosystem engineers

Terminalia arjuna in the Cauvery riverine ecosystem can be referred as 'Ecosystem Emgineers" as it modifies the physical environment by releasing resources to be used by other species. The activities of many organisms provide habitat that would not otherwise be available, often by means of disturbance to the physical habitat. Because of structural alterations they support many organisms and are often referred to as ecosystem engineers (Jones *et al.*, 1994). *Terminalia arjuna* stabilizes river banks, trap sediments, increases nutrient availability in the top soil so

Figure 3. Species *Terminalia arjuna* growing along the banks of River Cauvery

as to provide a competitive advantage for adventitive forbs and grasses with higher nutrient requirements than their native counter parts.

Figure 4. Species *T. arjuna* with its interlocking root system

The interlocking root system of this tree reduces the efficiency of rivers to withstand flood events and the butresses roots of this species are effective soil binders. Thus play a significant role in modifying the physical environment in ways that release resources for other species. Flood is a regular event in the downstream of River Cauvery, *Terminalia arjuna* act as barrier against erosion and stabilizes river bank in the riparian forest. It is the lone species along the riparian corridor acting as an emergent layer with good amount of canopy contributing to maintenance of micro climatic conditions viz., soil moisture and nutrirents. It is also necessary for the survival of the other evergreen species such as *Olea dioca, Syzygium sp, Madhuca neriifolia, Madhuca latifolia* etc during the seedling and sapling stage in the lower riparian stretch. The laden and gravels retained between the roots of *T.arjuna* retains soil moisture required for vegetation establishment and also provides a new substrates for the colonization of riparian plants. This species with good canopy cover limits the establishment or invasive from the adjacent scrub and dry land harboring *Canthium sp, Alangium salviforum, Acacia catechu* etc., as potential dominants. Thereby competition with semi-evergreen species is avoided. Hence absence of this species along the riparian corridor might cause a major change in the riparian vegetation structure and composition.

9. Resource providers

Terminalia arjuna acts as resource provider, as the leaves and flowers of this species falling into the water form diet for a number of fishes. The tree-lined river bank also provides shelter and shade to fish. Shade also keeps the growth of water weeds in balance, and regulates the temperature of water. The smooth coated otter (*Lutra perspicillata*) categorized as 'vulnerable' by 2004 IUCN Red List in the Cauvery Wildlife Sanctuary (CWS), needs a healthy aquatic ecosystem with plenty of fish. The shade provided by trees along the water's edge help to promote fish abundance with obvious benefits for the otter. Besides, gaint trees of *Terminalia arjuna* in the riparian zone act as a good potential nesting sites for bees and numerous bats which roost during day time. The bats play an important role as pollinators and seed dispersal agents. The riparian vegetation in the middle reaches of the Cauvery river is fragmented by various types of anthropogenic pressures resulted in shrinkage of several endemic species in riparian zone (Sunil et al., 2010). As the larger forks and branches of *Terminalia arjuna* provides a habitat for natural pollinators like honey bees, bats etc., influences the chance of recovery of native species in the fragmented patches of the riparian buffer in the middle reaches of the river. The huge canopy offered by *T.arjuna* species forms a thick patches in the riparian zone serving as important buffers in the semi-arid ecosystem, enable to provide a vital links to sensitive wildlife species such as *Ratufa macroura* (grizzled giant squirrel), an IUCN Red listed –near threatened species (Baskaran et al., 2011) which demands thick canopy cover along the riparian zones for breeding and feeding purpose (Joshua and Johnsingh, 1994).

Figure 5. *T. arjuna* acting as the roosting sites for the bats during the day time

10. Control of Invasive species in riparian zones

Riparian habitats are more susceptible to exotic species invasion due to the nutrient rich laden sediments and periodic flooding followed by hydrochory (Pyse and Prach 1994; Gregory and Naiman 2000). Invasion of non-native species in the riparian zone constitutes most serious threats to the biodiversity through the displacement of native plants (Shigenari and Izumi 2004). The Cauvery river in the lower reaches is is surrounded by dry deciduous to scrub type forests, and moist deciduous to semi-evergreen type trees along the river bank. Since the riparian zone stands distinctly here by harboring moist deciduous to semi-evergreen type vegetation, during dry season they assumes a very significant place for wildlife (Natta et al. 2003) particularly to the otters and wide elephant herds found in the sanctuary. But, the riparian vegetation here stands in high risk areas, as there is a chance of invasion of several pioneer species resided in the adjoining dry deciduous and scrub type vegetation into the riparian areas (Manjunath, 2001). Some of the fragmented corridors in riparian forest has already witnessed the invasion of scrub type species by lessening the native riparian species (Sunil et al., 2011).

Riparian species demands shade and moisture in soil in the early stages of their germinations. Huge canopy offered by *Terminalia arjuna* provide sufficient shade and holds soil moisture during the germination stage of riparian tree species. Some native species which supports avifaunal abundance such as *Ixora bracheata, Syzygium cumini, Syzygium jambose, Diospyros melanoxylon* and *Madhuca latifolia* resembles healthy association to the keystone species *Terminalia arjuna*. Also, it checks the growth of pioneer species in the riparian zone, thereby competition with riparian and semi-evergreen species harbored in riparian zone is avoided. Decline in native species such as *Syzygium cumini, Syzygium jambose, Madhuca sp* along the river bank might lead to the decline of natural source of leaves, twigs, fruit and insects that underpins the aquatic food web (Lovett *et al.*, 2007). Hence, canopy species like *T.arjuna* is much inevitable in this region where their absence might cause a major change in the riparian vegetation structure and composition which inturn affects the aquatic ecosystem in the region.

Figure 6. A and B. Seedlings of *Syzygium cumini, Ixora bracheata, Madhuca latifolia* and *Dalbergia latifolia* growing in area under canopy of *Terminalia arjuna* species.

11. Social significance of *Terminalia arjuna* in Cauvery riverine ecosystem

The primary uses of Cauvery river are providing water for irrigation, household consumption, industries and the generation of electricity (Varunprasath and Daniel, 2010). Over 90% of the river water is abstracted for irrigation. Population density in Cauvery is perhaps among the highest in the world (350 people/ km²; Smakhtin et al., 2006) indicating that potential for human disturbance is inevitable along the basin. The watershed regions of the Cauvery river is strongly affected by water stress in recent years (Ferdin, et al., 2010). Besides meeting industrial and agricultural needs, drinking water demands from the two major urban centres namely Bangalore (6th largest city in India) and Mysore with a millions population is increasing at an faster rate. The river being completely dependent on the monsoon for replenishment, the amount of water the Cauvery can provide to the various users varies with the fluctuating strength of the monsoon rainfall (Ferdin, et al., 2010). Providing clean water and improving the chemical quality of waters for both human consumption needs and ecosystem health have become important policy goals in the worldwide. Management of riparian vegetation is one strategy to achieve these goals. *Terminalia arjuna* is one of the key species in the Cauvery river to fulfill the strategy to maintain the river quality healthier. The widespread rootmat of this species protect the waterway from erosion and pollutants entering the river. It acts has a natural wall along the river bank resists soil erosion during flooding thereby avoiding the water loss due to the bank widening. Keeping increasing water scarcity and flood disaster in the lower reaches during monsoon, conservation and management of *Terminalia arjuna* in the upper reaches helps to reduce flood velocities and increase the further flow towards lower reaches, thereby maintaining the river water healthier.

Acknowledgements

We thank University Grant Commission for providing financial assistance, Karnataka State Forest Department in for extending the permission to carry out the studies and helping in field work.

Author details

B.C. Nagaraja*, C. Sunil and R.K. Somashekar

*Address all correspondence to: nagenvi@gmail.com

Department of Environmental Sciences, Bangalore University, Bangalore, India

References

[1] Amitha Bachan, KH. 2003. *Riparian Vegetation along the middle and lower zones of the Chalakkudy River*, Kerala, India. Limnological Association of Kerala, Iringalakkuda.

[2] Baskaran, N., Senthilkumar. K and Saravanan, M. 2011. A new site record of the Grizzled Giant Squirrel Ratufa macroura (Pennant, 1769) in the Hosur forest division, Eastern Ghats, India and its conservation significance. Journal of Threatened Taxa 3(6): 1837–1841.

[3] Blom, CWPM., Bogemann, GM., Laan, P., van der Sman, A.J.M., van de Steeg, H.M. and Voesenek, LACJ. 1990. Adaptation to flooding in plants from river areas. Aquatic Botany 38: 29 - 47.

[4] Braatne, JP., Rood, SB., Simons, RK., Gom, LA., Canali, JE. 2002. Ecology of Riparian Vegetation of the Hells Canyon Corridor of the Snake River: Field Data, Analysis and Modeling of Plant Responses to Inundation and Regulated Flows. Technical Report Appendix E.3.3-3. Copyright © 2003 by Idaho Power Company.

[5] Darveau, M., Beauchesne, P., Belanger, L., Huot, J. and LaRue, P. 1995. Riparian forest strips as habitat for breeding birds in boreal forest. Journal of Wildlife Management 59: 67–78.

[6] Ferdin, M., Gorlitz, S., Schworer, S. 2010. Water Stress in the Cauvery Basin, South India - How current water management approaches and allocation conflict constrain reform. ASIEN 117: 27-44

[7] Ferreira, LV. 1997. Effects of the duration of flooding on species richness and floristic composition in three hectares in the Jau National Park in floodplain forests in central Amazonia. Biodiversity and Conservation 6: 1353 -1363.

[8] Gregory, SV., Swanson, FJ., W. A. McKee, and Cummins, KW. 1991. An Ecosystem Perspective of Riparian Zones. Bioscience, 41:540 - 551.

[9] Gregory HW, Naiman RJ. 2000. Vulnerability of riparian zones to invasion by exotic vascular plants. Plant Ecol. 148(1): 105–114.

[10] Gundersen, P., Lauren, A., Finer, L., Eva Ring, Koivusalo, H., Saetersdal, M., Weslien, JO., Sigurdsson, BD., Hogbom, L., Laine, J., Hansen, K. 2010. Environmental Services Provided from Riparian Forests in the Nordic Countries. Ambio: 39:555–566.

[11] Hylander, K. 2006. Riparian zones increase regional species richness by harboring different, not more, species: Comment. Ecology 87: 2126–2128

[12] Jayaram, KC. 2000. (Ed) *Kaveri Riverine System: An Environmental Study.* The Madras Science Foundation., Chennai. Pp. 1-6.

[13] John L. Sabo, Ryan Sponseller, Mark Dixon, Kris Gade, Tamara Harms, Jim Heffernan, Andrea jani, Gabrielle Katz, Candan Soykan, James Watts and Jill Welte. 2005. Riparian zones increase regional species richness by harboring different, not more, species. Ecology, 86 (1): 56 - 62.

[14] Jones, CG., Lawton, HJ. and Shachak, M. 1994. Organisms as ecosystem engineers. Oikos 69: 373 – 386.

[15] Joshua, J. and Johnsingh, AJT. 1994. Ecology of the endangered Grizzled Giant Squirrel (*Ratufa macroura*) in Tamil Nadu, South India, Report. Wildlife Institute of India.

[16] Kelsey, KA. and West, SD. 1998. Riparian wildlife. In: *River Ecology and Management. Lessons from the Pacific Coastal Ecoregion.* ed. R.J. Naiman and R.E. Bilby, Springer. pp. 235–258. New York.

[17] Klement Tockner and Stanford AJ. 2002. Riverine flood plains: present state and future trends. *Environmental Conservation.* 29 (3): 308–330

[18] Knopf, FL. and Samson, FB. 1994.+ Scale perspectives on avian diversity in western riparian ecosystems. *Conservation Biology.* 8: 669–676.

[19] Lakra, WS., Sarkar, UK. and Gopalakrishnan, A. 2010. Threatened Fresh Water fishes of India. Army Printing Press, Lucknow, India. pp 16- 24.

[20] Lovett S, Price P, editors. 2007. Principles for riparian lands management. Canberra (ACT): Land and Water Australia.

[21] Malanson, GP. 1993. Riparian landscapes. Cambridge studies in ecology. New York (NY): Cambridge University Press.

[22] Manjunath. 2001. "Management plan of Cauvery Wildlife Division" Kanakapura. Karnataka Forest Department.

[23] Menges, ES. and Waller, DM. (1983). Plant strategies in relation to elevation and light in floodplain herbs. The American Naturalist. 122: 454 - 473.

[24] Munzbergova, Z. and Ward, D. 2002. Acacia trees as keystone species in Negev desert ecosystems. Journal of Vegetation Science 13: 227–236.

[25] Naiman, RJ., H. Decamps, and M. Pollock. 1993. The role of riparian corridors in maintaining regional biodiversity. Ecological Applications. 3: 209-212.

[26] Naiman, RJ., Decamps, H. and McClain, ME. 2005. River Ecology and Management: Lessons from the Pacific Coastal Ecoregion. Springer-Verlag, New York.

[27] Natta, AK., Sinsin, B., Van der Maesen, LJG. 2003. Riparian forests and biodiversity conservation in Benin (West Africa). Paper submitted to XII World Forestry Congress, 2003, Quebec City, Canada.

[28] Nilsson, CM., Gardfjell, M. and Grelsson, G. 1991. Importance of hydrochory in structuring plant communities along rivers. Canadian Journal of Botany. 69: 2631-2633.

[29] Parson, A. 1991. The Conservation and Ecology of Riparian Tree Communities In the Murray Darling Basin, NSW A Review' NSW NPWS Hurstville.

[30] Plieninger, T., Pulido, FJ. And Konold, W. 2003. Effects of land-use history on size structure of holmoak stands in Spanish dehesas: implications for conservation and restoration. Environmental Conservation 30: 61–70.

[31] Pyse, P. and Prach K. 1994. How important are rivers for supporting plant invasions? In: De Waal L, Child LE, Wade PM, Brock JH, editors. Ecology and management of invasive riverside plants. New York (NY): Wiley. p. 23–31.

[32] Risser, PG. 1990. The Ecological Importance of land water ecotones' [in Naiman, R.J. and Decamps H. (eds) "The Ecology and Management of Aquatic - Terrestrial Ecotones'] UNESCO, Paris; Parthenon Publishing Group. Man and the Biosphere Series

[33] Shenoy, K. 2005. Otters in River Cauvery, Karnataka. New Delhi:Wildlife trust of India.

[34] Shigenari, M., Izumi, W. 2004. Invasive alien plant species in riparian areas of Japan: the contribution of agricultural weeds, revegetation species and aquacultural species. Global Environ Res. 8(1):89–101.

[35] Skagen, SK., Melcher, CP., Howe, WH. and Knopf, FL. 1998. Comparative use of riparian corridors and oases by migrating birds in Southeast Arizona. Conservation Biology. 12: 896 – 909.

[36] Smakhtin, V., Arunachalam, M., Behera, S., Chatterjee, A., Das, S., Gautam, P., Joshi, GD., Kumbakonam G. Sivaramakrishnan, KG. and Unni, KS. 2006. Developing the procedures for assessment of ecological value and condition of Indian Rivers in the context of Environmental Water Demand (pp. 9-10, 13). Retrieved 8 March 2008 from http://nrlp.iwmi.org/ pdocs/DReports/phase_01/18.

[37] Sunderraj, SFW. and Johnsingh, AJT. 2001. 'Impact of biotic disturbance on Nilgiri langur habitat, demography and group dynamics.' Current Science 80 (3): 428–436.

[38] Sunil, C., R.K. Somashekar, RK. And B.C. Nagaraja, BC. 2011. Impact of anthropogenic disturbances on riparian forest ecology and ecosystem services in Southern India, International Journal of Biodiversity Science, Ecosystem Services & Management.

[39] Sunil C, Somashekar RK, Nagaraja BC. 2010. Riparian vegetation assessment of Cauvery River Basin of South India. Environ Monit Asses. 170(1–4):545–553.

[40] Tews, J., Brose, U., Grimm, V., Tielborger, MC., Wichmann, MC., Schwager, M. and Jeltsch, F. 2004. Animal species diversity driven by habitat heterogeneity / diversity: the importance of keystone structures. Journal of Biogeography 31:79-92.

[41] Tiker, GM. and Evans, MI. 1997. Habitats for birds in Europe: A conservation strategy for the wider environment. Birdlife Conservation Series 6. Cambridge, UK: Birdlife International

[42] Tockner, K. and Stanford, JA. 2002. Riverine flood plains: present state and future trends. Environmental Conservation 29: 308-330.

[43] Varunprasath. and Daniel, AN. 2010. Comparison Studies of Three Freshwater Rivers (Cauvery, Bhavani and Noyyal) in Tamilnadu, India. Iranica Journal of Energy and Environment 1 (4): 315-320.

[44] Walter, T., Umbricht, M. and Schneider, K. 1998. Datenbank zur Fauna der Auen. FAL, Reckenholz, Switzerland [www document]. URL http://www.admin.ch/sar/fal/aua/

[45] Warner, RF. 1983. Channel Changes in Sandstone and Shale Reaches of the Nepean River, NSW' in Nanson, G.C. and Young, R.W. (eds) 'Ecology and Management of Riparian Zones' Marcoola QLD Occasional Paper Series LWRRDC.

Chapter 6

Evolution, Biodiversity and Ecology in Microbial Communities: Mathematical Modeling and Simulation with the "Haploid Evolutionary Constructor" Software Tool

Sergey A. Lashin, Yury G. Matushkin,
Alexandra I. Klimenko, Valentin V. Suslov and
Nikolay A. Kolchanov

Additional information is available at the end of the chapter

http://dx.doi.org/10.5772/58302

1. Introduction

Life on Earth exists in the form of interacting communities of two types: populations and ecosystems. Populations maintain reproduction and Darwinian selection. Substance and energy exchange both with the environment and between populations is one of the major functions of ecosystems, which, as a result, enables the necessary conditions for organism reproduction [1–3]. Consequently, it is reasonable to consider any evolutionary process both in terms of population and in terms of ecosystems. Therefore, the evolutionary success of any given genotype carrier is related not only to fixation in population, but also the influence of such a population on functioning of the ecosystem. It is impossible to understand patterns of certain species' evolution without considering the trophic structure of the ecosystem, as a part of which they exist. An ecosystem enables the sparing use of environmental resources by setting up various linear or branched trophic chains [4]. When several linear chains or their branches integrate, a trophic cycle can arise, which enables reproduction of one or another resource [5–7]. On the other hand, an unusable resource – a metabolic cul-de-sac – can arise in ecosystems, which is buried or removed by the flow [8–10].

Symbiotic relationships are widespread in ecosystems, and especially specific to prokaryotic communities. These communities often have taxonomical diversity and are characterized by a complicated trophic chain with a certain extent of closeness. Specific traits of

prokaryotes, such as a pinotrophy diet, small cell size and low speed, lead to an organism's inability to escape from the surrounding environment quickly, and to the "biogenic desert" problem [11], when organisms that inhabit the center of a high-density population have to go hungry due to lack of nutrients. Therefore, the characteristic feature of prokaryotic ecosystems is **metabolism integration** – from the close association of single organisms of different species [12] to spatially-structured, trophically highly-closed ecosystems of meromictic [13] and soda lakes [14], and, after all, global biogeochemical cycles of the biosphere [15]. The large majority of prokaryotes in nature exist as a part of communities with a complicated structure – bacterial mats or biofilms, whose common metabolite pool often forms a complete cycle [15,16]. The presence of such cycles optimizes community members' metabolism. Trophic rings, within which metabolic products of one species or strain are used or can in certain conditions be used for food by others, are revealed in bacterial biofilms and in complex metabolic graphs, reconstructed on the basis of metagenomic projects [17,18].

Close bacterial association in biofilms enhances the probability of horizontal gene transfer between different bacteria, which enable them to obtain new features [19–21]. The importance of the horizontal transfer is supported by recently discovered natural vectors for cloning exogenes in prokaryotic genomes – integrons and superitegrons [22,23]. Therefore, most prokaryotes exist as a part of communities. A wide variety of prokaryotes are unable to grow in pure culture (uncultured prokaryotes) confirms this [24]. The evolution of such highly-integrated communities has its qualitative specific factors and cannot be reduced to the evolution of distinct populations composing them. Thus, while the reproduction rate of prokaryotes is exceptionally high, an experimental study of prokaryotic evolution is difficult, as it requires the study of the whole prokaryotic community. Accordingly, the mathematical modeling of evolutionary processes, adjusted for different types of trophic interactions, spatial distribution of organisms, genetic structure of populations, speciation, different reproduction schemes, environmental influence and other factors, is one of the main methods for the study of the evolutionary process. The modeling of evolution is one of the primary challenges of 21st century biology, mathematics and computer science.

Traditional approaches to evolutionary and population process modeling include methods of population dynamics [25,26] and methods of population genetics [27,28]. Population dynamics modeling methods describe population size changes through time subject to environmental conditions, trophic interactions between populations and other features, but as a rule, population genetic structure changes cannot be studied using these methods. As to population genetics methods, they are generally based on methods of the probability theory and mathematical statistics [29–31]. They allow for studying the evolution of population genetic structure, but do not provide means for the modeling of the population dynamics process in detail.

Further development of modeling and simulation methods led to so-called "hybrid" methods [32,33], which allow us to investigate changes both in the size and genetic structure of a population simultaneously. However, it should be noted that the large majority of these methods (like the other population genetics methods) are focused on diploid organism population modeling (generally, with sexual reproduction).

Methods of "portrait" (individual-oriented, agent-based) modeling [34–36] bridge the gap between individual characteristics and community structure, evolving as a consequence of some or other rules of interactions between individuals. These approaches are illustrative enough, but may have a high computational complexity (ideally, they require a detailed description of every individual in the population). Besides, most of them are static, i.e. during calculations their structure can be changed only within the limits predetermined by the developer. As a result, if modern computer capacity is sufficient for modeling of a population of diploid organisms with sexual reproduction (whereas the effective size of such populations is generally no more than 100-1000 individuals), then direct simulation modeling is often problematic for the modeling of a population of haploid organisms, particularly bacteria (the effective size of bacterial populations is 10^6-10^9 individuals).

With due consideration to the abovementioned requirements of the evolutionary and population process modeling tool, we previously developed a modeling method and software package "Haploid evolutionary constructor" (HEC) [37,38]. The HEC provides tools to simulate the functioning of a haploid organism population network, trophically linked with substrate-product relationships under the environmental effect. This modeling approach provides a means of simultaneously describing the prokaryotic community at various levels of its biological organization: genetic, metabolic, population and ecological, flexibly varying the degree of description detail at any level. During the model simulation, each population may vary both its size and genetic diversity due to selection and mutations. The key feature of this methodology is the ability to model such evolutionary and population processes that require an intense structure rearrangement of a model during simulation. Such processes contain, for example, horizontal genetic transfer and speciation. Besides, the methodology offers the possibility to describe the polymorphism of one or several genes in a population, where the number of alleles can be changed during the simulation process.

In summary, the HEC provides comprehensive study of the bacterial community model, analyzing the dynamics of changes in allelic frequency, the population size, the concentration of metabolites, the community trophic structure and its evolution, including stochastic genetic factors.

During model simulation via the HEC, the number of populations reached 300, with an approximate total 10^{20} individuals, the size of distinct population came to 10^{18}, with around $10^6 - 10^7$ different genotypes (up to 4000–15000 different genotypes within one population in case of multiple polymorphism of 10–15 genes).

2. Haploid Evolutionary Constructor methodology description

The HEC methodology provides modeling and simulation of biological and evolutionary processes in trophically linked communities of unicellular haploid organisms. Figure 1 shows the scheme of the main HEC objects and processes. The HEC simulates trophic interconnected haploid organism networks that are combined into populations according to genetic proximity and reside in one whole volume termed *environment*. Organisms may consume and *utilize* substrates and synthesize, and then *secrete* products into the environment, which inversely can be used by other organisms as substrates. Some substrates favorably affect population growth, others, on the contrary, may have an inhibiting effect. The efficiencies (as rate constants) of substrate utilization and product production are controlled by certain genes.

So-called *nonspecific* substrates are reproduced by the in-flow (Figure 1 shows scheme with one nonspecific substrate N_1). By contrast substrates running into the environment only through cell activity (synthesis and secretion) are known as *specific* (represented by S_i in Figure 1).

2.1. Environment

Environment is a bounded flow system of fixed volume V_{total}, containing all populations and substrates. The environment is also a mediator of relationships between populations (and substrates); the inflowing and outflowing processes of both substrates and cells are connected with it. The environment is characterized by following variables:

- V_{total} – environment capacity (in liters);

- $k_{kenv,flow}$ – flow rate (in % V_{total} per unit time);

- $N_{env,i}$ – concentration of the ith nonspecific substrate in the environment (in mM);

- $S_{env,i}$ – concentration of the ith specific substrate in the environment (in mM);

- $N_{flow,i}$ – concentration of the ith nonspecific substrate in the flow (in mM);

and the following processes:

substrates inflow into the environment – increases a nonspecific substrate concentration according to the flow rate and concentration of these substrates in the flow;

substrates outflow from the environment – reduces both nonspecific and specific substrate concentrations according to the flow rate;

the inflow/outflow of nonspecific substrates follows the formulas:

$$N_{env,i}(t+1) = N_{env,i}(t) + k_{env,flow} \cdot \left(N_{flow,i} - N_{env,i}(t) \right) \tag{1}$$

the specific substrates outflow follows the formula below (specific substrates flow into the environment associated with these substrates through cell synthesis as described below):

$$S_{env,i}(t+1) = S_{env,i}(t) \cdot \left(1 - k_{env,flow} \right) \tag{2}$$

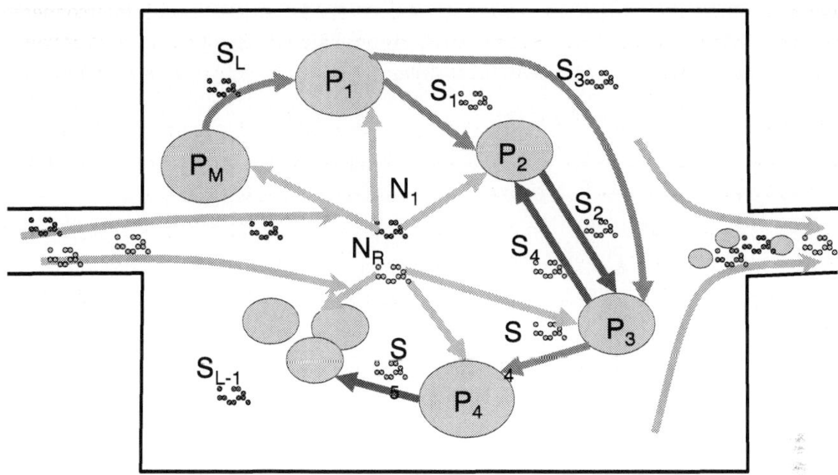

Figure 1. Scheme of HEC objects and processes. Blue circles represent populations P1, P2, ... PM. Groups of colored dots represent substrates: specific S_1, S_2, ... S_L and nonspecific N_1. Arrows between populations represent trophic relationships: the population, from which the arrow comes, produces the substrate, used by the population to which this arrow leads. Red arrows represent a substrate's activating effect, blue arrows – an inhibiting effect. The usage of a nonspecific substrate is indicated by orange arrows. The flow action is represented by the thick arrow in the lower left-hand corner of the figure. The bounded area represents the environment.

2.2. Population modeling

We consider populations to be a set of cells, that have common substrate using and producing properties. Cells are considered to belong to the same population (same species) if they:

1. utilize the same variety of nonspecific and specific substrates;

2. produce the same variety of specific products;

3. have the same trophic strategy;

4. have the same synthesis strategy.

We define basic terms and notions, used hereinafter as follows:

- **trait** – any given substrate synthesis or utilization rate constant. Every trait is considered as unambiguously controlled by one **gene**. In this particular case, the **gene** is considered a unit of inheritance;

- **allele** – a gene variant, i.e. a particular value of the corresponding constant;

- **individual's genotype** is the set of alleles, divided into five groups. The first group (c_i) characterizes the efficiency of specific substrate utilization (s_i), the second group (d_i) – substrate production rate, the third group (r_i) – efficiency of nonspecific substrate utilization, the fourth group (m_i) – efficiency of immunity against phages, and the fifth group (v_i) – phage (viruses) genes;

- **mutation** is a change of a corresponding trait value, which can be interpreted as a gene shifting into another state (allele).

Using the introduced terminology, the concept of a **monomorphic population** can be formed – a population of "genetically identical" cells, where all cells have corresponding genes represented by the same alleles. The genotype, common for all the cells of such population, is called the **monomorphic population genotype**.

In order for cells to use substrates for their reproduction and population size growth along with products synthesis, at first they have to get these substrates from environment. In the HEC, the process of substrate consumption is described by the particular step, where a cell's requirement of various substrates and the availability of substrates in the environment are taken into consideration. In case of any substrate deficiency, a competition for this substrate may occur either intrapopulation or interpopulation (when cells of several populations may use the same substrate at the same time). In case of substrate excess, the maximum amount of the substrate consumed by one cell is defined by the value of the substrate **consumption rate**. This value is species-specific and cannot be changed due to mutation (in this HEC version) – this may be equivalent to, for example, the size limit for one cell. Hence, monomorphic population is additionally characterized by the amount of substrate molecules consumed.

Trophic strategies is a term for the formulas and laws, describing population changes in a single generation depending on population size, the amount of consumed substrates, the flow rate, mortality rate and other factors. Examples of trophic strategies are illustrated by the equations below (other formulas may be also used, including those defined by the user):

$$\Delta P = F_1(\vec{N},\, \vec{S},\, \vec{R},\, \vec{C},\, P) = \sqrt{r_0 n_0(P) \bullet \sum_{i \in I_{consumed}} c_i s_i(P)} - k_{flow} \bullet P - k_{death} \bullet P^2 \tag{3}$$

$$\Delta P = F_2(\vec{N},\, \vec{S},\, \vec{R},\, \vec{C},\, P) = P \bullet \frac{\left(\dfrac{n_0/P}{K_{01(r_0)}}\right)^{\gamma_0}}{1 + \left(\dfrac{n_0/P}{K_{02(r_0)}}\right)^{\gamma_0}} \bullet \prod_{i \in I_{consumed}} \frac{\left(\dfrac{s_i/P}{K_{i1(c_i)}}\right)^{\gamma_i}}{1 + \left(\dfrac{s_i/P}{K_{i2(c_i)}}\right)^{\gamma_i}} - k_{flow} \bullet P - k_{death} \bullet P^2 \tag{4}$$

$$\Delta P = F_3(\vec{N},\, \vec{S},\, \vec{R},\, \vec{C},\, P) = a_{basal}(n_0) \bullet P - \sqrt{\sum_{i \in I_{consumed}} c_i s_i(P)} - k_{death} \bullet P^2 \tag{5}$$

where

$I_{consumed}$ – set of indices of substrates consumed;

n_0 – amount of nonspecific substrate, consumed by the cells of population from the environment (in proportion to the population size);

$\vec{N}$ – vector of specific substrates, consumed by the cells of the population from the environment (in proportion to the population size) values;

r_0 – utilization rate for the unique nonspecific substrate (trait, controlled by the corresponding gene);

$\vec{S}$ – vector of corresponding specific substrate utilization rates (traits, controlled by correspondent genes);

P – population size;

k_{flow} – flow rate in the environment ("washout" rate);

k_{death} – population mortality rate;

a_{basal} – "natural increase" of the population;

$\gamma,\ \gamma_0,\ \gamma_i$ – coefficients, describing the nonlinear nature of substrate influence on population growth;

K_{ij} – coefficients, describing the efficiency of substrate influence on population growth (depends on corresponding traits).

The equation [eq.3] governs the utilization process of several specific and one nonspecific substrate, where the latter is essential for cells – when it is not available, the population does not grow. Substrates have strong cooperativity. Besides, some substrates are able to compensate in some degree the lack of other necessary substrates, including a nonspecific substrate deficiency. The trophic strategy described in [eq.3], satisfies the **Rubel's law of replaceability of ecological factors** [39], and is called the **compensatory trophic strategy**.

The equation [eq.4] again governs the utilization process of several specific and one nonspecific substrate. Nevertheless, every substrate is essential. A deficiency of one substrate cannot be compensated by an excess others. The trophic strategy described by [eq.4] satisfies the ecological **Liebig's law of the minimum** [4] and called the **noncompensatory trophic strategy**.

The equation [eq.5] governs the utilization process of one nonspecific substrate coupled with the inhibiting effect of specific substrates on population growth. Besides, this effect is cooperative. This trophic strategy is called the **inhibitory trophic strategy**.

Similar to the concept of a monomorphic population, we define the concept of **polymorphic population**, which can be regarded as a set of monomorphic subpopulations. Cells in a polymorphic population have the same gene variety, while different cells can have different alleles of one or several genes. The polymorphic population is characterized by the "generalized genome" – a set of population's **genetic spectra**. The **genetic spectrum** is the distribution of allele occurrence frequencies in a population (for one gene) (Figure 2).

Mutation in terms of a genetic spectrum means the change of its profile (thus, the formation of a new allele is possible, Figure 3).

The change of the polymorphic population size is calculated according to the following scheme. The polymorphic population is split into many monomorphic subpopulations. Then,

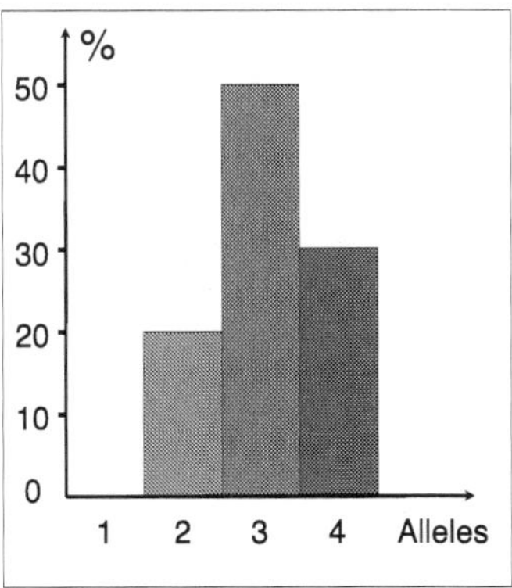

Figure 2. The genetic spectrum shows that the given trait value is 2 for 20% of individuals in this population, 3 for 50%, and 4 for 30%.

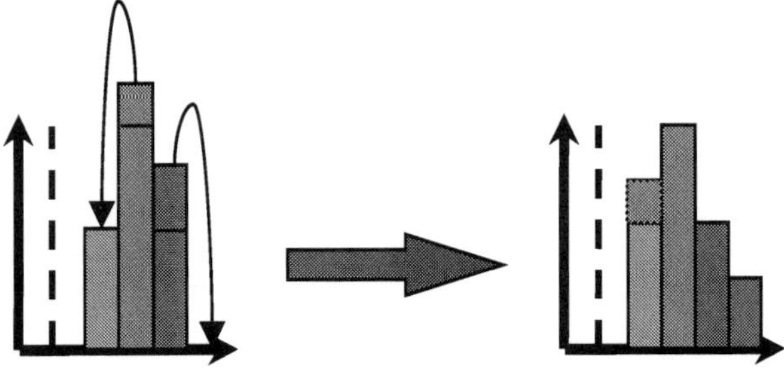

Figure 3. Change of genetic spectrum due to mutations.

the growth for each subpopulation is calculated according to formulas similar to the above-mentioned (eq.3-5). Next, the monomorphic populations are merged into a polymorphic population. The polymorphic population is split based on the proportion of every allele for all genes in the population genotype. The substrates consumed by the polymorphic population are divided proportional to the sizes of the single monomorphic populations. It noteworthy, that each single monomorphic population growth may differ markedly from the growth of other subpopulations. It depends on the monomorphic population genotype, population size and the amount of certain substrates consumed by the population. Consequently, the proportion of alleles in a population may change (which may be interpreted as the adaptation of the population to certain conditions).

2.3. Metabolism

When a cell synthesizes a product, that it can utilize itself, it is obvious that there is "no use" in secreting this product into the environment and then competing for it on "equal terms" with the other populations. For this reason, we consider two forms (states) of substrates in the modeling of the internal cell substrates. The first form are the substrates that are **"ready for utilization"**, the second form are **synthesized** substrates. The substrates of the first group are replenished by substrate consumption from the environment and through transition of synthesized substrates (if the cell can synthesize it). During reproduction and product synthesis, the substrates of the first group are expended. The substrates of the second group are replenished only by synthesis. A principal scheme of cell metabolism in a population, consisting of the stages of substrate consumption, utilization, synthesis, and secretion is illustrated by (Figure 4).

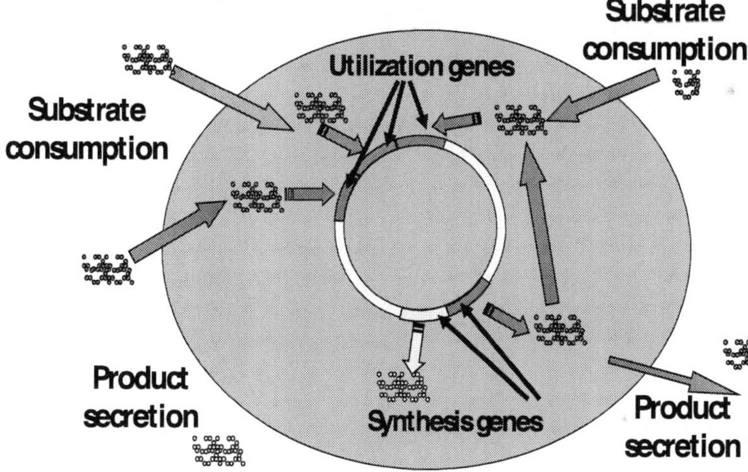

Figure 4. A scheme of trophic processes through the example of one cell. The cell utilizes three types of substrates and synthesises two types of substrates, one of which it can utilize itself. A synthesized substrate, which cannot be utilized by the cell (yellow) is comprehensively secreted into the environment, while the second substrate (red) is partly used by the cell itself (consequently, there is less or no secretion into the environment).

Product synthesis by the cells of a polymorphic population is described by the gene network model of metabolite synthesis, which we call the **synthesis strategy** (Figure 5).

An example of a formula describing a simple synthesis strategy:

$$\Delta s_i = P \cdot \sum_{j \in Spectr_i} d_{ij} s_i \left(P_j / P \right) \tag{6}$$

where

Δs_i – amount of synthesized i-type substrate;

d_{ij} – trait value in genetic spectra $Spectr_i$;

P – population size;

P_j – proportion of individuals having d_{ij} trait value in population (in the genetic spectra).

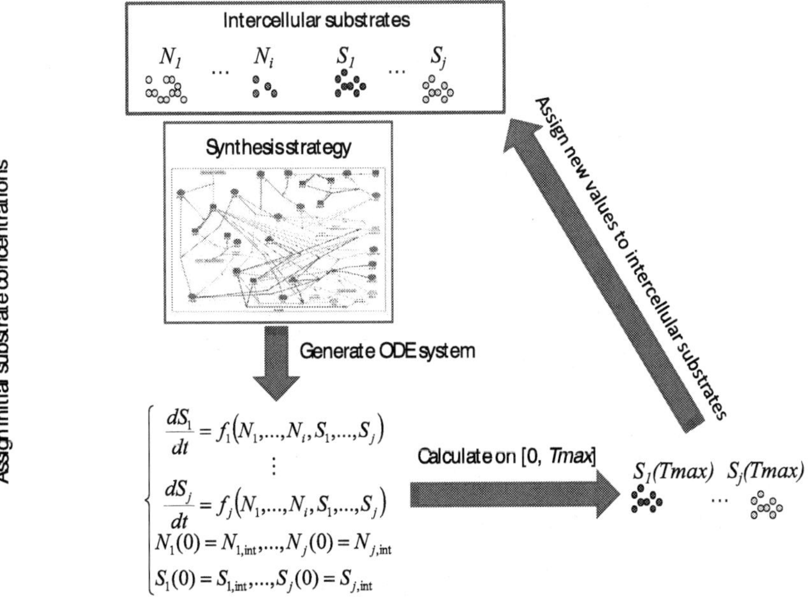

$$\begin{cases} \dfrac{dS_1}{dt} = f_1\!\left(N_1,\dots,N_i,S_1,\dots,S_j\right) \\[4pt] \quad\vdots \\[4pt] \dfrac{dS_j}{dt} = f_j\!\left(N_1,\dots,N_i,S_1,\dots,S_j\right) \\[4pt] N_1(0) = N_{1,\mathrm{int}},\dots,N_j(0) = N_{j,\mathrm{int}} \\[4pt] S_1(0) = S_{1,\mathrm{int}},\dots,S_j(0) = S_{j,\mathrm{int}} \end{cases}$$

Figure 5. Common pattern of synthesis strategies calculation.

2.4. Infection process modeling

To model a phage infection, the HEC provides extended objects of polymorphic populations: *polymorphic phage populations* and the *polymorphic population of infected cells* (the "normal" polymorphic populations are therefore regarded as "healthy").

The infection modeling includes the following phases: infection of the healthy cells through phage penetration from the environment into one part of the cell population, phage reproduction inside of the infected cells, and finally, the phages burst into the environment after the lysis of cells. The infected cells form polymorphic populations, further reproduction of which may follow a lytic or a lysogenic pathway. The lytic pathway means death (lysis) of the infected cells with synchronous phage formation and their transportation into the environment (the number of phages depends on their profusion). The lysogenic pathway means prophage formation and no phage transportation into the environment follows. At the same time the population of the infected cells reproduces like an ordinary polymorphic population in the HEC (i.e. according to the trophic strategy), acquiring immunity to that type of phage (Figure 6).

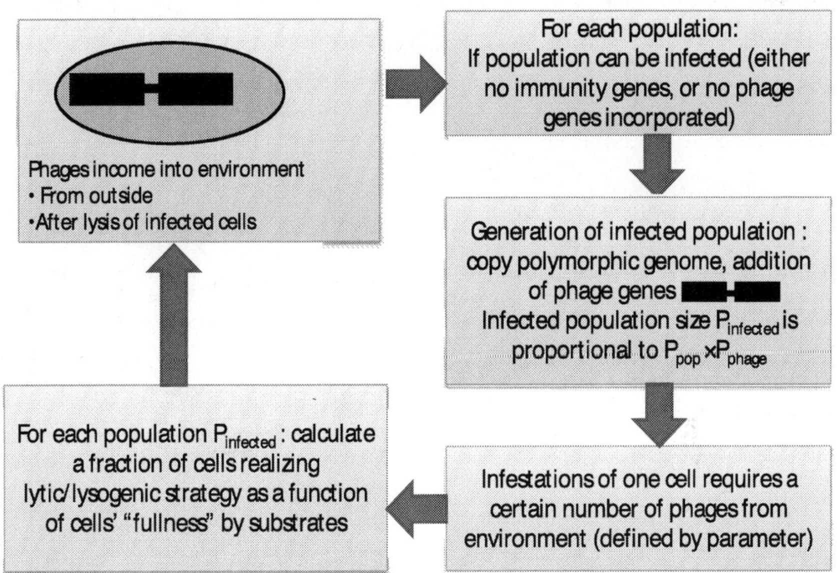

Figure 6. Scheme of phage infestation.

The choice of the lysogenic or the lytic scenario depends on the conditions of the cell population at the moment of infection: optimum conditions lead to a lytic type, pessimal conditions– to a lysogenic. In the latter case, a part of the population randomly switches to the lytic form if conditions improve, causing the death of this part and phage generation (Figure 7).

The polymorphism of phages, the formation of new strains owing to mutations and competition between strains are also described via the genetic spectra arithmetics. However, in the phage populations unlike prokaryotic populations, genes define a specific virulence (an ability to infect certain populations) and abundance (the number of copies per lysed cell).

3. Simulation of prokaryotic communities via the Haploid Evolutionary Constructor

Through the use of the HEC software tool, we have simulated a number of biological models of the functioning and evolution of unicellular haploid organism communities [37]. Inter alia, the correspondence of the modeling results to biological data as well as previously published mathematical models has been illustrated. We estimated the key parameters of the model, regarding cell size, number of substrates required for cell division and other factors based on *E.coli* cell information [40].

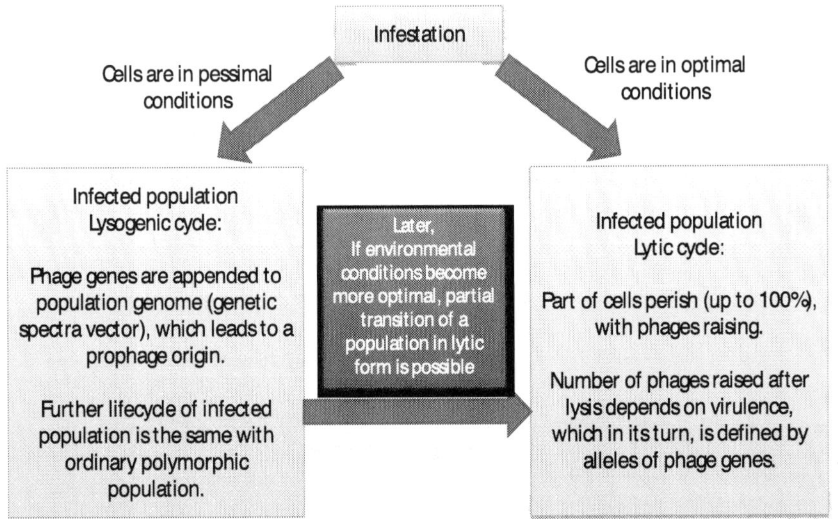

Figure 7. Scheme of the choice of the lysogenic or lytic scenarios.

To verify the modeling approach, a number of basic models were considered, and the obtained results confirmed correspondence to both classical models and experimental information [38]. Thus, the models examined proved the biological adequacy of the HEC and its applicability for a wide range of population, genetic and ecological problems.

3.1. Modeling of biodiversity dynamics and adaptability in bacterial communities

The evolutionary success of a biological system relates to the balance of two characteristics: its stability (ability to preserve its structure and homeostasis despite changes in the environment) and its adaptability –ability to preserve an evolutionary flexibility in response to uncontrolled environmental changes. Traditionally, communities that are more complex are considered to be more stable, but there can be exceptions. The study of model dynamics, when the population size growth of one species depends on the population size of others species in the community, confirms that increased complexity leads to the improvement of stability only when community connectivity increases at the same time.

We maintained a comparative simulation of stability dynamics, the adaptability and biodiversity of trophic closed communities with compensatory and non-compensatory metabolism (according to Rubel and Liebig respectively). The trophic strategies formulas representing these laws [eq.3,4] are mentioned in the HEC models description. The deficiency or low concentration of a nonspecific substrate in the environment leads to population extinction for both strategies. However, in case of compensatory nutrition strategy, the deficiency (a low concentration, but not a complete lack) of a nonspecific substrate in the environment can partly be compensated by the high concentration of specific substrates. This satisfies Rubel's law of replaceability. In case of non-compensatory nutrition strategy, the deficiency of any substrate cannot be compensated by the extra concentration of other substrates. This case satisfies Liebig's law of the minimum.

It is known that the change of conditions on Earth shows a certain cyclism. Its sources are geophysical and astronomical cyclic processes [41,42]. As a result, the input of matter and energy into the ecosystem tends to change. To keep the common value of this flow constant, the ecosystem must conduct an evolutionary search for new sources of matter and energy [6,43]. Hence, the "learning ability" of the ecosystem becomes a critical parameter. The horizontal gene transfer (HGT) in prokaryotic populations is a relatively easy way to carry out such "learning". The simulation showed that the trophic structure of a community imposes substantive restrictions on the benefits of the HGT – a long-term effect is possible only in case of HGT between populations, whose metabolisms are sufficiently rich and flexibly adjustable (compensatory trophism). For populations with simplified metabolisms (non-compensatory trophism), the HGT offers only localized advantages. This contradicts the assumption that the major trend of prokaryotes evolution is individual genome simplification, compensated by relationship amplification in the bacterial community. Such simplification in the long term leads to the death of the community.

The simulation showed dramatically better adaptivity of trophic rings with compensatory (TRC) trophism in comparison with non-compensatory ones. The fixation of beneficial mutations even in one TRC population improved stability of the entire system, significantly extending its lifetime, i.e. offering an additional chance to wait out the starvation, or even completely saved the TRS from extinction due to the metabolism optimization. Nevertheless, many taxa in the world are unable to compensate one resource with another on a broad scale. Why is the major trend of the evolution on Earth increasing biodiversity by means of progressive specialization, rather than the formation of biota based on several taxa-generalists? Let us assume that "learning ability" is a critical parameter for evolution within an ecosystem. Subsequently, in case of a low biodiversity level in an ecosystem with compensatory trophism, it is entirely possible that new sources of energy and matter will never be found. In fact, the probability of finding such sources is higher in ecosystems that preserve a high level of biodiversity until extinction. Then, if the value of the initial nutrient flow is recovered, a new source will be added, providing the ecosystem with resources for further progressive evolution. Therefore, in the long term Liebig's systems have an advantage over Rubel's. It should be noted that this advantage has, as with all evolutionary processes, a nondeterministic, stochastic nature, while the stability of the advantage is determined. Thus, when biodiversity is high, the system may die out for stochastic reasons. This matches paleontological information. A permanent rotation of hegemonic biotas, without sacrificing comparably small amounts of epibiotic ecosystems, accumulating virtually the entire range of biochemical activities (cyanobacterium tufts, alkalophilic communities), is observed in the fossil records.

3.2. Evolutionary trends of genome sophistication and simplification

One of the most important periods in the evolution of life on Earth is eukaryotic cell formation [44]. The comparative study involving genome-wide information for bacteria, archaea and eukaryotes suggest that the microbial eukaryotic domains could not be inherited from the ancestors of mitochondria and plastids, but were borrowed from other bacteria [45]. In such a case, the origin of eukaryotes is a consequence of the autonomism of one member of the complex synotrophic prokaryotic community through the looping back of major regulatory interactions. It is considered that such autonomism is the result of symbiosis of several types of prokaryotes, and it is fairly probable that the entire series of HGT between symbionts took place.

3.2.1. Genome amplification

By applying the HEC, we investigated the evolution of a community, which at the beginning of computations constituted a trophic ring, in which each of the three populations consumed and produced precisely one specific substrate. For example, the first population produced the substrate consumed by the second, and the second produced the substrate consumed by the third population, etc. (Figure 8).

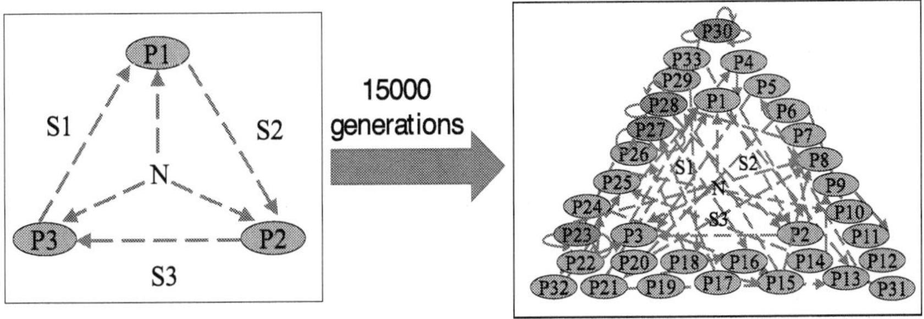

Figure 8. The trophic ring scheme consisted of three populations at start time (left) and after 15000 generations (right).

During calculations with a certain probability between cells of different populations (10^{-7} per generation per cell), HGT could take place. As a result, new populations, either consuming a larger number of specific substrates or producing a larger number of specific products would be formed. It was found, that in the longer term (after 10000–15000 generations), the population with the "most complete genome" (i.e. populations consuming and producing the maximum possible number of specific substrates in the given trophic system), or the population with an "almost complete genome" gain an extreme biomass advantage at the specified conditions. Namely, at modetare genome-length-penalty level values (0.01-0.05), and with stable poor environmental conditions (nonspecific substrate concentration in the inflow is around 10^{-4} mM, i.e. at the survival minimum for parent population cells) (Table 1). In the long-term, such populations replaced all other populations from the trophic system ("outsider" populations either died or reached a maximum number in the environment that hovered around 10-100 individuals) (Figure 9).

from: http://www.annualreviews.org/doi/abs/10.1146/annurev.genet. 38.072902.091216?journalCode=genet

[19] Ochman H, Lawrence JG, Groisman EA. Lateral gene transfer and the nature of bacterial innovation. Nature [Internet]. Macmillian Magazines Ltd.; 2000 May 18 [cited 2013 Nov 5];405(6784):299–304. Available from: http://dx.doi.org/10.1038/35012500

[20] Shestakov S. On the Early Biological Evolution from the Viewpoint of Genomics. Paleontol. J. [Internet]. 2003 [cited 2013 Nov 5];(6):50–7. Available from: http://www.maik.ru/abstract/paleng/3/paleng6_3p609abs.htm

[21] Gogarten JP, Townsend JP. Horizontal gene transfer, genome innovation and evolution. Nat. Rev. Microbiol. [Internet]. Nature Publishing Group; 2005 Sep [cited 2013 Oct 30];3(9):679–87. Available from: http://dx.doi.org/10.1038/nrmicro1204

[22] Rowe-Magnus DA, Guerout A-M, Mazel D. Bacterial resistance evolution by recruitment of super-integron gene cassettes. Mol. Microbiol. [Internet]. 2002 Mar [cited 2013 Nov 5];43(6):1657–69. Available from: http://doi.wiley.com/10.1046/j. 1365-2958.2002.02861.x

[23] Ilyina TS. Bacterial Superintegrons, a Source of New Genes with Adaptive Functions. 2006;42(11):1294–302.

[24] Amann R, Ludwig W, Schleifer K. Phylogenetic identification and in situ detection of individual microbial cells without cultivation. Microbiol. Rev. [Internet]. 1995 Mar 1 [cited 2013 Nov 5];59(1):143–69. Available from: http://mmbr.asm.org/content/ 59/1/143.short

[25] Pielou EC. An introduction to mathematical ecology [Internet]. 1969 [cited 2013 Nov 5]. Available from: http://books.google.ru/books/about/An_introduction_to_mathematical_ecology.html?id=apprpg3cdfgC&pgis=1

[26] Murray JD. Mathematical Biology®: I. An Introduction, Third Edition. 3rd ed. Antman SS, Mardsen JE, Sirovich L, Wiggins S, editors. New York: Springer; 2002. p. 551.

[27] Ewens W. Mathematical population genetics: I. Theoretical introduction. 2004 [cited 2013 Nov 5]; Available from: http://www.google.com/books?hl=ru&lr=&id=twXIyX-yod2MC&oi=fnd&pg=PR7&ots=DMOTg871XN&sig=8mpfOFr83jewa-bR1OKPCr3a28s

[28] Neuhauser C. Mathematical Models in Population Genetics [Internet]. Handb. Stat. Genet. Third Ed. John Wiley & Sons, Ltd; 2004 [cited 2013 Nov 5]. Available from: http://onlinelibrary.wiley.com/doi/10.1002/9780470061619.ch22/summary

[29] Weir BS. Genetic data analysis. Methods for discrete population genetic data. Sinauer Associates, Inc. Publishers; 1990 [cited 2013 Nov 5]; Available from: http://www.cabdirect.org/abstracts/19900180990.html;jsessionid=D364ACD3FCDB6360579FC4C81317E39F

[30] Balloux F. EASYPOP (Version 1.7): A Computer Program for Population Genetics Simulations. J. Hered. [Internet]. 2001 May 1 [cited 2013 Nov 5];92(3):301–2. Available from: http://jhered.oxfordjournals.org/content/92/3/301.short

[31] Excoffier L, Laval G, Schneider S. Arlequin (version 3.0): an integrated software package for population genetics data analysis. Evol. Bioinform. Online [Internet]. 2005 Jan [cited 2013 Oct 31];1:47–50. Available from: http://www.pubmedcentral.nih.gov/articlerender.fcgi?artid=2658868&tool=pmcentrez&rendertype=abstract

[32] Moskaleichik FF. Normal and Adverse Genetic Processes in Subdivided Populations of the Island Type: Computer Simulation. Russ. J. Genet. [Internet]. 2004 Aug [cited 2013 Nov 5];40(8):938–44. Available from: http://link.springer.com/10.1023/B:RUGE.0000039729.14117.e0

[33] Semovski S., Verheyen E, Sherbakov D. Simulating the evolution of neutrally evolving sequences in a population under environmental changes. Ecol. Modell. [Internet]. 2004 Aug [cited 2013 Jan 31];176(1-2):99–107. Available from: http://linkinghub.elsevier.com/retrieve/pii/S0304380003005593

[34] DeAngelis DL, Gross LJ. Individual-based models and approaches in ecology:populations, communities and ecosystems. Chapman & Hall; 1992 [cited 2013 Nov 5]; Available from: http://www.cabdirect.org/abstracts/19931977135.html

[35] Grimm V, Railsback SF. Individual-based Modeling and Ecology [Internet]. 2005 [cited 2013 Nov 5]. Available from: http://www.google.ru/books?hl=ru&lr=&id=12MvUbMeog8C&pgis=1

[36] Grimm V, Berger U, Bastiansen F, Eliassen S, Ginot V, Giske J, et al. A standard protocol for describing individual-based and agent-based models. Ecol. Modell. [Internet]. 2006 Sep [cited 2013 Jan 29];198(1-2):115–26. Available from: http://linkinghub.elsevier.com/retrieve/pii/S0304380006002043

[37] Lashin S a, Suslov V V, Kolchanov N a, Matushkin YG. Simulation of coevolution in community by using the "Evolutionary Constructor" program. In Silico Biol. [Internet]. 2007 Jan;7(3):261–75. Available from: http://www.ncbi.nlm.nih.gov/pubmed/18415976

[38] Lashin S a, Matushkin YG. Haploid evolutionary constructor: new features and further challenges. In Silico Biol. [Internet]. 2012 [cited 2013 Jan 31];11(3-4):125–35. Available from: http://www.ncbi.nlm.nih.gov/pubmed/22935966

[39] Rübel E. The replaceability of ecological factors and the law of the minimum. Ecology [Internet]. 1935 [cited 2013 Nov 5]; Available from: http://www.jstor.org/stable/10.2307/1930073

[40] Sundararaj S, Guo A, Habibi-Nazhad B, Rouani M, Stothard P, Ellison M, et al. The CyberCell Database (CCDB): a comprehensive, self-updating, relational database to coordinate and facilitate in silico modeling of Escherichia coli. Nucleic Acids Res. [In-

ternet]. 2004 Jan 1 [cited 2013 Nov 5];32(Database issue):D293–5. Available from: http://nar.oxfordjournals.org/content/32/suppl_1/D293.short

[41] Bennett K. Milankovitch cycles and their effects on species in ecological and evolutionary time. Paleobiology [Internet]. 1990 [cited 2013 Nov 5]; Available from: http://marineecology.wcp.muohio.edu/climate_projects_04/glacial_cycles/web/pdf/MilankovitchCycles.pdf

[42] Dobretsov N, Kolchanov N, Suslov V. On important stages of geosphere and biosphere evolution. Biosph. Orig. Evol. [Internet]. 2008 [cited 2013 Nov 5]; Available from: http://link.springer.com/chapter/10.1007/978-0-387-68656-1_1

[43] Levchenko VF, Starobogatov YI. Origin of biosphere. J. Evol. Biochem. Physiol. [Internet]. 2010 May 20 [cited 2013 Jan 31];46(2):213–26. Available from: http://www.springerlink.com/index/10.1134/S0022093010020122

[44] Markov A, Kulikov A. Origin of Eukaryota: Conclusions Based on the Analysis of Protein Homologies in the Three Superkingdoms. Paleontol. J. [Internet]. 2005 [cited 2013 Nov 5]; Available from: http://www.maikonline.com/maik/showArticle.do?pii=S0031030105040015&lang=en

[45] Markov A V., Kulikov AM. Homologous Protein Domains in Superkingdoms Archaea, Bacteria, and Eukaryota and the Problem of the Origin of Eukaryotes. Biol. Bull. [Internet]. 2005 Jul [cited 2013 Nov 5];32(4):321–30. Available from: http://link.springer.com/10.1007/s10525-005-0108-0

[46] Cavalier-Smith T. Principles of Protein and Lipid Targeting in Secondary Symbiogenesis: Euglenoid, Dinoflagellate, and Sporozoan Plastid Origins and the Eukaryote Family Tree, 2. J. Eukaryot. Microbiol. [Internet]. 1999 Jul [cited 2013 Nov 5];46(4):347–66. Available from: http://doi.wiley.com/10.1111/j.1550-7408.1999.tb04614.x

[47] Olson M V. When less is more: gene loss as an engine of evolutionary change. Am. J. Hum. Genet. [Internet]. 1999 Jan [cited 2013 Nov 5];64(1):18–23. Available from: http://www.pubmedcentral.nih.gov/articlerender.fcgi?artid=1377697&tool=pmcentrez&rendertype=abstract

[48] Ochman H. Genes Lost and Genes Found: Evolution of Bacterial Pathogenesis and Symbiosis. Science (80-.). [Internet]. 2001 May 11 [cited 2013 Nov 5];292(5519):1096–9. Available from: http://www.sciencemag.org/content/292/5519/1096.short

[49] Jacob F, Monod J. Genetic regulatory mechanisms in the synthesis of proteins. J. Mol. Biol. [Internet]. 1961 [cited 2013 Nov 5];3(3):318–56. Available from: http://www.sciencedirect.com/science/article/pii/S0022283661800727

[50] Ebeling W, Schimansky-Geier L, Romanovsky YM. Stochastic Dynamics of Reacting Biomolecules [Internet]. 2002 [cited 2013 Nov 5]. Available from: http://www.google.ru/books?hl=ru&lr=&id=ieVmkC6tY0MC&pgis=1

[51] Riznichenko GY. Mathematical models in biophysics [Internet]. 2012. Available from: http://www.biophysics.org/Portals/1/PDFs/Education/galina.pdf

[52] Tchuraev RN. On a stochastic model of a molecular genetic system capable of differentiation and reproduction of the initial state. Biometrical J. [Internet]. 1980 [cited 2013 Nov 5];22(2):189–94. Available from: http://doi.wiley.com/10.1002/bimj.4710220212

[53] Tchuraev R., Stupak I., Tropynina T., Stupak E. Epigenes: design and construction of new hereditary units. FEBS Lett. [Internet]. 2000 [cited 2013 Nov 5];486(3):200–2. Available from: http://www.sciencedirect.com/science/article/pii/S0014579300023000

[54] Lashin S a., Suslov V V., Matushkin YG. Comparative Modeling of Coevolution in Communities of Unicellular Organisms: Adaptability and Biodiversity. J. Bioinform. Comput. Biol. [Internet]. 2010 Jun [cited 2013 Jan 31];08(03):627–43. Available from: http://www.worldscientific.com/doi/abs/10.1142/S0219720010004653

[55] Lashin S a., Matushkin YG, Suslov V V., Kolchanov N a. Evolutionary trends in the prokaryotic community and prokaryotic community-phage systems. Russ. J. Genet. [Internet]. 2011 Dec 8 [cited 2013 Jan 31];47(12):1487–95. Available from: http://www.springerlink.com/index/10.1134/S1022795411110123

Diversity and Abundance of Fish Larvae Drifting in the Madeira River, Amazon Basin: Sampling Methods Comparison

R. Barthem, M.C. da Costa, F. Cassemiro,
R.G. Leite and N. Silva Jr.

Additional information is available at the end of the chapter

http://dx.doi.org/10.5772/57404

1. Introduction

The Amazon River and its main tributaries are known as downstream migratory way of eggs and larvae of many migratory fishes [1, 2, 3]. The young fish drift may occur during all year but the great abundance is associated with the flood pulse pattern, generally with the rising of the river level [4, 5]. The eggs, larvae and juveniles may drift for hundred or thousand kilometers and repopulate the stocks in the down river portion [6; 7; 8; 9].

The maintaining of the larval drift process is fundamental for the conservation of a great number of migratory fishes as well as for the maintaining of local economy and food security. The migratory species are the most important fisheries resource of the Amazon Basin and they are responsible for about 87% of the total landing in 66 ports of the main Amazon cities [10]. The fishery production, in special the fishery of migratory species, maintains an important sector of the Amazon economy, generating about US$130 million yr^{-1} and more than 160 thousand jobs [11]. The fish consumption rate by Brazilian Amazonian population is 24 kg annual per capita or 343 thousand ton/year, almost four times of the fish consumption rate of the Brazilian population [12].

One of the biggest threats to the drift movement of young fish drift movement are large dams, which interrupt the river connection and reduce or change the natural flood pulse pattern [13, 14]. There were few large hydroelectric dams in the Amazon basin, but this situation is changing by the construction of several hydroelectric dams in large tributaries of the Amazon River [15]. The most recent dams are two hydroelectric power plants nearby Porto Velho city, in the State of Rondônia, Brazil, which are damming the largest tributary of the Amazon River, the Madeira River.

The Madeira River Basin occupies 1.4 million km² or 20% of the Amazon Basin, which 50% belongs to Bolivia, 40% to Brazil and 10% to Peru. The Andes headwater is in Bolivia and Peru and the lowland portion and the Brazilian Shields headwater are in Brazil. The Madeira River at the frontier border of Bolivia and Brazil is characterized by a sequence of waterfalls, which separate the Llanos de Mojo savanna in Bolivia, the largest floodplains area of the Madeira River, and its lowland portion [16]. The Jirau and Santo Antonio hydroelectric power plants are being built above of the low portion in order to generate energy from the waterfalls of the Madeira River. The reservoirs of both dams should occupy an area around 530 km², which will flood only the Brazilian territory. The highest difference between the upper and lower limits of both reservoirs is about 50 m (Environmental Impact Assessment (EIA) of Madeira River, http://www.ibama.gov.br/licenciamento/index.php).

In order to assess the impact of the dams over the ichthyoplankton drifting along the Madeira River, it was established by the hydroelectric company a sampling protocol to collect periodically along the river before closuring the dam and, then in the reservoir, just below the dam. One methodology adopted was used to compare the larval density estimated from samples taken at different depths along the river transect [2, 3, 4, 5, 23]. In addition, it was developed a methodology in order to estimate the larval flux in different cross section of the river. This methodology was adapted from the direct methods for measuring suspended-sediment discharge in rivers [25]. The utilization of both methods will help to decrease the gear selectivity of the nets and methodologies and to compare the results of each one.

The aim of this study is to compare the selectivity of three sampling methods over the species composition and developmental stage of fish, as well as characterize the drift movement of ichthyoplankton in the waterfalls stretch of the Madeira River, considering the flood seasonal and spatial variation, before the closure of the Jirau Dam. The present study is a reference to the future impact assessment of the ichthyoplankton drift pattern related to the establishment of power plants.

2. Study area

The fish eggs, larvae and juveniles were collected monthly from October 2009 to September 2012 in two sampling sites in the Madeira River. The locations of those sites were defined by the limits of the planned reservoir of the Jirau Hydroelectric Power Plant (HPP). The sample sites were established in the upper and downriver limits of the planned reservoir in order to compare future modifications of the drift fish larvae patterns. At upriver site is the confluence of Abunã and Madeira River (9° 40′ S-65° 26′W), and at downriver site is the Jirau Hydroelectric Power Plant (9° 15′ S-64° 38′W). Between these two sampling sites, there are four perennial rapids (Pederneiras or Tamborete, Paredão, Jirau and Caldeirão do Inferno) and five seasonal rapids (Machado, Prainha, Vai Quem Quer, Dois Irmãos and Embaúba). Along this stretch there is the biggest rapid, known as Jirau waterfall, which is located about 13 km upside of Jirau HPP (Figure 1).

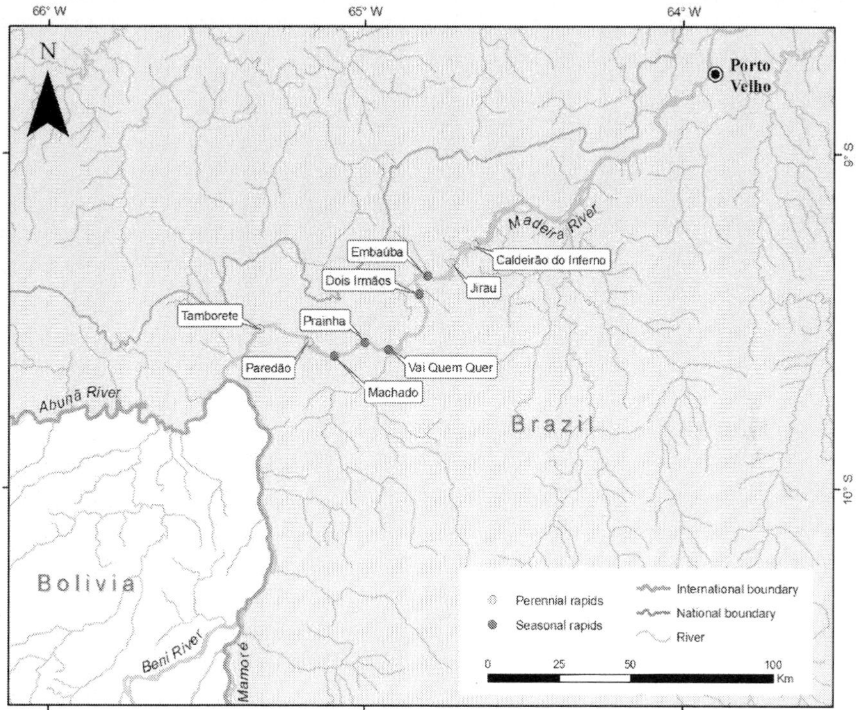

Figure 1. Perennial and seasonal rapids between the two sampling sites in the Madeira River.

The river's length between two sampling sites is about 140 km, river's width ranges from 300 to 1200 m, and altitude varies from 72 to 92 m. The flooding area in this stretch is very narrow if compared to the flooding area of the Beni or Mamoré River in upriver of the Jirau's Reservoir. The annual precipitation is about 2,650 mm, with intensive rainfall from December to April, while the driest period goes from May to November. The difference between high and low water level, in general, goes from 10 to 12 m [16]. The monthly mean of freshwater discharge ranges from 34,512 m³/s, in March, and 5,959 m³/s, in September, showing an annual mean of 19,222 m³/s (River Discharge Database-Sage: http://www.sage.wisc.edu).

3. Sampling and data analysis

Ichthyoplankton samples were obtained in two perpendicular transects in the Madeira River located at two sampling sites. In the Abunã site, at upriver, was established one transect above and other below of the confluence with the Abunã River and Madeira River. At downriver site, the positions of transects were established upside and downside of the Caldeirão do Inferno rapid, where the Jirau Hydroeletric Power Plant (Jirau HPP) is being built (Figure 2).

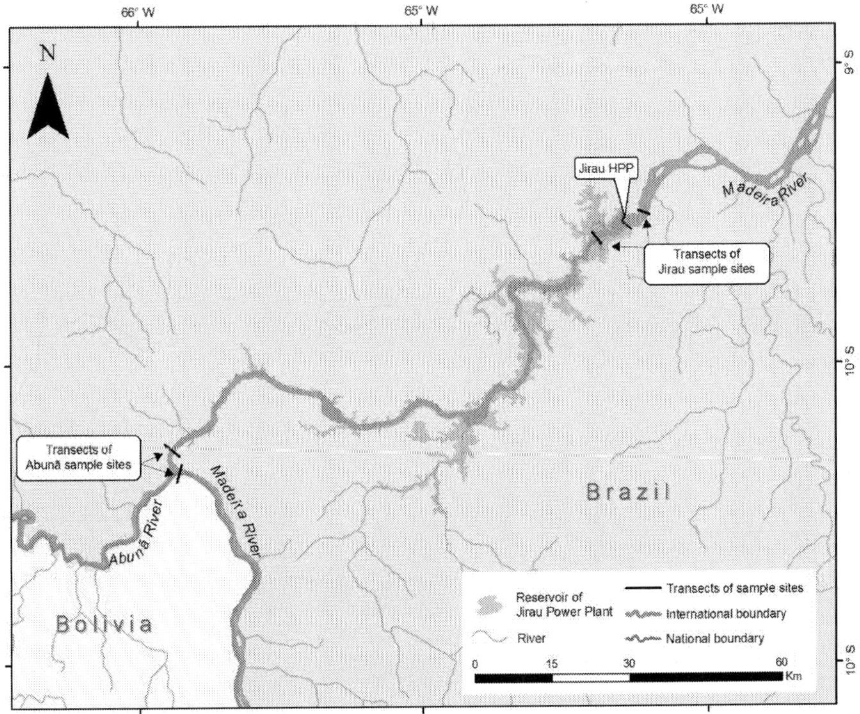

Figure 2. Four transects where ichthyoplankton samples are collected in reservoir area of the Jirau HPP, in Madeira River Basin.

The quantitative samples of ichthyoplankton were obtained during the daylight by towing an ichthyoplankton net from an outboard motor boat equipped with an electric winch. During sampling, the boat's position was maintained by the motor, which was moving upstream with the same speed of the river. Nocturnal samples were not collected because was assumed that the ichthyoplankton drift does not vary along day periods, as reported for rivers that show similar physicochemical characteristics of Madeira River [2].

Two different nets were used to collect ichthyoplankton in the Madeira River. The first one, nominated as larval net, has a conical-cylindrical format, with an aperture diameter of 0.5 m (0.1935 m²), 1.5 m length, and a mesh size of 0.5 mm [17]. This type of net has a collector cup used to open and close the net at different depth. The second one, nominated as juvenile net, was developed to collect large larvae and juvenile fish. It has a square frame in the open mouth, with 1 m^2 of area, 2 m of length, and 5 mm of mesh size. The volume of water filtered was measured by a flowmeter set up in the mouth of each net.

It was tested two different sampling methods to estimate the abundance of the ichthyoplankton drift in the Madeira River. The first used the point sampling method [18] and fixed five sites sampling in a linear transect to take samples in margins, main channel and zones in-between the main channel and the margins [3, 5]. The second used the integrating sampling method [18], which applied the systematic sampling method to define the sites sampling along of transect. The Point sampling with Larval net - PL - collects two samples in different depth at the water column. The remote open/close system set up in the mouth of the larval net ensured the net to collect samples at one meter from surface and at 70% of the total depth. The time spent in each sample was five minutes, and the distance from the margin ranged from 5 to 20 m. The Integrating sampling using Larval net – IL - and Juvenile net – IJ - takes one integrating samples of the water column, from the surface to the bottom and back. The velocity of net moving down and up was constant and determined by an electric winch. The distance of sampling sites was about 100 m each other and, at least, 20 m of the margin. The number of spots in each transect was related with width of the river.

The physical and chemical characteristics of the Madeira River's water were monitored by measuring dissolved oxygen (mg/l), pH, conductivity (µS/cm), temperature (°C) and turbidity (ntu) in the surface and at 70% of the bottom depth of each sample spot. The estimative of the freshwater discharge of the Madeira River was based on the river level of the Abunã site. The river level and the river discharge of the Abunã station was supplied by fluviometric station located in the district of Abunã, in the city of Porto Velho / RO (UTM 20L 240534 and 8926519), operated by the Geological Survey of Brazil (CPRM).

The ichthyoplankton samples were fixed in a 4% formalin solution and then they were counted and identified considering the taxonomy and the developmental stages. The developmental stages considered in this study were egg, larva yolk, pre-flexion, flexion, post-flexion and juvenile [19, 20]. The fish larvae were identified according to [21, 17].

The larval density was determined for each sample considering the number of ichthyoplankton and the volume of water filtered (larvae/m^3). The larval flux (larvae/s) of the cross section of the river was estimated multiplying the average of the larval density of the transect by the diary discharge (m^3/s). It was not considered the discharge of the small tributaries along the study area due to this minimal contribution to the overall freshwater discharge. The larval flux variation was analyzed by Factorial ANOVA, which tested the effect of the sampling method, river site and ichthyoplankton development stage. Pearson coefficient (r) was used to test the relationship between the variables.

4. Hydrology and environment variables

The freshwater discharge of the Madeira River is characterized by an annual unimodal cycle defined by four phases: low, rising, high and falling water level. The low water level phase is when the discharge is minimal and the river beach is exposed, usually between August and November, while the high water level phase is when the discharge is high and the river floods the marginal areas, it occurs generally between February and May. Other months are considered transition phases when the water level is rising or falling. The rising water level phase is when begins the rainy season and the river discharge starts to increase, between December and January, and the falling water level phase is when the river discharge decreases and the flood retreats, between June and July. The diary discharge median estimated for the study period was 15 thousand m³/s, which was used to separate the low and high discharge phases. Interannual variations of the hydrological cycle were observed during the period studied, with a short and intensive high discharge phase in 2010 (<160 days and >39 thousands m³/s) in relation to other years (>190 days and <36 thousands m³/s) (Figure 3).

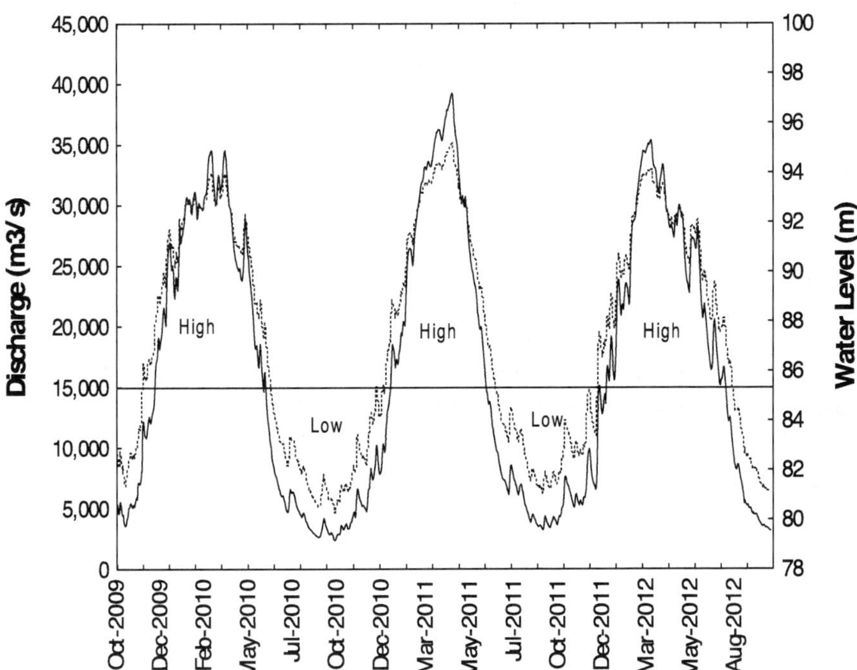

Figure 3. The hydrological cycle of the Madeira River Basin obtained at the Hydrological Station of Abunã, at the up-river of the study area (discharge- continuous line; water level- dotted line).

The Madeira River is a muddy river that receives a large amount of sediments rich in mineral salt, which comes from Andean region. Due to this, the monthly averages of the conductivity and pH are in general higher than the other rivers of the Amazon Basin, varying from 69 to 131 µS/cm and 6 and 8, respectively. The conductivity and pH values were inversely related to the discharge, and the highest values occurred during the low discharge period ($r_{conductivity}$= -0.74, r_{pH}= -0.46; p<0.01, n=36). The mean turbidity is also high in relation to the other rivers, varying from 84 to more than one thousand NTU, with the highest turbidity value occurring in the beginning of the high discharge period (Pearson correlations: r= 0.63, p<0.01, n=36). The dissolved oxygen mean was relatively high, which can be associated with the aeration process caused by the movement of water in the rapids. The monthly average ranged from 5.7 to 8.8 mg/l and there was an inverse relationship with the discharge (r= -0.41, p<0.05, n=36). The high river discharge is associated with the flooding of the wetland and decomposition of large amount of organic material, which consumes oxygen of the aquatic environment (Table 1).

Month	Conductivity	Temperature	Oxygen	Ph	Turbidity
1	73	27.5	8.6	7.03	1.137
2	69	27.1	7.0	6.83	1.129
3	73	27.0	5.7	6.94	961
4	78	27.6	6.7	7.10	564
5	80	27.6	6.7	6.69	401
6	74	26.9	7.6	6.75	298
7	90	27.2	8.9	6.08	205
8	107	26.4	8.8	7.88	180
9	131	28.1	8.6	8.07	84
10	121	29.4	7.8	7.98	411
11	105	29.4	7.5	7.48	320
12	78	27.9	7.1	7.28	806
Mean	89.61	27.67	7.59	7.17	551.57
SD	22.68	1.14	2.20	0.87	421.58

Table 1. Average monthly of the physicochemical parameters obtained in the Madeira River.

5. Ichthyoplankton abundance

During the three years, 4,148 individuals were collected by 432 samples of ichthyoplankton realized monthly in four transects, with an average of 9.5 samples by transect and method. The number of samples were similar for each combination of sampling methods and nets, however there were more samples in downriver (Jirau: 56%) than in upriver (Abunã: 44%). This difference in samples number was due to the river width, which is related to the number of sampling sites for integrating sampling method. A total of 21,665 larvae (99%) and 282 eggs (1%) were collected. The point sampling (PL) method was more efficient at collecting eggs than the integrating sampling methods. The PL collected 53% of the eggs, followed by the integrating sampling method with juvenile net (IJ), that collected 29%, and the integrating sampling method with larvae net (IL), that caught the remains 17% of the eggs. The PL method also collected more larvae (61%) than the other methods and the IL method collected more larvae (34%) than the IJ method (5%). The number of eggs was similar for the both sites, and in the upriver (Abunã) showed twofold more larvae than in downriver (Jirau), even at downriver showing more samples (Table 2).

Sites	S&N	Eggs				Larvae & Juveniles		
Upriver (Abunã)	IJ	53	19%		631	3%	
	IL	19	7%	49%	4,316	20%	66%
	PL	65	23%		9,369	43%	
Downriver (Jirau)	IJ	30	11%		489	2%	
	IL	30	11%	51%	2,979	14%	34%
	PL	85	30%		3,881	18%	
Total		282	100%		21,665	100%	

Table 2. Number of eggs, larvae, juveniles collected in up, downriver places, and the combination of sampling methods and nets: IJ: Integrating sampling with juvenile net; IL: integrating sampling with larval net; and PL: point sampling with larval net.

The development stage dominant in all samples was pre-flexion (66%), followed by flexion (19%) and post-flexion (5%). This stage composition was similar for the both sites, but different for the net types, where more than 95% of the ichthyoplankton caught by IL and PL were in flexion or early stages and 99% of the ichthyoplankton caught by IJ was in flexion or older stages. Considering each development stage, about 2/3 of all larvae in larval yolk or pre-flexion stages were caught by the PL method, whilst 49% of all larvae in post-flexion stages and 71% of juveniles were caught by the IJ method. Only nine small fish in adult stage were collected during this study (Table 3).

Stage	Methods And Nets			Sites		Total	
	IJ	IL	PL	Upriver	Downriver		
Unknown	22	403	318	423	320	743	3%
Larval Yolk	7	376	790	910	263	1,173	5%
Pre-Flexion,	7	4,599	9,713	9,830	4,489	141319	66%
Flexion	396	1,667	2,073	2,368	1,768	4,136	19%
Post-Flexion	522	220	319	647	414	1,061	5%
Juvenile	159	30	35	134	90	224	1%
Adult	7	0	2	4	5	9	0%
Total	1,120	7,295	13,250	14,316	7,349	21,665	100%

Table 3. Composition of the larvae collected considering the development stages and in relation to the up and downriver places and the combination of sampling methods and nets (IJ: integrating sampling with juvenile net; IL: integrating sampling with larval net; and PL: point sampling with larval net).

The abundance index considered the average of the larval density and the estimative of the larval flux in the cross section of four transects along the Madeira River. Peaks of larval density and flux were observed during the rising discharge and the beginning of the high discharge phases (December to March) and a decreasing in the next months. However, larval flux was minimal from June to November while larval density showed some peaks during this period

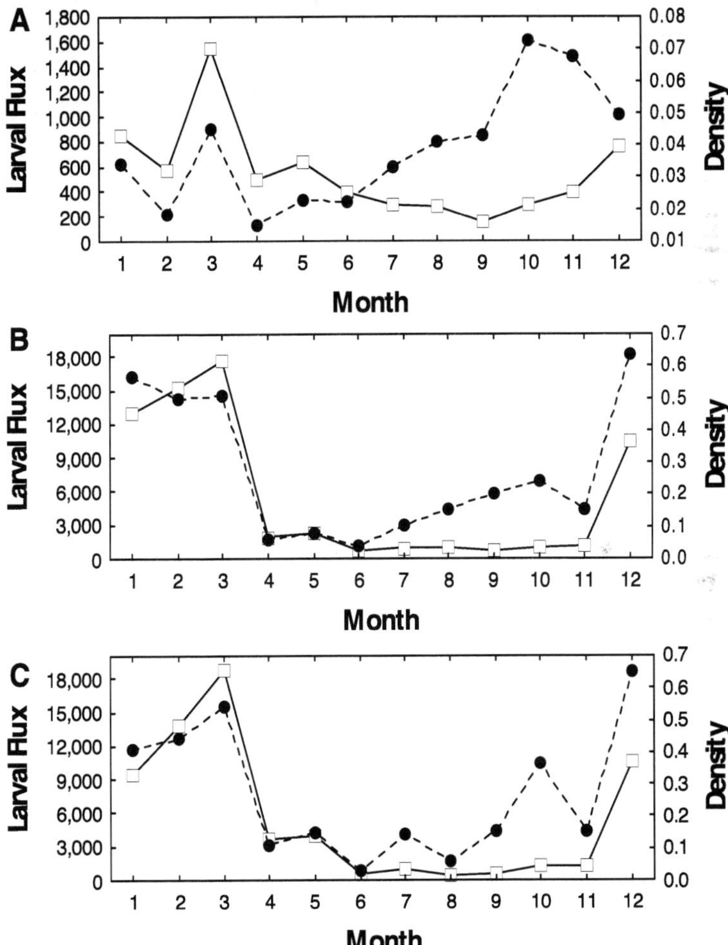

Figure 4. Monthly variation average of the larval flux (larvae/s) (□) and larval density (larvae/m³)(●) estimated for (A) integrating sampling with juvenile net-IJ; (B) integrating sampling with larval net-IL and (C) point sampling with larval net-PL. Months: 2-5 (February-May) high discharge; 6-7 (June-July) falling discharge; 8-11 (August-November) Low discharge; 12-1 (December-January) rising discharge.

(Figure 4). The relationship between the two abundance indexes for the three methods is presented in Figure 5. The highest correlation value between larval density and flux was observed for the IL method (r^2=0.86) and the lowest values was observed for the IJ method (r^2=0.23). The correlation between IL and PL methods was higher for larval flux (r^2=0.77) than for larval density (r^2=0.61) (Figure 5). The mean composition of the larval flux by larval stages of 144 samples shows the importance of IJ and IL methods for the juvenile abundance estimative and the IL and PL for the abundance estimative of the early stages (Table 4).

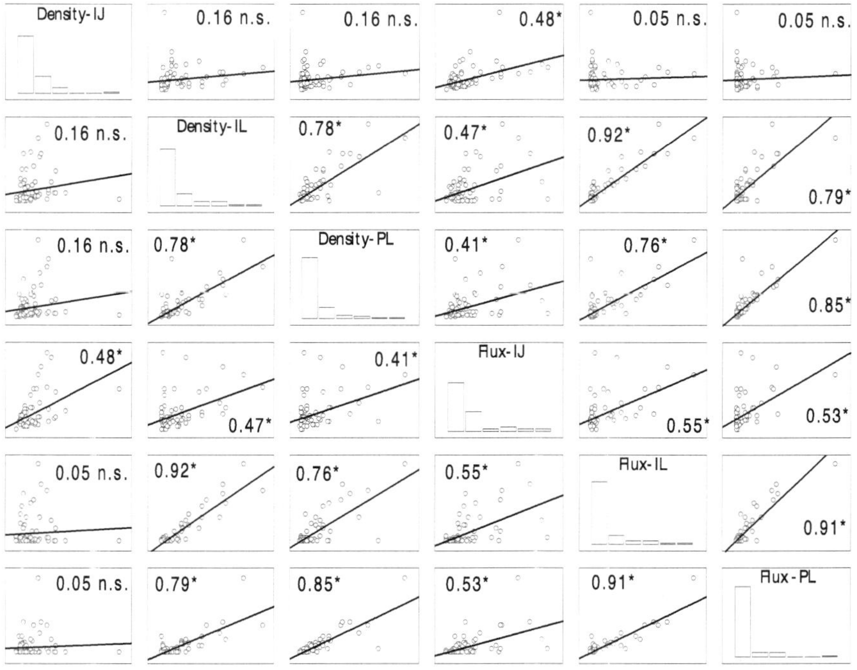

Figure 5. Abundance indexes plots for the three sampling methods and the respective Person's coefficient r (* p<0.01, n=144; n.s: no significant). Abundance indexes: Density (larvae/m³) and Flux (larvae/s). Methods: IJ: integrating sampling with juvenile net; IL: integrating sampling with larval net; and PL: point sampling with larval net.

The seasonal and spatial variation of the larval flux was analyzed only for IL and PL data. The IJ data was not considered due its low capacity to detected larvae in early stages and to assure the assumptions for homogeneity of variance. An ANOVA-two-way was performed to analyze the variation in larval composition considering the square root of the larval flux as dependent variable and the sampling method (IL and PL), the up and downriver sites and five development stages (larval yolk, pre-flexion, flexion, post-flexion and juvenile) as independent variables (Table 5). The assumptions for homogeneity of variance was met according to the

Stage	IJ	IL	PL	Total
Larval Yolk	12 (1%)	685 (55%)	539 (44%)	1,236 (100%)
Pre-Flexion	13 (0.2)%	3,675 (48%)	3,996 (52%)	7,684 (100%)
Flexion	217 (7%)	1,507 (52%)	1,175 (41%)	2,898 (100%)
Post-Flexion	344 (19%)	677 (38%)	758 (43%)	1,779 (100%)
Juvenile	198 (41%)	201 (42%)	82 (17%)	481 (100%)
Total	822 (5%)	7,261 (48%)	6,944 (46%)	15,027 (100%)

Table 4. Average and percent of the development stage composition for each method considering larval flux (larvae/s). Methods: IJ- integrating sampling with juvenile net; IL- integrating sampling with larval net; and PL-point sampling with larval net.

Levene's test (F=0.31, p>0.05). The larval flux estimated by PL method is significantly lower than the IL method for all stages (Method: F=7.53, p<0.01), especially for juvenile stage, but the interactive effects of sampling method with the other factors was not significant (Method-Site, Method-Stage, Method-Site-Stage: p>0.05). The larvae of the pre-flexion stage were more abundant than the other stages (Stage: F=26.10, p<0.01). The up and down river sites was not a significant factor (Site: p>0.05), but the interactive effect with development stage was significant (Site-Stage: p<0.05) indicating larvae in pre-flexion stage is higher in upriver than downriver, but the other stages presented similar values in both sites (Tukey HSD test, p<0.01) (Table 5).

	SS	d.f.	MS	F	p
Intercept	580306.0	1	580306.0	689.5715	<0.0001**
Method	6340.9	1	6340.9	7.5349	0.0062**
Site	1882.5	1	1882.5	2.2369	0.1352
Stage	87861.8	4	21965.5	26.1013	<0.0001**
Method-Site	23.4	1	23.4	0.0278	0.8676
Method-Stage	891.3	4	222.8	0.2648	0.9006
Site-Stage	8308.0	4	2077.0	2.4681	0.0436*
Method-Site-Stage	2193.1	4	548.3	0.6515	0.6260
Error	637050.2	757	841.5		

Table 5. Result of Factorial ANOVA testing the methods (IL and PL), river site (up and downriver) and development stage (larval yolk, pre-flexion, flexion, post-flexion and juvenile) factors over the square root of the larval flux. S.S. = sum of squares; d.f. = degrees of freedom; M.S. = mean square; F = F statistic; p = significance level. One asterisk means significant at 0.05 level and two asterisks means significant at 0.01.

6. Larval diversity

Eggs and larvae identification depends on the integrity of the sampled larvae as well as on previous ontogeny studies. High diversity of the Amazon fish and the paucity of ichthyo-plankton ontogeny studies become the larvae identification a challenge. Despite of this, just 0.3% of the larvae and juveniles collected were completely unknown and only 8% of them were identified at the order level, the lowest possible level. The most larvae (92%) was identified at family level and the identification at genus and species level was possible in 30% and 13% of the total larvae, respectively. The Characiformes and Siluriformes larvae were dominant in the samples and represented about 98% of the total larvae collected. The PL net method collected almost twofold more Characiformes than Siluriformes larvae, while IJ method collected almost fourfold Siluriformes than Characiformes larvae. The larvae composition collected by IL method did not show a large discrepancy among the most important taxonomic groups, even so Characiformes were more abundant than Siluriformes larvae (Table 6).

Curimatidae, Auchenipteridae and Pimelodidae were the most abundant families that totalizing about 2/3 of the total larvae. The Curimatidae and Auchenipteridae families represented the half of the larvae collected by the PL method and the Pimelodidae family represented 2/3 of the collected by IJ method. The caught of three families aforementioned were similar for the IL method. The Characidae family was the fourth in the rank of abundance, representing about 12% of the larval composition collected by the three methods. The larvae of Prochilodontidae, Anostomidae and Hemiodontidae together represented 12% of the total larvae. The sampling methods using larval net collected 95% or more of the larvae of Characidae, Prochilodonti-dae, Anostomidae and Hemiodontidae families, which they represented together about ¾ of all larvae caught. Other families represented less than 5% of the total (Table 6).

Less than 5% of the larvae of Curimatidae, Auchenipteridae, Anostomidae, Prochilodontidae and Hemiodontidae families, the most abundant families, were identified at the genus level. Nonetheless, 94% of the Characidae larvae and 83% of the Pimelodidae were identified at the genus level. The larvae of *Brycon, Mylossoma, Triportheus* and *Piaractus* genera were the most abundant among the Characidae family. Each one represented more than 5% of all Characidae larvae and together represented 93% of the Characidae larvae and 11% of all larvae, but few of them were identified at the species level. On the other hand, most larvae of the four most abundant genera of the Pimelodidae family were identified at the species level. *Pinirampus, Brachyplatystoma, Pimelodus* and *Pseudoplatystoma* represented each one more than 5% of all Pimelodidae larvae and together represented 71% of Pimelodidae larvae and 13% of all larvae.

The Curimatidae, Auchenipteridae and Anostomidae families are diversified and they were represented in the study area by 7 genera and 21 species, 18 genera and 21 species and 6 genera and 13 species, respectively. For the other side, the diversity of Prochilodontidae and Hemio-dontidae families are low and they have only two genera with three species and two genera with six species in the area, respectively [22].

[22] identified three species of *Brycon* (*B. amazonicus, B. falcatus* and *B. melanopterus*), two species of *Mylossoma* (*M. aureum* and *M. duriventre*) and three species of *Triportheus* (*T. albus, T.*

	Order	Total		IJ		IL		PL	
Order	Unknown	63	0,3%	0	0%	29	0.4%	34	0.3%
	Characiformes	12,485	58%	231	2%	4,005	55%	8,249	62%
	Siluriformes	8,806	41%	825	74%	3,182	44%	4,799	36%
	Perciformes	189	1%	20	2%	49	1%	120	1%
	Clupeiformes	69	0%	12	1%	21	0.3%	36	0.3%
	Gymnotiformes	53	0%	32	3%	9	0.1%	12	0.1%
Family	Unknown	1,834	8%	27	2%	832	11%	975	7%
	Curimatidae	5,891	27%	9	1%	1,676	23%	4,206	32%
	Auchenipteridae	3,965	18%	11	1%	1,272	17%	2,682	20%
	Pimelodidae	3,960	18%	717	64%	1,582	22%	1,661	13%
	Characidae	2,650	12%	134	12%	880	12%	1,636	12%
	Prochilodontidae	1,482	7%	1	0,1%	495	7%	986	7%
	Anostomidae	749	3%	1	0,1%	231	3%	517	4%
	Hemiodontidae	276	1%		<0.1%	101	1%	175	1%
	Cynodontidae	251	1%	71	6%	70	1%	110	1%
	Sciaenidae	188	1%	20	2%	48	1%	120	1%
	Doradidae	116	1%	67	6%	20	0.3%	29	0.2%
	Trichomycteridae	92	0.4%	5	0.4%	34	0.5%	53	0.4%
	Cetopsidae	61	0.3%	3	0.3%	16	0.2%	42	0.3%
	Heptapteridae	45	0.2%	3	0.3%	16	0.2%	26	0.2%
	Apteronotidae	27	0.1%	22	2%	2	<0.1	3	<0.1
	Callichthyidae	15	0.1%	34	1%	2	<0.1	5	<0.1
	Loricariidae	15	0.1%	16	0.1%	7	0.1%	7	0.1%
	Engraulidae	13	0.1%	9	1%	3	<0.1	1	<0.1
	Pristigasteridae	11	0.1%	3	0.3%	2	<0.1	6	<0.1
	Erythrinidae	8	<0.1	2	0.3%	4	0.1%	2	<0.1
	Gasteropelecidae	8	<0.1	0	<0.1	2	<0.1	6	<0.1
	Sternopygidae	7	<0.1	6	1%	0	0%	1	<0.1
	Clupeidae	1	<0.1	0	0%	0	0%	1	<0.1
	Total	21,665	1	1,120	100%	7,295	100%	13,250	100%

Table 6. Larvae composition caught by the combination of methods and nets: IJ- integrating sampling with juvenile net; IL- integrating sampling with larval net; and PL- point sampling with larval net.

angulatus and *T. auritus*) in the study area. The present study identified the two species of *Mylossoma* and only two species of Triportheus (*T. angulatus* and *T. auritus*) and none of *Brycon*. Considering the shortage of larvae identified at specie level, it is not possible to discuss the composition of those genera. On the other side, all *Pinirampus* larvae and 99% of the *Brachyplatystoma* larvae were identified at specie level. *Pinirampus pirinampu* was the only specie of the genus. Five *Brachyplatystoma* species were identified, being *B. filamentosum*, *B. rousseauxii* and *B. capapretum* the most abundant, totalizing 90% of the genus. The other two species in the larval collection were *B. platynemum* and *B. juruense*. Two *Brachyplatystoma* species, *B. tigrinum* and *B. vaillantii*, were present in the local fish collection [22] but they are not present in the present larvae samples. Nevertheless, only 56% of the *Pimelodus* and 11% of the *Pseudoplatystoma* genus were identified at specie level. *Pimelodus blochii* was the main specie identified of the genus and only one individual represented *P. altissimus* was in all samples. The two species of *Pseudoplatystoma* present in the area were in the larvae samples, *P. punctifer* and *P. tigrinum*, but most larvae were not identified at specie level. The larval net caught the majority of the larvae of those species, with the exceptions of *P. blochii*, *B. rousseauxii* and *B. capapretum*, which the juvenile net caught more than 50% of the larvae of each specie.

The majority (c.a. 95%) of the larvae of the most families were in pre-flexion or earlier stages, whilst Characidae and Pimelodidae presented the larvae in advanced stage. The most Characidae larvae were found in pre-flexion (63%) and flexion (35%) stages, while Pimelodidae larvae were found in flexion (69%) and post-flexion (23%) stages. However, the development stage composition was different for each species or genus. *Mylossoma* spp., *Triportheus* spp. and *Piaractus brachypomus* presented 90% or more of the larvae in pre-flexion stage, while *Brycon* spp. showed developed larvae, with 87% in flexion stage. The majority (2/3 or more) of *Pseudoplatystoma* spp., *Zungaro zungaro*, *Pinirampus pirinampu*, *Brachyplatystoma filamentosum* and *B. capapretum* larvae were in flexion stage, and more than a half of *Pimelodus blochii* and *B. rousseauxii* were in post-flexion stage.

The larvae drift in the river occurred mainly when the discharge is increasing. Some families concentrated the drifting movement in just a few months and others remained drifting during all year, indicating the length of the reproductive season. The average larval flux estimated by the three samples method in the two sites was compared in order to understand the general drifting pattern of the most abundant taxa. The IL and PL methods were efficient to identify the great variation of the larval flux for most of the families and IJ detected satisfactorily the annual flux variation only for Characidae and Pimelodidae.

The Curimatidae and Prochilodontidae families showed the shortest reproductive season between December and April, but strongly concentrated in March. The larvae of Auchenipteridae and Anostomidae families remained drifting in the river for a longer period, between November and May, with peak in January and March. The general pattern of the larval drifting of the Characidae and Pimelodidae families is also intensive in the period of rising discharge and less intensive in the low discharge months. However, the results of the three methods showed some differences, especially in April and May, when the IL method pointed the end of reproduction season and the PL method still detects an intensive flux of larvae (Figure 6). *Brycon* spp. and *Piaractus brachypomus* showed a shorter spawning season, during the rising discharge months (December-January), and *Mylossoma* spp. extended the reproduction beyond the period of the rising discharge (December-March). *Triportheus* spp. presented a delayed drifting larval movement in comparison with the other species. The peak was detected in March by the IL method and in May by the PL method, both moments in the high discharge period (Figure 7).

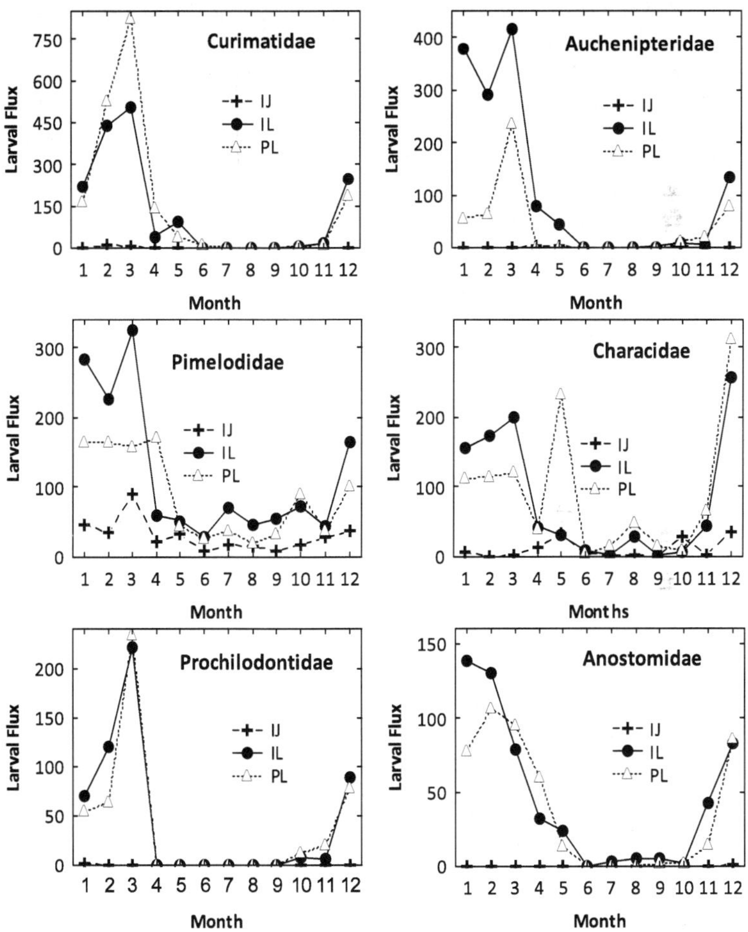

Figure 6. Monthly average of the larval flux (larvae/s) of main families: IJ- integrating sampling with juvenile net; IL- integrating sampling with larval net; and PL- point sampling with larval net. Months: 2-5 (February-May) high discharge; 6-7 (June-July) falling discharge; 8-11 (August-November) low discharge; 12-1 (December-January) rising discharge.

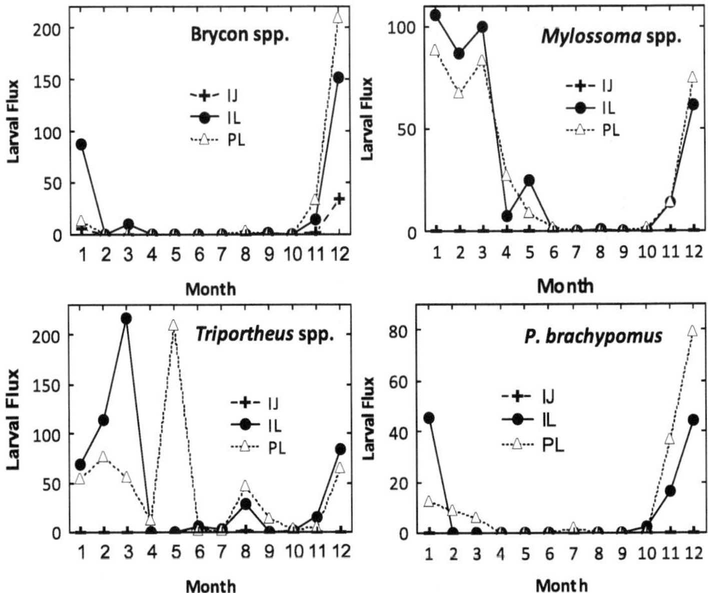

Figure 7. Monthly average of the larval flux (larvae/s) of the main species of Characidae family: IJ- integrating sampling with juvenile net; IL- integrating sampling with larval net; and PL- point sampling with larval net. Months: 2-5 (February-May) high discharge; 6-7 (June-July) falling discharge; 8-11 (August-November) low discharge; 12-1 (December-January) rising discharge.

The drifting movement of the most abundant Pimelodidae species was longer than the Characidae species. In addition, the IJ method presented an expressive flux estimative when compared with the other groups, in special for the *Brachyplatystoma rousseauxii* and *Pimelodus blochii*. Despite of large amount of larvae of each species, drifting in the channel during the rising and or high discharge months, the movement of larvae did not stop during the low discharge period suggesting an uninterrupted spawning season. Larvae of *Pseudoplatystoma* spp. and *P. blochii* were scarce during the falling and low discharge, but the larvae of *Pinirampus pirinampu* or *Brachyplatystoma* spp. were abundant in that period. The different sampling methods also presented expressive differences in the larval flux of some Pimelodidae species. *P. blochii*, *B. capapretum* and *B. rousseauxii* presented isolated peaks of larval flux detected by only one method (Figure 6).

7. Discussion

The point sampling method carried out at fixed habitats along of transects is a good way to compare the larval abundance or composition in different depths or habitats [2, 3, 4, 5, 23]. However, when the aim is to assess the impact of the river modification over the ichthyo-

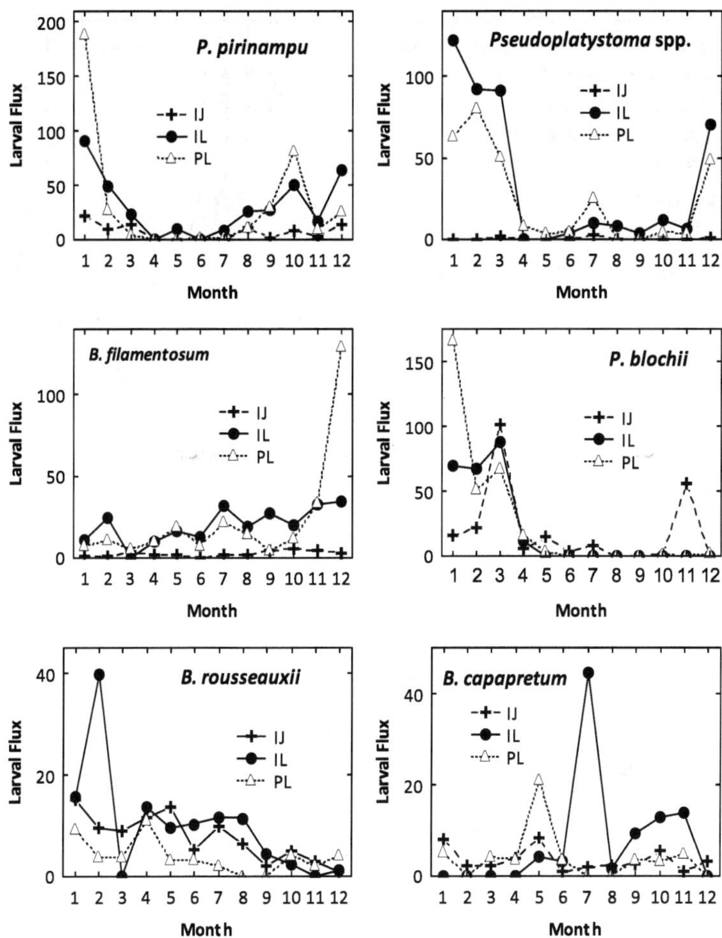

Figure 8. Monthly average of the larval flux (larvae/s) of the main species of Pimelodidae family: IJ- integrating sampling with juvenile net; IL- integrating sampling with larval net; and PL- point sampling with larval net. Months: 2-5 (February-May) high discharge; 6-7 (June-July) falling discharge; 8-11 (August-November) low discharge; 12-1 (December-January) rising discharge.

plankton drifting process, it is necessary to develop a methodology adapted to estimate the larval flux in natural or dammed river cross sections.

The three methods caught few eggs in relation to total of ichthyoplankton, and it was similar for the up and downriver sites. [4] also found few eggs in the ichthyoplankton composition in the Solimões River (Table 2). They hypothesized that the eggs have a short residence time, less than 16 hours, and as the spawning occurs mainly during the dusk, the eggs had hatched before the sampling moment. The point sampling and the larval net caught more individuals than the integrating sampling and the juvenile net. However, the juvenile net caught mainly larvae in post-flexion stages and juveniles when compared to samples collected by larval net (Table 3 and 4). Using juvenile net is appropriated to sample ichthyoplankton of species that go through the study area in advanced development stages, like *Pimelodus blochii* and *B. rousseauxii*.

The abundance indexes, larval density and flux considered in this study, pointed an increasing of larvae and juveniles abundance during the rising discharge and the beginning of the high discharge phases (December to March), and a decreasing in the next months. It was also observed by [4], in Solimões River, and by [3], in the Madre de Dios River. However, the larval flux index was minimal from June to November while larval density index presented some expressive peaks during May to November. The relationship of both indexes was low, although had been statistically significant, for the IJ method and high for the IL and PL methods (Figure 5). The larval density is affected by the dilution effect, which occurs during the rising and high discharge, and by the concentrating effect, that occurs during the low water discharge. In that way, the larval density is a measure more accurate for the drift inversely proportional to the water discharge if the larval flux is constant. The larval flux is a more accurate measure to assess the changes of the larval drift in modified rivers. This measure is not biased by the discharge moment or by the width or depth of the river cross section.

The composition of the larval flux in relation to the larval development stages of each method (Table 4) indicated the IL method as the less selective in relation to the development stages of the fish when they are drifting in the river. The IJ method is selective for juveniles and underestimated the abundance of all larvae in early phases and the PL method is selective for larvae and underestimated the abundance of juvenile phase.

The larval flux estimated by the IL method was higher than the larval flux estimated by the PL method and they were similar in the upper and downriver sites, in spite of the number of larvae caught in the upriver was two times of the number caught downriver (Table 2 and 5). The two methods estimated the larval flux in pre-flexion stage as bigger than the larval flux in other stages and the larval flux in pre-flexion stage were more abundant in upriver than downriver, while the larval flux of the other stages were similar in both sites (Table 5). The upriver abundance of larvae in early stages may suggest a more intensive spawning activity more intensive upriver of the study area.

The identification of the larvae in rivers with high diversity is a challenge. Most larvae was identified at family level and it was possible to identify only few genera or species, mostly of the Characidae and Pimelodidae family. The larvae of Siluriformes and Characiformes orders as well as Curimatidae, Auchenipteridae, Pimelodidae, Characidae, Prochilodontidae, Anostomidae and Hemiodontidae families represented together 95% or more of all larvae caught during this study. These results do not corroborate the [4] results, in which the most abundant larvae found in the Solimões River were of the Clupeiformes, Characiformes and Perciformes orders. The small amount of larvae of Siluriformes order must be due to the differences in methodology. Most of the samples in the Solimões River were collected on the surface, while in the present study samples covered all water column. However, the difference in the abundance of Clupeiformes and Perciformes larvae in the Solimões River and Madeira River is difficult to be explained only by the methodology and the environmental difference of both rivers must be considered. The selectivity of the method for different species was very clear, while IJ method was very selective for Siluriformes larvae, in particular for the Pimelodidae family; PL method was more selective for the Characiformes larvae, in particular for the Curimatidae and Auchenipteridae families (Table 6). For most families, larvae were found in pre-flexion or earlier stages, except for Characidae and Pimelodidae that showed larvae in more advanced stages, such as *Brycon* spp., *Pimelodus blochii* and *B. rousseauxii*.

The larvae drift indicated the reproductive period of the fish. The IL and PL methods identified a short annual reproductive period for Curimatidae and Prochilodontidae families, between December and April, and a prolonged annual reproductive period for Auchenipteridae and Anostomidae families, between November and May. However, the three methods showed an almost continuous reproductive period for Pimelodidae and Characidae families, with some differences in the moment of the most intensive reproductive peak (Figure 6).

The larval flux estimated by IL and PL methods showed a short annual reproductive period, between December and January, for *Brycon* spp. and *Piaractus brachypomus,* and an annual prolonged reproductive period for *Mylossoma* spp., between December and March. Both methods indicated also a short annual reproductive period for *Triportheus* spp., but IL method indicated March as the reproductive period and PL method indicated May (Figure 8). This difference may be related with the selectivity of the different methods for the three species of *Triportheus* in the area. Thereby, it is necessary to identify the effect of the methods in the evaluation of the larval abundance of those species.

The different methods presented peaks of reproduction activity in different months for *Brachyplatystoma* and *Pimelodus* species (Figure 8), which must be biased by the few number of larvae sampled. The IJ method collected mainly larvae of *P. blochii* and *B. rousseauxii,* and its larval flux showed a biannual short reproductive period for *P. blochii,* in March and in November, and a continuous reproductive period for *B. rousseauxii,* more intensive in the January and decreasing until December. *P. blochii* is considered a species that shows short annual reproductive period, which occurs in the beginning of rainy season [24]. The *P. blochii* larval flux peak occurred in March, at the end of the rainy season, and in November, at the beginning of the rainy season. As the studied area receives the flow discharge of the Beni and Mamoré Rivers, it must be investigated if the two larval peaks found in the area (Figure 8) have originated at the different basins upriver. *B. rousseauxii* reproduces in Andes foothill and the eggs and larvae drift the river toward to Amazon estuary [9]. Studies in Madre de Dios River in Peru, considered the main tributary of the Beni River, showed that this area is a spawning area for *B. rousseauxii* and the reproductive period is prolonged, and spawning period is concentrated at the high water period [5]. The larval flux peak of *B. rousseauxii* larval is in January (Figure 8) and the presence of larvae in advanced developmental stages is in accordance with the drifting movement from the Andes foothill during the high discharge.

Finally, the impact assessment of the new infrastructures projects in the Amazon depends of the data quality obtained before the impact. The present study discussed the effect of the sampling method in the evaluation of the larval drift pattern in the Madeira River. The result is background knowledge for the future studies to assess the impact of the Jirau hydroelectric power plant to the larval drifting in the Madeira River.

Acknowledgements

We thank Energia Sustentável do Brasil S.A., responsible for the construction, maintenance, operation and sale of energy to be generated by the HPP Jirau, for financial support and for providing the data analyzed in the present study as well as Systema Naturae Consultoria Ambiental Ltda., company responsible for implementing the Conservation Program of the Fish Fauna in the area of influence of HPP Jirau. We would like to extend our thanks to all professionals involved with the Subprogram of the Ichthyoplankton, especially André Almeida Uchôa, Andréa Souza Leão, Antônio Cléber Nunes Ferreira, Camila Afonso dos Santos Rosa, Cibelle Mendes Cabral, Cláudia Milena Siqueira Lopes, Fernanda Capuzo Santiago, Ivan VianaTibúrcio, Josmara dos Passos Carvalho, Lívia Naves de Moraes, Luciana Fugimoto Assakawa, Marcio Lima Santos, Marcos Paulo dos Santos Fonseca, Marina Granai, Suzana Silva Peres Rodrigues and Thiago Piassa.

Author details

R. Barthem[1], M.C. da Costa[2], F. Cassemiro[2], R.G. Leite[3] and N. Silva Jr.[2,4]

1 Department of Zoology, Museu Paraense Emilio Goeldi, Belém, Brazil

2 Systema Naturae Consultoria Ambiental Ltda., Goiânia, Brazil

3 Department of Aquatic Biology and Fish Ecology, Instituto Nacional de Pesquisas da Amazônia, Manaus, Brazil

4 Pontifícia Universidade Católica de Goiás, Goiânia, Brazil

References

[1] Pavlov DS, Nezdoliy VK, Urteaga AK, Sanches OR. Downstream Migration of Juvenile Fishes in the Rivers of Amazonian Peru. Journal of Ichthyology 1995; 35(9): 753-767

[2] Araujo-Lima, CARM, da Silva, VV, Petry, P, Oliveira E.C. & S.M.L. Moura, 2001. Diel variation of larval fish abundance in the Amazon and rio Negro. Brazilian Journal of Biology. 61(3) 357-362.

[3] Cañas CM, Pine WE. Documentation of the temporal and spatial patterns of Pimelodidae catfish spawning and larvae dispersion in the Madre de Dios River (Peru): insights for conservation in the Andean-Amazon headwaters. River Research Application 2011; 27:602-611.

[4] Araujo-Lima CARM., Oliveira C. Transport of larval fish in the Amazon. Journal of Fish Biology1998. 53(Supplement A) 297-306

[5] Cañas CM, Waylen PR. Modelling production of migratory catfish larvae (Pimelodidae) on the basis of regional hydroclimatology features of the Madre de Dios Basin in southeastern Peru. Hydrolical Processes 2011. DOI: 10.1002/hyp.8192.

[6] Goulding M, Carvalho ML. Life history and management of the tambaqui (Colossoma macropomum, Characidae): An important Amazonian food fish. Revista Brasileira de Zoologia, São Paulo 1982; 1:107-133.

[7] Carvalho JL, Merona B de. Estudos sobre dois peixes migratórios do baixo Tocantins, antes do fechamento da barragem de Tucuruí. Amazoniana 1986; 9(4):595-607.

[8] Ribeiro MCLB, Petrere Jr M. Fisheries ecology and management of the jaraqui (Semaprochilodus taeniurus, S. insignis) in central Amazon. Regulated Rivers: Research and Management 1990; 5: 195-215

[9] Barthem R., Goulding M. The Catfish Connection: Ecology, Migration, and Conservation of Amazon Predators. New York, Columbia University Press; 1997.

[10] Barthem RB, Goulding M. An Unexpected Ecosystem: The Amazon revealed by the fisheries. Amazon Conservation Association (ACA) - Missouri Botanical Garden Press. Lima, Peru; 2007.

[11] Almeida O, Lorenzen K, McGrath D. The commercial fishing sector in the Regional Economy of the Brazilian Amazon. In: R. Welcomme and Peter. T. (Org.). Proceedings of the Second International Symposium on the Management of Large Rivers for Fisheries. 1 ed. Bangkok: FAO-Regional Office for Asia and the Pacific/RAP Pulication2004; 2 15-24

[12] Isaac VJ, Almeida MC. El consumo de pescado en la Amazonía brasileña. COPESCAALC Documento Ocasional 2011; 13: 43.

[13] Humphries P, Lake PS., Fish larvae and the management of regulated rivers. Regulated Rivers: Research Management 2000; 16:421–432.

[14] Agostinho AA., Pelicice FM, Gomes LC. Dams and the fish fauna of the Neotropical region: impacts and management related to diversity and fisheries. Brazilian Journal Biology 2008; 68(4, Suppl.): 1119-1132.

[15] Tollefson J. A struggle for power. Nature 2011; 479: 160-161 doi: 10.1038/479160a.

[16] Goulding M, Barthem R, Ferreira EJG. The Smithsonian Atlas of the Amazon. Washington: Smithsonian Institution; 2003.

[17] Leite RG. Cañas C, Forsberg B, Barthem R, Goulding M. Larvas dos Grandes Bagres Migradores. Instituto Nacional de Pesquisas da Amazônia (INPA) - Asociación para la Conservación de la Cuenca Amazónica (ACCA). Lima, Peru; 2007.

[18] Lewis WM. Saunders III JF. Two new integrating samplers for zooplankton, and water chemistry. Archiv für Hydrobiologie 1979; 85(2):244-249.

[19] hlstrom EH, Moser HG. Eggs and larvae of fishes and their role in systematic investigations and in fisheries. Revue des Travaux de l'Institut des Pêches Maritimes 1976; 40(3-4) 379-398

[20] Snyder DE. Terminologies of intervals for larval fish development. In: J. Boreman (Ed.) Great Lakes Fish Eggs and Larval Identification. U.S. Fisheries and Wildlife 1976: 41-58.

[21] Araujo-Lima CARM. Larval development and reproductive strategies of Amazonian fishes. PhD Thesis. Oban, Argyll, Scotland, University of Sterling; 1990.

[22] Torrente-Vilara G. Heterogeneidade ambiental e diversidade ictiofaunística do trecho de corredeiras do rio Madeira, Rondônia, Brasil. PhD Thesis. Manaus, Instituto Nacional de Pesquisas da Amazônia (INPA)/Fundação Universitária do Amazonas (FUA); 2009

[23] Bialetzki A, Sanches PV, Cavicchioli, M, Baumgartner, G, Ribeiro, RP, Nakatani K.. Drift of ichthyoplankton in two channels of the Paraná River, between Paraná and Mato Grosso do Sul states, Brazil. Brazilian Archives of Biology and Technology 1999. http://www.scielo.br/pdf/babt/v42n1/v42n1a08.pdf (accessed 18 September 2013).

[24] Marcano D, Cardillo E, Rodriguez C., Poleo G, Gago N, Guerrero H.Y. Seasonal reproductive biology of two species of freshwater catfish from the Venezuelan floodplains. General and Comparative Endocrinology 2007; 153(1-3): 371-377.

[25] Carvalho, N. de O., Filizola Jr., N. P., Santos, P. M. C. dos, and Lima, J. E. F. W., 2000 - Guia de práticas sedimentométricas. Brasília: ANEEL: 154 p.

Chapter 8

Arbuscular Mycorrhizal Fungi and their Value for Ecosystem Management

Andrea Berruti, Roberto Borriello, Alberto Orgiazzi,
Antonio C. Barbera, Erica Lumini and
Valeria Bianciotto

Additional information is available at the end of the chapter

http://dx.doi.org/10.5772/58231

1. Introduction

Arbuscular Mycorrhizal Fungi (AMF) are a group of obligate biotrophs, to the extent that they must develop a close symbiotic association with the roots of a living host plant in order to grow and complete their life cycle [1]. The term "mycorrhiza" literally derives from the Greek *mykes* and *rhiza*, meaning fungus and root, respectively. AMF can symbiotically interact with almost all the plants that live on the Earth. They are found in the roots of about 80-90% of plant species (mainly grasses, agricultural crops and herbs) and exchange benefits with their partners, as is typical of all mutual symbiotic relationships [2]. They represent an interface between plants and soil, growing their mycelia both inside and outside the plant roots. AMF provide the plant with water, soil mineral nutrients (mainly phosphorus and nitrogen) and pathogen protection. In exchange, photosynthetic compounds are transferred to the fungus [3].

Taxonomically, all AMF have been affiliated to a monophyletic group of fungi, i.e. the Glomeromycota phylum [4]. They are considered to be living fossils since there is evidence that their presence on our planet dates back to the Ordovician Period, over 460 million years ago [5]. Investigations on AMF taxonomy began in the nineteenth century with the first description of two species belonging to the genus *Glomus* [6]. Since that date, many Glomeromycotan species, genus and families have been discovered and characterized by means of traditional approaches based on the phenotypic characteristics (mainly spore morphology). Molecular DNA sequencing-based analyses have recently contributed to a great extent by shedding light on a previously unseen and profound diversity within this phylum [7].

Nevertheless, an open debate on the phylogeny of AMF, and in particular concerning some taxonomical groups, is still puzzling scientists [8–10] (Figure 1). Besides a general disagreement about the number of families and genera (Figure 1), what emerges from reference [8] is that Gigasporales are considered to be a separate order from Diversisporales. This is different from what has been reported in the tree on the right side of Figure 1, which was presented in reference [9], and supported by the recent reference [10].

Functionally, AMF form the so-called arbuscular mycorrhizae with plant roots. The most typical AMF structure, which also gives the name to this group of fungi, is the arbuscule (Figure 2). This structure, whose shape recalls that of a small shrub, forms inside the root cortical cells by branching in several very thin hyphae. In this way, the surface area, where the nutritional exchanges between the plant and fungus take place, is maximized. Fungal hyphae that grow between root cortical cells are able to produce other AMF structures, such as intercellular hyphae and vesicles (Figure 2). All these structures that grow inside the plant roots represent the intraradical phase of the fungus. Hyphae also grow outside the plant roots, and generate a network that extends over long distances and explores the soil beyond the nutrient depletion zone that normally characterizes the area surrounding the roots. At the end of the AMF life cycle, or in response to particular environmental conditions, spores (Figure 2) of variable size (up to 400 µm), depending on the species, are produced in the roots and/or in the soil. These, along with external explorative and running hyphae, represent the extraradical phase of the fungus. The synergic action of the intra-and extraradical phases is responsible for the ecological significance of the AMF, a soil-root-living key group of organisms [3].

1.1. The ecological roles of AMF

Arbuscular mycorrhizal fungi have a high relevance in many ecosystem processes. Since they can be found in many different plant species, they can provide their favorable services to almost all terrestrial ecosystems, from grasslands to forests, deserts and agroecosystems [11]. AMF can play several roles in such environments. The most agriculturally significant and frequently investigated one, from both the ecological and physiological points of view [12], is their positive effect on plant nutrition and, consequently, on plant fitness. In particular, they play a pivotal role in helping the plant uptake phosphorus from the soil [13]. Without AMF, it is rather difficult for the plant to absorb this macroelement from the soil, since it is mainly available in its insoluble organic or inorganic form. Besides phosphorus, AMF can also translocate water and other mineral nutrients (in particular nitrogen) from the soil to the plant. These nutritional exchanges are bidirectional. As a consequence, particularly efficient symbiotic associations have been demonstrated to stabilize through unknown mechanisms, with the plant selecting the most cooperative fungal partners and vice versa [14]. The AMF-inducible recovery of plant nutritional deficiency can inevitably lead to an improvement in plant growth, with a potential positive impact on productivity. Needless to say, AMF have attracted a great deal of interest from the agricultural world over the years [15].

AMF are also responsible for other services that favour the plants they colonize: (a) they positively affect plant tolerance towards both biotic (e.g., pathogens) and abiotic stresses (i.e., drought and soil salinity) by acting on several physiological processes, such as the production

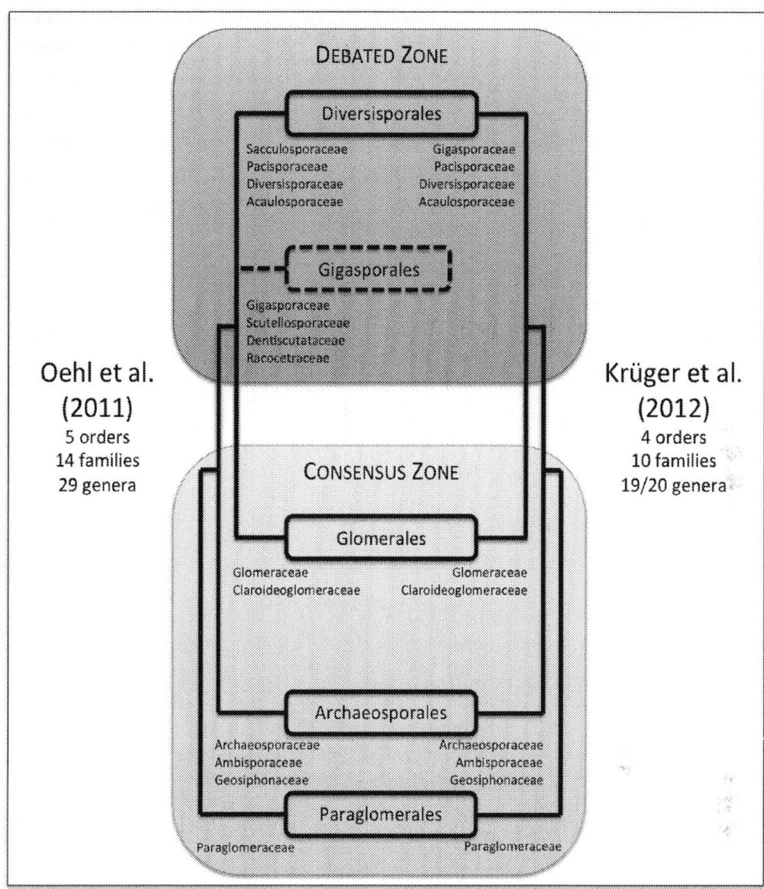

Figure 1. A schematic representation of two recently published and partly controversial phylogenetic trees of the Glomeromycota phylum (reference [8] for the tree on the left side and [9] for the tree on the right side). The one published in reference [8] was based on molecular (SSU, ITS, partial LSU rDNA, and partial β-tubuline gene) and morphological analyses (spore wall structures, structures of the spore bases and subtending hyphae, germination, and germination shield structures). The tree published in reference [9] was based on concatenated SSU rDNA consensus sequences (ca 1.8 kb).

of antioxidants, the increment of osmolyte production or the improvement of abscisic acid regulation [16,17], and the enhancement of plant tolerance to heavy metals [18]; (b) they help plants become established in harsh/degraded ecosystems, such as desert areas and mine spoils [19]; (c) they increase the power of phytoremediation (the removal of pollutants from the soil by plants) by allowing their host to explore and depollute a larger volume of soil [20,21]. Another crucial ecological role played by AMF is their capacity to directly influence the diversity and composition of the aboveground plant community. Several studies have

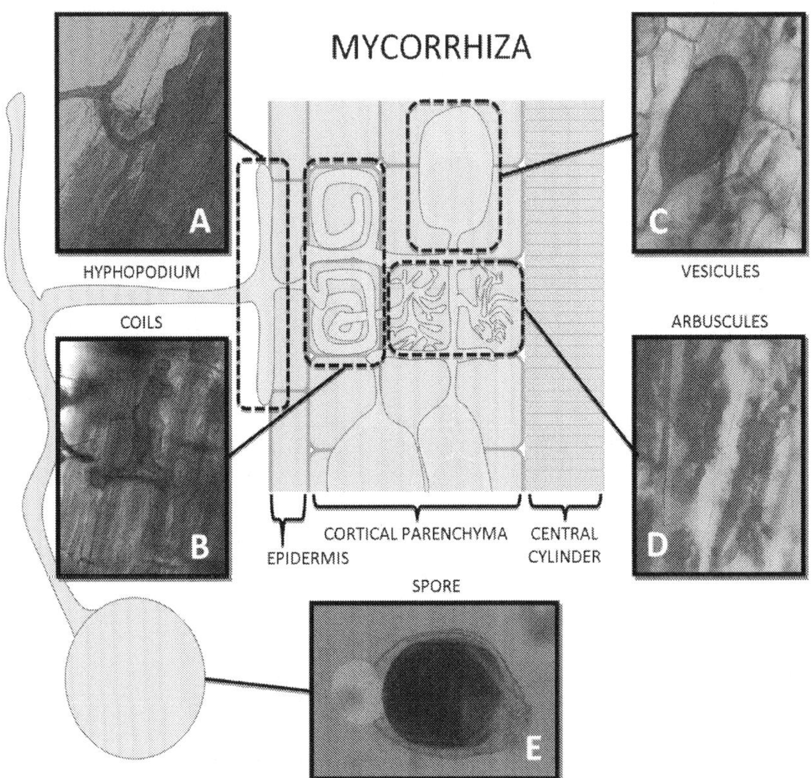

Figure 2. Extraradical and intraradical phases of AMF growth. The spore (Figure E, a *Scutellospora sp.* spore stained with Melzer's reagent and squeezed) germinates in the bulk soil and approaches the root of a host plant. The fungus penetrates through a hyphopodium (Figure A, stained with 0.1% cotton blue encountered in *Camellia japonica* L. roots) and develops intracellular coils, extracellular vesicles and intracellular arbuscules (Figures B, C, D) within cortical parenchyma, without entering the central cylinder where the vessels are.

confirmed that plant species richness can be altered not only by climatic and edaphic factors, but also by soil microbial assemblages [22–24]. The underlying mechanism is not completely understood, but could be related to the promotion of seedling establishment of secondary plant species [25]. Nevertheless, on some occasions, AMF can also negatively affect the diversity and growth of plants, which is particularly significant for the management of weeds [26]. Last but not least, AMF play a critical role in soil aggregation, thanks to their thick extraradical hyphal network, which envelops and keeps the soil particles compact. It has been suggested that glycoproteins (glomalin and glomalin related proteins) secreted by AMF into the soil could exert a key role in this process [27,28]. These proteins are exuded in great quantities into the soil, and could have implications on carbon sequestration. This potential capability of AMF is likely to contribute to a great extent to the soil ecosystem carbon dioxide (CO_2) sequestration process. This aspect has led to the recognition of the importance of this group of organisms in processes related to climate change mitigation [29].

All the services offered by arbuscular mycorrhizal fungi confirm the need to study and describe all their features, including their biology, ecology, taxonomy, phylogeny and biodiversity. Over the years, several techniques have been developed to reach this goal: a brief history is reported in section 1.2.

1.2. Methods used in the study of AMF

This group of organisms has a constraining characteristic that makes their study very complex: as obligate symbionts, they cannot be cultivated in vitro, away from their host plant. The development of an artificial system that is capable of going beyond this barrier dates back to the 1980s, when in vitro transformed carrot roots were successfully colonized by AMF species [30]. Thanks to this method, the study of arbuscular mycorrhizae became easier and many researches on both physiology and genetics became possible [31,32]. Over the last two decades, many molecular and physiological mechanisms involved in the symbiotic process between plants and AMF have been discovered and described, thanks to the increasing innovations and opportunities offered by molecular biology. For example, it is now known how the infectious process of AMF arises, and many of the involved genes have been identified successfully [33].

Molecular biology has also revolutionized the analysis of the biodiversity of AMF, making it easier and more accurate to characterize the AMF community composition of large quantities of samples from many different ecosystems, from prairies to savannas, and from grasslands to forests (Table 1). The first studies on the diversity and distribution of AMF were mainly focused on the identification of the species that colonize the roots of a given plant in a given environment [34]. This was mainly due to the deficiency in the available investigation techniques, as they were primarily based on spore surveys and intraradical fungal structure morphological identification by means of microscopy. Such morphological identification surveys were time consuming and often lacked accuracy, since many species were easily confused with others. The situation changed radically when the use of DNA-based techniques became common, and the extraction of DNA from plant tissue was reduced to a few relatively easy steps that could be reproduced in any laboratory [35,36]. The load bearing principle is simple: by sequencing a specific DNA region, it is possible to univocally identify the corresponding AMF [37]. So far, the most used DNA target regions for AMF identification are located on the ribosomal genes (Small and Large ribosomal Subunits – SSU and LSU, respectively – and the Internal Transcribed Spacers – ITS1 and ITS2), as they show a rate of variability that is sufficient to discriminate between AMF species/isolates [9]. All this has led to the current era of molecular identification of AMF species [10]. Next-Generation Sequencing (NGS) tools represent a further step forward for biodiversity surveys of all organisms [38], including AMF. Over the last few years, the number of NGS-based AMF biodiversity studies has increased, while the spectrum of the target environments has broadened [39]. Furthermore, new primer pair sets for the specific amplification of AMF DNA sequences, capable of providing higher accuracy and a comprehensive coverage of the whole Glomeromycota phylum, have been developed [40]. Nowadays, AMF assemblages are no longer studied only in plant roots, but also in the bulk soil [41–43]. The main result obtained from the application of NGS to the study of AMF biodiversity has been the discovery of an unpredictable diversity within the Glomeromycota phylum [39]. However, this series of innovative molecular tools has introduced a new issue, that is, the continuously increasing number of unidentified AMF DNA sequences from environmental samples with no correspondence whatsoever to sequences of known species [44]. This has naturally made scientists aware of the fact that the number of AMF species could be larger than expected. However, it is not reliable to have new species described on just the basis of short DNA sequences obtained by means of NGS tools. Instead, for each new suggested taxon, a series of steps needs to be followed to characterize the morphotype, the functional traits, and the ecological role offered when present in combination with other organisms in a given environment. Therefore, NGS tools cannot be considered as complete replacements of the traditional methods of identification and description of new species. The combined approach is still necessary to shed light on such a key group of organisms and to make them available for agricultural application and, more in general, for other practices useful for the wellbeing of humankind [45].

1. Reference	2. Year	3. Method	4. Target region	5. Studied compartment	6.Ecosystem	7. AMF sequences	8. OTUs
[39]	2013	Clon-seq/NGS	SSU	Plant root	Tropical, subtropical, temperate and boreal forests, subtropical and temperate grasslands, tropical and subtropical deserts and shrublands, and polar tundras (Africa, Asia, Oceania, Europe, North and South America)	2353/22391	204
[46]	2013	NGS	SSU	Soil	Prairie (Cananda)	1335521	120
[47]	2013	NGS	SSU	Plant root and Soil	Temperate forest (Estonia)	35738	76
[48]	2013	Clon-seq	SSU	Plant root	Mediterranean semi-arid soils (Spain)	467	30
[49]	2013	Clon-seq	SSU	Soil and plant root	Prairie (USA)	232	13
[43]	2012	NGS	SSU	Soil	Forest (Estonia)	13320	37
[50]	2012	NGS	SSU	Soil	Arable field (China)	59611	70

1. Reference	2. Year	3. Method	4. Target region	5. Studied compartment	6. Ecosystem	7. AMF sequences	8. OTUs
[51]	2012	NGS	SSU	Soil	Prairie - Chernozem (Cananda)	7086	33
[52]	2012	NGS	LSU	Plant root	Grassland (Denmark)	82511	32
[42]	2012	Clon-seq	SSU/LSU	Soil and plant root	Arable field (Italy)	427/364	20/23[a]
[53]	2012	Clon-seq	SSU	Plant root	Alpine meadow ecosystem (China)	4452	38
[54]	2011	NGS	SSU	Plant root	Broadleaf, mixed broadleaf and coniferous forests, botanical gardens, greenhouse	65001	73
[55]	2011	NGS	SSU	Plant root	Grassland, wood and heath (UK)	108245	70
[56]	2011	Clon-seq	SSU	Plant root	Hardwood forest (USA)	1598	17
[41]	2010	NGS	SSU	Soil	Mediterranean soils (Italy)	2815	19/80[a]
[57]	2010	Clon-seq	SSU	Soil and plant root	Vineyard (Italy)	681	37
[58]	2009	Clon-seq	SSU	Plant root	Woodland (UK)	617	33/37[b]
[59]	2009	Clon-seq	SSU	Plant root	Mediterranean semi-arid soils (Spain)	1443	21
[60]	2009	NGS	SSU	Plant root	Boreal forest (Estonia)	111580	47
[61]	2008	Clon-seq	LSU	Soil and plant root	Arable field (Italy)	183	8
[62]	2008	Clon-seq	SSU	Plant root	Boreal forest (Estonia)	911	26/27[c]
[63]	2008	Clon-seq	SSU	Plant root	Arable field (Mexico)	213	16
[64]	2008	Clon-seq	SSU	Plant root	Serpentine soils (USA)	1249	19
[65]	2008	Clon-seq	SSU	Plant root	Arable field (Sweden)	115	8
[66]	2007	Clon-seq	ITS	Soil, plant root and spores	Meadow (Germany)	180	>18
[67]	2007	Clon-seq	SSU	Rhizoids	Liverworts (World-wide)	150	10

1. Reference	2. Year	3. Method	4. Target region	5. Studied compartment	6.Ecosystem	7. AMF sequences	8. OTUs
[68]	2007	Clon-seq	LSU	Soil and plant root	Arable field (France)	246	12
[69]	2007	Clon-seq	SSU	Plant root	Grassland (Sweden)	185	19
[70]	2007	Clon-seq	ITS	Plant root	Volcanic desert (Japan)	205	11
[71]	2006	Clon-seq	SSU	Plant root	Polluted soils (Italy)	115	12
[72]	2005	Clon-seq	SSU	Plant root	Warm-temperate deciduous forest (Japan)	394	5
[73]	2004	Clon-seq	SSU	Plant root	Wetland (Germany)	546	35
[74]	2004	Clon-seq	LSU	Plant root	Grassland (Denmark)	158	11
[75]	2004	Clon-seq	ITS	Plant root	Pasture (UK)	30	10
[76]	2004	Clon-seq	SSU	Plant root	Grassland (Japan)	200	8
[77]	2004	Clon-seq	SSU	Plant root	Grassland (UK)	606	9
[78]	2003	Clon-seq	ITS	Plant root	Afromontane forests (Ethiopia)	92	20
[79]	2003	Clon-seq	SSU	Plant root	Boreal forest (Estonia)	16	6
[80]	2002	Clon-seq	SSU	Plant root	Seminatural grassland (UK)	88	24
[81]	2002	Clon-seq	SSU	Plant root	Woodland (UK)	232	13
[82]	2002	Clon-seq	SSU	Plant root	Tropical forest (Republic of Panama)	1536	18/23[d]
[83]	2002	Clon-seq	SSU	Plant root	Tropical forest (Republic of Panama)	558	18
[84]	2001	Clon-seq	SSU	Plant root	Arable field (UK)	303	8
[36]	1999	Clon-seq	SSU	Plant root	Seminatural woodland (UK)	141	6/8[e]
[85]	1998	Clon-seq	SSU	Plant root	Woodland (UK)	253	6/10[b]

a-taxa obtained with different primer sets; b: taxa obtained at different study sites; c-taxa obtained from forest ecosystems of different ages and management intensities; d-taxa obtained from roots of different plant species; e-taxa obtained at different sampling times.

Table 1. The table shows an overview of DNA-based studies on the diversity of Arbuscular Micorrhizal (AM) fungal communities. For each study, the following are reported in sequence: 1. Reference, 2. Year of publication, 3. Used method (Clon-seq=cloning and sequencing; NGS=next generation sequencing), 4. Studied DNA region (SSU=Small Subunit; LSU=Large Subunit, ITS=Internal Transcribed Spacer), 5. Compartment from which the DNA was analyzed, 6.

Ecosystem from which the samples were collected, 7. Number of DNA sequences and 8. OTUs (Operational Taxonomic Units) from arbuscular mycorrhizal fungi.

2. The impact of humans on AMF biodiversity

Most human activities have an arguable impact on the physical and biological aspects of soil. As mentioned before, AMF are among the most widespread soil microorganisms, and each human activity that has an impact on soil, such as agricultural practices, therefore has a side effect on them. These practices, alone or in combination, exert an enormous selective pressure on AMF that shapes their community structure and evolution by modifying several of their biological features, such as sporulation strategy, resource allocation and spatial distribution [86]. As in natural ecosystems, AMF are also present and active in agricultural ecosystems, where they colonize several major arable crops (sorghum, maize, wheat and rice). Many studies have indicated that AMF diversity, effectiveness, abundance and biodiversity decline in agroecosystems subjected to high input practices [41,42]. Modern intensive farming practices that implement deep and frequent tillage, high input inorganic fertilization and pesticide use are evidently a particular threat to AMF. This is surely a drawback for agriculture, since the more AMF biodiversity losses, the fewer AMF functional traits the host plant can benefit from. On the other hand, the activity and diversity of AMF, following conversion from conventional to organic farming, have not yet been investigated thoroughly. However, the available data seem to indicate that AMF respond positively to the transition to organic farming through a progressive enhancement of their activity [87]. Even though it is difficult to discriminate between the effects that different agricultural treatments exert on AMF communities, they are here considered separately, and their role in shaping AMF communities will be analyzed.

2.1. Tillage: A conventional practice detrimental to AMF

One of the most ancient and representative agricultural techniques is tillage. Tillage has played a crucial role in the evolution and technological development of agriculture, particularly for food production. The benefits produced by tillage include a better conservation of water and soil fertility, the abatement of weeds and the preparation of a suitable seedbed. To fulfill these tasks, the undisturbed soil is mechanically manipulated in an effort to modify the physical characteristic of the soil and eliminate weeds. The physical, chemical and biological effects of tillage on the soil can be both beneficial and negative, depending on the methods that are used. The inappropriate use of tillage techniques can therefore have a dramatic impact on the soil structure and on soil microorganism community assemblage. It is possible to identify different tilling levels, ranging from a very low impact, "No-tillage", to a high impact, "conventional tillage". A continuum of intermediate conditions lies in between these two extreme situations, e.g. varying frequency and intensity of the plowing.

The mechanical soil disturbance experienced by AMF in tilled agricultural soils has no equivalent in natural ecosystems. This is why tillage has been widely recognized to be one of the principal causes of the modification of the AMF communities that colonize plant roots in agricultural fields [88]. Mycorrhizal diversity, at a family level [88], and the timing of root colonization [89] can be affected negatively. As a consequence, the effectiveness of AMF [90] is likely to be reduced. Periodically repeated mechanical soil disturbance destroys the extra-radical mycelial network formed by AMF. This very complex underground structure can reach lengths of up to some tens of meters in one gram of soil [91], and represents a soil "highway" for nutrient transport. For this reason, it is often claimed to be closely correlated to biodiversity, biomass production and the functioning of plant communities [22,25,92].

An ecological shift in AMF communities is particularly noticeable when frequently and infrequently tilled agroecosystems are compared [42,63,88,93]. This is probably due to the different tolerance to hyphal disruption among the different AMF species [94,95]. Although AMF species can colonize plants from spores, this process often requires a certain amount of time. Faster root colonization can be reached in the presence of a viable and well-structured underground mycelial network that facilitates AMF proliferation and speeds up plant root penetration [96]. On the other hand, AMF species differ greatly in their capacity to restart colonization from fragmented mycelium or root fragments [97]. Intense tillage could be a factor that favors those AMF species that are more able to proliferate from fragmented hyphae or root fragment [98], and could therefore determine a shift in AMF community assemblages. A clear example of this is the large presence of Glomeraceae species found in tilled soil all over the world [99]. AMF species belonging to this group are able to randomly connect hyphae in close proximity after disruption, a condition that can easily be found in disturbed soil. This allows these species to proliferate more easily and to rapidly become dominant over slow-growing AMF. The members of the Gigasporaceae family, for example, use spores as the main source of root colonization, but do not regrow from hyphal fragments [97].

2.2. Fertilization

Another agricultural practice that has major ecological fall-outs is chemical fertilization. This practice is often claimed to be fundamental in improving the growth performance of plants, but it is sometimes abused. In addition to the environmental drift and the possible pollution of underground water reservoirs, the presence in the soil of high levels of fertilizer dramatically alters the interaction between plants and microbial communities. The central role of arbuscular mycorrhizae in plant nutrition makes them very susceptible to changes in soil nutrient availability. Generally, in a nutrient-rich environment, a plant can directly uptake enough nutrient from the soil, without the "catering" service provided by the AMF partners. As a result, the dependency of plants on their AMF partners gradually diminishes, and AMF community richness and diversity decline [42,53,100,101]. It is thought that fertilization can alter the performance of this symbiosis, making microbial partners costly, and even parasitic [102]. It has been hypothesized that the enrichment of soil resources, due to high input fertilization, could lead to a reduction in plant allocation to roots and mycorrhizas [103], and an accumulation of nutrient resources in epigeous plant sinks [104]. A reduction in host plant resource allocation to the fungal partners can therefore result in a decrease in AMF root colonization [105], and an increase in fungal competition for limited C resources. Moreover, this reduction in host nutrient availability is thought to shift the competitive balance between microbes, favoring more aggressive, antagonistic microbial genotypes [106–108]. This change in competitive balance can alter the evolution of the functional traits of AMF by reprogramming AMF to reduce their allocation to structures devoted to nutrient exchange (arbuscules and coils), and increase their allocation to internal storage and growth structures (vesicles and intraradical hyphae) [103,109,110]. This is likely to result in an incremented presence of highly competitive AMF which, on the other hand, will be less beneficial to the host crop [111].

Particular AMF taxa have been found to be more sensitive than others to specific fertilization conditions [42,50,53,65,93,112]. This is probably due to the different taxon-related ability of the AMF taxa to manage nutrient absorption. For instance, *Acaulospora* species have been demonstrated to be very effective in P uptake, and in the transfer to the host plant, compared to Glomeraceae species [113]. In line with these findings, Acaulosporaceae species have been considered to decrease to a great extent under high input P fertilization [50]. The same thing has been observed for Gigasporaceae in N-enriched soils [50,103]. On the other hand, Glomeraceae species, such as *Rhizophagus intraradices*, are able to cope well with nutrient rich environments [50,53].

2.3. Crop rotation

The choice of crop and rotation made by the farmer has a crucial impact on AMF communities. Even though AMF are commonly recognized as generalist symbionts that show the ability to interact with different plant species, some plant-fungus combinations can perform better than others. The choice of the partner is not univocal, but is believed to be driven by a reciprocal reward mechanism between the two symbionts involved [14]. This means that both the plant and the AMF communities can exert an important role in modifying the community composition of the partner [22,23]. Thus, different cultivation practices that involve a variation in plant diversity, such as monoculture, fallow and crop rotation, could show different and profound effects on AMF community assemblages.

Monoculture can be highly deleterious for AMF communities, and result in a significant reduction in mycorrhizal root colonization [114] and mycorrhizal diversity [115,116]. The effect of continuous monocropping, especially when crops that are not highly dependent on AMF-mediated nutrition (e.g. wheat) are used, favors the selection and proliferation of less cooperative and more aggressive fungal symbionts. These are likely to enact similar behavior to parasitism [102,106]. In addition, intensive tillage treatments, which are necessary in the case of monoculture practices, can overly disperse fungal propagules, thus allowing fewer AMF isolates to dominate the community profile. The dominion of AMF species with a poor mutualistic attitude could be toned down by alternating the cultivation of plant species that are less dependent on AMF with 'break crops', such as *Brassica* [117] or legumes [118]. The former is a non-mycorrhizal crop that can therefore act as an inhibitor of the dominant AMF species proliferation. The latter represent the opposite approach, since legumes are AMF-dependent crops that favor the overall propagation of AMF communities. This is the fundamental principle of crop rotation, a practice that can exert a control function that prevents particular AMF from dominating the soil matrix. Hence, crop rotation has the potential of driving AMF communities to be less parasitic [86]. It has been experimentally demonstrated that crop rotation promotes higher AMF diversity [115,119], and can reshape AMF communities derived from agricultural fields to be more diverse and similar to the ones detected in natural ecosystems [87].

3. AMF biodiversity restoration

Agricultural fields, degraded lands and the so-called "third landscapes" are all soil environments in which humans have had an impact on the ecological balances, by unchaining a series of inevitable ecosystem alterations. Therefore, the restoration of such balances should be a necessity. Owing to their role in the promotion of plant health, soil nutrition improvement and soil aggregate stability, AMF are primary biotic soil components that, when missing or impoverished, can lead to a less efficient ecosystem functioning. The presence of a high degree of AMF biodiversity is in fact typical of natural ecosystems and indicates good soil quality [120]. Consequently, a process that aims at the re-establishment of the natural level of AMF richness is a pivotal step towards the restoration of the ecological balances. As previously mentioned, the cultivation practices adopted for major crops include anthropic inputs that can impact AMF occurrence and/or diversity. Of these, the use of fertilizers and pesticides also has an adverse impact on production costs, and should be reconsidered due to the heightened social concern about the corresponding environmental drift [121]. As a consequence, the need to benefit from AMF as a biofertilizer, with a view to sustainable agriculture, is becoming increasingly urgent. An appropriate management of these symbiotic fungi would lead to a great reduction in chemical fertilizer and pesticide inputs, a key target for growers facing a crisis, and having to deal with a more environmentally aware clientele. Two main strategies are possible to achieve this goal: the direct re-introduction of an AMF pool (referred to as "inoculum") into the target soil, or the selective management of the target ecosystem. These strategies can be selectively adopted when a population of AMF propagules of low effectivity is present, or when the indigenous AMF are absent or very low. This means that the AMF restoration process is suitable for different purposes, e.g. greenhouse and open-field cultivation, and even in helping the rehabilitation of degraded lands.

3.1. AMF inoculation and the role of enterprises

The re-introduction of AMF into soils that are impoverished in belowground biodiversity is a complex strategy, but it can be very rewarding. Unfortunately, the production of AMF inoculum on a large-scale is very difficult using the techniques currently available. The main obstacle to the production of an AMF inoculum lies in their peculiar symbiotic behaviour, the AMF compulsorily requiring a host plant for growth. This means that AMF are propagated through cultivation with the host plant, and this usually requires time-demanding protocols and cumbersome infrastructures. The maintenance of AMF reference collections requires methodologies that are rather different from those used for other microbial collections and inoculum production. Unlike non-obligate symbionts, the production of AMF inoculum requires the control and optimization of both host growth and fungal development. Thus, these propagation techniques involve high costs that are not apparently competitive with fertilization-related costs. The impossibility of rapidly assessing AMF colonization on the host plant, together with the complexity of AMF species identification, also contribute to the pitfalls of inoculum agricultural usability. Moreover, the management of the high amount of inoculum necessary for extensive use is very challenging. It has been suggested that AMF is more suitable for plant production systems that involve a transplant stage, as inoculation is carried out more easily, and smaller quantities of inoculum are needed. At a first glance, establishing an open-field, large-scale inoculation treatment would seem technically impractical and economically prohibitive. However, once AMF biodiversity has been restored, AMF-friendly practices, such as fall cover cropping [122], can be put in place in order to help the AMF persist. If no detrimental agricultural practices are carried out, the biodiverse mycelial network will remain unaltered and infective in the future. For example, in revegetation schemes, it would be totally impractical to restore an entire degraded land, which often appears as a highly extended surface, through inoculation. A particular approach must be considered when it is necessary to face these situations. First, the ability of specific cover crop mixtures and even target indigenous plant species to elevate the native AMF inoculum has to be taken into account as a potentially successful selective management tool to aid the recovery of desertified ecosystems [123]. However, since ecosystem functioning is supported by a close liaison between the aboveground plant diversity and belowground AMF diversity [22], the excessive loss of AMF propagules in degraded ecosystems could, in some cases, preclude either natural or artificial revegetation. For this reason, an inoculation step may also be needed. Although it would be too laborious and expensive to re-introduce AMF and cover plants into entire lands, a smaller-scale approach should be adopted. Taking inspiration from the idea of creating the so-called "fertility islands" [124], only small patches of cover plants could be inoculated with AMF. This could lead, in time, but with reduced costs, to the re-establishment of a mycelial network that would also be able to allow native plant species to quickly recover the nutrient impoverished land.

Hence, AMF restoration would only represent an initial cost and, if soil AMF persistence is favoured, this cost could be subjected to amortization over the years. This makes the application of AMF particularly attractive since, as already demonstrated [125,126], it could provide considerable savings for growers and for degraded land recovery projects, in comparison to conventional fertilization. It is important that the end-users cultivate a portion of their crop without inoculum in order to assess the cost-effectiveness and the beneficial effects on plant fitness due to AMF inoculation [127]. Growers are starting to understand the significance of sustainable agricultural systems, and of reducing phosphorus inputs using AMF inocula, especially in the case of high value crops, such as potted ornamental plants. These crops can easily be regarded as the result of organic crop farming, and be sold at a premium price to an eco-friendly orientated consumer class. However, the absence of solid inoculation practices still represents a problem, and applied research should therefore be focused on defining the best inoculum formulation strategies [128] and imparting know-how to the growers.

Since large-scale AMF production is impractical for growers, the significance of AMF has not been ignored by the commercial sector, and many AMF-based inocula are nowadays available for sale. AMF inoculum production began in the 1980s and flourished in the 1990s. Nowadays, several companies produce and sell AMF inocula. In recent years, these products have come under increasing scrutiny by scientists and end-users. Most manufacturers advertise their products by pointing out their suitability for a wide range of plants and environmental conditions. Unfortunately, their promises made about these products and the results seen are too often worlds apart. This has led to radical generalisations, both positive and negative, about the efficacy of the currently available products. The problem is that success, in terms of root colonization and plant response, is unpredictable since no plant does best with the same AMF mix [129]. In terms of fungal content, the manufacturer's tendency is to introduce a more or less biodiverse mix of AMF. Some companies have chosen the approach of single formulations, while others produce a range of differently shaped products for their target end-users. Glomeraceae species are usually used, but also Gigasporaceae, Scutellosporaceae and Acaulosporaceae families are gradually being introduced to commercial inoculum production. These few used species can be routinely propagated for spore applications, are found in association with a large variety of host plants and are geographically distributed all over the world.

Great problems arise in formulating the inoculum product in its most suitable state for the market. In the coming years, it is likely that greater regulation and controls will be introduced concerning the production and selling of AMF inocula. In Europe, the regulation of these products varies from country to country, with some having very strict regulations, while others are less demanding. In North America, Canada, for instance, considers AMF inocula to be only supplements and not fertilizers. In the USA, registration may fall either to the fertilizer or the pesticide sectors, depending on the supposed action of the formulated AMF inoculum. However, in most countries, AMF are no longer considered dangerous for human or animal health, and no infectivity or toxicity tests are therefore necessary. Normally, an application for registration has to be filled in and a series of meticulous information needs to be attached to the registration request. These data should also be reported on the inoculum label, and should include the list of all the ingredients and their concentrations, a detailed taxonomic description of the AMF, the isolate's history, the geographic origin and distribution, some literature on the beneficial effects of the isolate, a list of possible contaminants, an official safety data sheet, information about the producer, the number of viable AMF propagules or the percentage of colonization expected on reference plants after a known quantity is inoculated, the list of recommended plant hosts, the suggested soil conditions for inoculum effectiveness, the recommended application method/dosage, the suggested storage conditions, the expiration date and information on the manufacturing processes. Other information regarding previous tests performed with different soil, and which confirms the climatic conditions and the beneficial effect of the inoculum should also be added in order to highlight the reliability of the product and to help direct the consumer. Preventing over-regulation will be crucial in assisting the development of SMEs (Small and Medium Enterprises), and in helping refresh the market with this eco-friendly biotechnological tool.

In order to allow the AMF inoculum market to develop, scientists should define a series of 'best practices' that could be adopted by these SMEs to solve serious issues related to their product quality. One of these issues arises from the need to control the biological composition of the product, especially for the possible presence of pathogens, but above all to assess its quality in terms of AMF composition. Being obligate symbionts, AMF are non-axenically culturable, while only a few can be monoxenically cultured. Therefore, an inoculum is produced above all using a containerized-culture, either in greenhouses, growth chambers, or in fields, and, as a result, cannot be completely free from external microorganisms. There is increasing awareness of the risk of pathogens, and many concerned producers are even making use of agrochemicals in an attempt to avoid contamination of their product. Others have instead decided not to include host root residues in their formulation, in order to avoid pathogen carry-over. Alternatively, surface sterilization of the incorporated colonized roots can be introduced without affecting the viability of the AMF propagules [130]. As far as quality control in terms of AMF composition is concerned, it is essential to verify whether the product effectively has the potential described on the label. With AMF, in order to confirm the fungal identity, such an assessment can be done through morphological identification of the spores [131,132]. Unfortunately, this technique requires a great deal of labor and there are very few experts in the world that are able to conduct a reliable identification solely on the basis of spore morphology [133]. Quick and user-friendly molecular techniques have been developed to detect AMF strains from complex matrices, such as soil [41,42] and AMF inocula [129,134]. The discrimination of AMF, on the basis of these techniques, relies almost completely on the sequencing of the ribosomal genes, the genetic region on which the AMF phylogenesis was constructed (4), and is still under debate [8–10]. Molecular techniques also allow the inoculated isolates to be reliably traced inside the host plant and their persistence in the soil to be established [135]. The use of Realtime qPCR and specific primers appears to be a very promising tool for the tracing of AMF isolates and their quantification in the host roots after application [136]. A recent study has even used laser microdissection to qualitatively monitor the arbuscule formation in *Camellia japonica* L., after inoculation with a highly biodiverse AMF inoculum [134]. Such a quality control is very important to exclude poor quality or defective AMF inocula from the market.

3.2. Key steps and current techniques for inoculum production

The actual inoculum propagation and formulation process entails a series of key steps that are crucial for the good quality of the final product. The most determining aspect of inoculum formulation is the choice of the AMF content. As mentioned before, the tendency is to introduce a mix of several AMF into commercial inocula. The most scientifically investigated AMF isolate, i.e. *Rhizophagus irregularis* DAOM197198 [137], is also one of the most frequently used for commercial inoculum formulation. This species is a very generalist symbiont that can colonize a large variety of host plants, survive long-term storage, is geographically distributed all over the world and, last but not least, adapts well to both in vivo and in vitro propagation. These characteristics make this isolate of *R. irregularis* suitable to be a premium component of commercial inocula. As previously mentioned, several other AMF that mainly belong to Glomeraceae species, but also to Gigasporaceae, Scutellosporaceae, and Acaulosporaceae families, are gradually being introduced into commercial inoculum production. It is important to notice that AMF are sometimes marketed as consortia that contain ectomycorrhizal fungi, saprophytic fungi and plant growth-promoting rhizobacteria (PGPR), in order to increase the product potential for plant protection and production. The proper choice of the inoculum AMF content is unfortunately constrained by a lack of knowledge on the specificity of the relationships between a specific AMF strain and a particular crop, and on the compatibility and competition of the AMF strains for niches in the soil environment [128]. When AMF are examined as a community, there is abundant evidence that fungal growth rates can be host- and niche-specific. In reference [60], it has been suggested that partner specificity in AM symbiosis may occur at an ecological group level of both the plant and fungal partners. In [14], it has been demonstrated how reciprocal "rewards" stabilize cooperation between the host-plant and the fungus, thereby enforcing the best symbiotic combinations. Thus, the best way of finding the most cooperative and specific AMF isolates for the formulation of more targeted inocula is to directly screen what nature offers, by fathoming out the naturally occurring symbiotic combination set. For example, some AMF species are commonly recognized to be more stress tolerant than others, and are usually found in stressed and polluted soils [18,138]. Native AMF from areas affected by osmotic stresses can potentially cope with salt stress in a more efficient way than other fungi [139]. Thus, it is preferable to take this into account when "tuning" an inoculum to a particular kind of degraded/stressed soil and in order to avoid failure of the revegetation process [140,141]. Optimal benefits will only be obtained from inoculation after a careful selection of the favorable host/niche/fungus combinations. For this reason, natural or semi-natural ecosystems, in which the desired host plant is well established, represent a valid source of naturally selected AMF. However, this highly selective inoculum formulation requires time and hard work. An intriguing approach would be to formulate a series of highly biodiverse inocula, including several AMF species/strains of different geographical/environmental origin, which would be capable of offering benefits to multiple host plants under different environmental conditions, thus making researchers switch from looking for a superstrain to formulating a superinoculum.

AMF can use a number of different types of propagules to colonize new roots with different degrees of efficiency [142]. These are components of the extraradical and intraradical phase of AMF. The extraradical phase comprises spores and a mycelium that forms the hyphal network. Several fungal structures, inside both living and dead root fragments, can represent a source of inoculum [143]. Vesicles, in particular, have been shown to be very infective [97]. Considering that a number of different propagule types exist, it is of primary importance to determine the most eligible and user-friendly to be adopted as inoculum sources. Unfortunately, this is more complex than may be expected, since different AMF taxonomical ranks differ in their ability to propagate from a given propagule. As already mentioned, for instance, it seems that propagation through mycelial fragmentation may be more important for species of the Glomeraceae family, whereas spore germination may be the preferential type of propagation for species in other families (e.g. Gigasporaceae). In reference [144], the authors tested the establishment of a biodiverse community of AMF in a pot culture using different sources of inoculum from the field. They found that spores were successful in establishing most species of Acaulosporaceae, Gigasporaceae and Scutellosporaceae, whereas Glomeraceae species were only dominant when root fragments or soil cores were used. It is important to consider that these different propagation strategies can also reflect on the potential agricultural use of a particular AMF inoculum.

Once the AMF content has been selected, pure monospecific cultures are normally obtained from a single spore, or a small piece of colonized root fragment, or mycelium collected directly from field plants, or obtained from AMF collection cultures. The AMF propagule spreads and colonizes the root apparatus of the host plant, and the subsequent pot-culture generations lead to the production of high quantities of AMF inoculum. Several organizations throughout the world have research culture collections (The International Culture Collection of VA Mycorrhizal Fungi, INVAM; The Banque Européenne des Glomales, BEG; The Canadian National Mycological Herbarium, DAOM; The Canadian Collection of Fungal Cultures, CCFC; The non-profit Biological Resource Center ATCC; The Glomeromycota In Vitro Collection, GINCO; NIAS, National Institute of Agribiological Science) and provide users with reliable AMF propagules to start propagation. Moreover, detailed information on species origin and distribution, spore morphology, and molecular biology and biochemistry are often provided by these organizations. The common purpose of these available AMF collections is to provide a stock source of pure and reliable material for fundamental and applied research use.

A pivotal step during AMF inoculum propagation is the choice of an adequate host plant. The criteria required for the host plant are its high mycorrhizal dependency and potential, i.e. its capacity of being highly colonized by a high number of AMF species, and its inclination to promote growth and sporulation, its suitability to grow under growth chamber or greenhouse conditions and its production of an extensive root system with a high number of fine feeder roots in a short time. A series of plants are commonly recognized as actual AMF "trap" plants, due to their mycorrhizal dependency and lack of specificity, and they are routinely used as host plants during propagation. These include clover (*Trifolium* spp.), plantains (*Plantago* spp.), ryegrass (*Lolium perenne* L.), the tobacco plant (*Nicotiana tabacum* L.), leek (*Allium porrum* L.), Sudan grass (*Sorghum bicolor* (L.) Moench), corn (*Zea mays* L.) and bahia grass (*Paspalum notatum* Flugge).

Pasteurization, steaming and/or irradiation are necessary to avoid contamination of the growing media. The use of a well-aerated substrate is also recommended. The manufacturer must provide the customer who intends to introduce the AMF inoculum to a target plant with basic information and assistance concerning its chemical and physical characteristics, such as nutrient content, pH and salinity. In particular, when elevated quantities of inoculum are used in agricultural fields, or in a pot-culture, controlling the nutrient content is of crucial importance, as it might lead growers to rethink their normally adopted fertilization practices. Conventionally, inoculum formulation processing consists of sieving the substrate and chopped roots of the trap plant in order to retrieve AMF propagules that can be included in the inoculum. This means that the carry-over of a certain amount of nutrients to the final product is unavoidable. Nevertheless, if trap plant pots are not over-fertilized, as it should be during inoculum formulation, the nutrient content will be negligible. A solution to the problem could be the laborious approach of completely separating the spores, mycelium and colonized trap plant root fragments from the used growing media. These substrate-free propagules could then be mixed with an inert-like carrier at a desired rate. The amendment of the inoculum should be compatible with the AMF, almost inert and only serve to support mycorrhizal development. Optimum P and N, but also other macroelement levels, have to be tuned to specific plant–AMF combinations, as mentioned in the previous section, in order not to reduce AMF propagation and diminish plant dependency on mycorrhization after inoculation. Other edaphic factors, such as pH, salinity, soil temperature, moisture and soil aeration, should also be controlled to optimize AMF inoculation. Since the inadequacy of the nutrient composition dramatically affects AMF development, conventional soil analyses should be performed on the formulated inoculum, in independent official laboratories, as a quality control step. This way, the manufacturer will be provided with a certificate that guarantees the customers the validity of the data reported on the label and, therefore, enhances the quality of the inoculum. During experimental tests on the beneficial effects of inoculants, researchers often adopt an important practice in order to be able to differentiate between the effects of the inoculum carrier and the AMF portion, i.e. the use of a sterilized inoculum as a control, the so-called "mock" inoculum [145]. This practice of including a non-inoculated and a "mock" inoculated control should be considered by end-users who are willing to assess the eventual beneficial effect of AMF inoculation.

A few alternatives to the pot-culture method are available, regarding inoculum production and formulation. Other soilless culture systems, such as aeroponics and hydroponics, enable the production of pure clean spores and maximize growing conditions for the host plant [146]. Aeroponic inoculum production has long been scientifically validated [147,148], and could soon reach massive commercialization levels. Root-organ monoxenic culture is another method that allows the successful large-scale propagation of AMF which can be used directly as an inoculum. Unfortunately, the protocol for this method of propagation is not easily adjustable to all AMF strains. So far, several dozens of AMF species and strains have been propagated in vitro with the right synthetic growth medium and growth conditions. This type of culture consists of AMF inoculated excised roots (often *Daucus carota* L.) that have acquired the ability to uncontrollably proliferate, without the epigeous portion, after transformation with an *Agrobacterium rhizogenes* Conn. strain. This method of propagation does not require high specialization, and facilitates the control of AMF strain purity. As mentioned before, it is suitable for large-scale production, as a massive number of spores (several thousand), mycelium and colonized roots [149] can be obtained from one Petri dish in just 4 months, and from the consecutive subcultures [150]. AMF propagated with this technique have been shown to successfully re-colonize plant roots [151,152]. A possible further advantage of the AMF inoculum production process could be the use of bioreactors with liquid transformed root-organ cultures aimed at the large-scale propagation of AMF [153]. These tools may become suitable for commercialization in the near future and will lead to reduced labor and enhanced automation. However, as the AMF are produced in association with transformed roots, the product will only be intended for research use and may not be used for open-field inoculation.

The final product could become available on the market as a powder or granular substrate made from mixed inert-like materials, such as peat, compost, vermiculite, perlite, quartz sand, micronized zeolite and expanded clay, where colonized root fragments (1-5 mm long), spores and hyphal networks are uniformly distributed. Liquid inocula, dedicated to horticultural use, obtained from a hydroponic culture, or from a spore/mycelium suspension in a liquid carrier, represent a possible alternative final product [154]. As a final step before commercialization, the AMF composition should be characterized in order to control inoculum purity and to trace the inoculated strains. This prevents poor quality inocula from being put on the market.

The storage methodologies should preserve a product's high and consistent quality, and be simple and inexpensive at the same time. AMF viability and efficiency can be maintained for several months at room temperature (20-25°C), but the inocula must be kept in their packaging and must be partially dried. The main inconvenience that could occur during the storage period is that spores can sometimes become dormant, thus decreasing germination rates drastically [155]. However, a cold-storage period could be used to break dormancy [156]. Longer-term storage of liquid or dry inocula could be conducted at 5°C for both in vivo and in vitro propagated AMF [127]. Research culture collections are often stored using more sophisticated and expensive preservation techniques. These include the maintenance of monospecific inocula on living host plants (with regular molecular checks regarding the AMF identity), or alginate bead mediated encapsulation-drying and cryopreservation [157,158].

4. Perspectives

Future research in this field will have to concern the formulation of AMF isolate collections, with comprehensive information on host-preference, edaphic and climatic adaptation, and stress and disturbance tolerance. This will help manufacturers address their product towards different uses, including agricultural use, as well as new fields of application, such as the green architecture of urban sites [159]. At the same time, farmers will have to begin asking for assistance from experts in the field when introducing AMF to their cropping systems. Scientists should also carry out large-scale multi-location field trials, and conduct cost-benefit analyses, in order to increase awareness among the end-users of AMF inocula.

By 2050, global agriculture will have the task of doubling food production in order to feed the world [160]. At the same time, dependence on inorganic fertilizers and pesticides must be reduced. For these reasons, significant advances in AMF research are needed to allow their stable use in agriculture. Their application and synergistic combination with other functionally efficient microbial consortia that include PGPR (Plant Growth Promoting Rhizobacteria), saprophytic fungi and other helper microorganisms [161], will help farmers develop a more sustainable cropping system.

Acknowledgements

Our work was financially supported by the following institutions: Piemonte Region (ECO-FLOR and PRO-LACTE projects), Alcotra (FIORIBIO2 project), and EU (PURE project). The authors would like to thank Dr. Valentina Scariot for her coordinating work in the ECOFLOR project and Lucia Allione for her support in the funding management of the projects.

Author details

Andrea Berruti[1], Roberto Borriello[1], Alberto Orgiazzi[2], Antonio C. Barbera[3], Erica Lumini[1] and Valeria Bianciotto[1]

1 National Research Council, Plant Protection Institute – Turin UOS, Torino, Italy

2 European Commission, Joint Research Centre, Institute for Environment and Sustainability, Ispra (VA), Italy

3 DISPA, University of Catania, Catania, Italy

References

[1] Parniske M. Arbuscular mycorrhiza: the mother of plant root endosymbioses. Nature Reviews Microbiology 2008;6:763–75.

[2] Wang B, Qiu Y-L. Phylogenetic distribution and evolution of mycorrhizas in land plants. Mycorrhiza 2006;16:299–363.

[3] Bonfante P, Genre A. Mechanisms underlying beneficial plant–fungus interactions in mycorrhizal symbiosis. Nature Communications 2010;1:48.

[4] Schüßler A, Schwarzott D, Walker C. A new fungal phylum, the Glomeromycota: phylogeny and evolution. Mycological Research 2001;105:1413–21.

[5] Simon L, Bousquet J, Levesque RC, Lalonde M. Origin and diversification of endo-mycorrhizal fungi and coincidence with vascular land plants. Nature 1993;363:67–9.

[6] Tulasne LR, Tulasne C. Fungi nonnulli hipogaei, novi v. minus cogniti auct. Giornale Botanico Italiano 1844;2:55 – 63.

[7] Schüßler A, Walker C. 7 Evolution of the "Plant-Symbiotic" Fungal Phylum, Glomeromycota. In: Pöggeler S, Wöstemeyer J, editors. Evolution of Fungi and Fungal-Like Organisms, Springer Berlin Heidelberg; 2011, p. 163–85.

[8] Oehl F, Sieverding E, Palenzuela J, Ineichen K, da Silva GA. Advances in Glomeromycota taxonomy and classification. IMA FUNGUS 2011;2:191–9.

[9] Krüger M, Krüger C, Walker C, Stockinger H, Schüßler A. Phylogenetic reference data for systematics and phylotaxonomy of arbuscular mycorrhizal fungi from phylum to species level. New Phytologist 2012;193:970–84.

[10] Redecker D, Schüßler A, Stockinger H, Stürmer SL, Morton JB, Walker C. An evidence-based consensus for the classification of arbuscular mycorrhizal fungi (Glomeromycota). Mycorrhiza 2013:1–17.

[11] Öpik M, Moora M, Liira J, Zobel M. Composition of root-colonizing arbuscular my-corrhizal fungal communities in different ecosystems around the globe. Journal of Ecology 2006;94:778–90.

[12] Smith SE, Smith FA. Roles of Arbuscular Mycorrhizas in Plant Nutrition and Growth: New Paradigms from Cellular to Ecosystem Scales. Annual Review of Plant Biology 2011;62:227–50.

[13] Smith SE, Jakobsen I, Grønlund M, Smith FA. Roles of Arbuscular Mycorrhizas in Plant Phosphorus Nutrition: Interactions between Pathways of Phosphorus Uptake in Arbuscular Mycorrhizal Roots Have Important Implications for Understanding and Manipulating Plant Phosphorus Acquisition. Plant Physiol 2011;156:1050–7.

[14] Kiers ET, Duhamel M, Beesetty Y, Mensah JA, Franken O, Verbruggen E, et al. Recip-rocal Rewards Stabilize Cooperation in the Mycorrhizal Symbiosis. Science 2011;333:880–2.

[15] Wagg C, Jansa J, Schmid B, van der Heijden MGA. Belowground biodiversity effects of plant symbionts support aboveground productivity. Ecology Letters 2011;14:1001–9.

[16] Van der Putten WH. Plant Defense Belowground And Spatiotemporal Processes In Natural Vegetation. Ecology 2003;84:2269–80.

[17] Ruiz-Lozano JM, Porcel R, Azcón C, Aroca R. Regulation by arbuscular mycorrhizae of the integrated physiological response to salinity in plants: new challenges in phys-iological and molecular studies. J Exp Bot 2012;63:4033–44.

[18] Leyval C, Joner EJ, Del Val C, Haselwandter K. Potential of arbuscular mycorrhizal fungi for bioremediation. Mycorrhizal Technology in Agriculture, from Genes to Bio-products 2002:175–86.

[19] Requena N, Perez-Solis E, Azcón-Aguilar C, Jeffries P, Barea J-M. Management of In-digenous Plant-Microbe Symbioses Aids Restoration of Desertified Ecosystems. Appl Environ Microbiol 2001;67:495–8.

[20] Griffioen W a. J. Characterization of a heavy metal-tolerant endomycorrhizal fungus from the surroundings of a zinc refinery. Mycorrhiza 1994;4:197–200.

[21] Göhre V, Paszkowski U. Contribution of the arbuscular mycorrhizal symbiosis to heavy metal phytoremediation. Planta 2006;223:1115–22.

[22] Van der Heijden MGA, Klironomos JN, Ursic M, Moutoglis P, Streitwolf-Engel R, Boller T, et al. Mycorrhizal fungal diversity determines plant biodiversity, ecosystem variability and productivity. Nature 1998;396:69–72.

[23] Scheublin TR, Van Logtestijn RSP, Van Der Heijden MGA. Presence and identity of arbuscular mycorrhizal fungi influence competitive interactions between plant spe-cies. Journal of Ecology 2007;95:631–8.

[24] Facelli E, Smith SE, Facelli JM, Christophersen HM, Andrew Smith F. Underground friends or enemies: model plants help to unravel direct and indirect effects of arbuscular mycorrhizal fungi on plant competition. New Phytologist 2010;185:1050–61.

[25] Hart MM, Reader RJ, Klironomos JN. Plant coexistence mediated by arbuscular mycorrhizal fungi. Trends in Ecology & Evolution 2003;18:418–23.

[26] Veiga RSL, Jansa J, Frossard E, van der Heijden MGA. Can Arbuscular Mycorrhizal Fungi Reduce the Growth of Agricultural Weeds? PLoS ONE 2011;6:e27825.

[27] Rillig MC, Mummey DL. Mycorrhizas and soil structure. New Phytologist 2006;171:41–53.

[28] Bedini S, Pellegrino E, Avio L, Pellegrini S, Bazzoffi P, Argese E, et al. Changes in soil aggregation and glomalin-related soil protein content as affected by the arbuscular mycorrhizal fungal species Glomus mosseae and Glomus intraradices. Soil Biology and Biochemistry 2009;41:1491–6.

[29] Cheng L, Booker FL, Tu C, Burkey KO, Zhou L, Shew HD, et al. Arbuscular Mycorrhizal Fungi Increase Organic Carbon Decomposition Under Elevated CO2. Science 2012;337:1084–7.

[30] Mugnier J, Mosse B. Vesicular-arbuscular mycorrhizal infection in transformed root-inducing T-DNA roots grown axenically. Phytopathology 1987;77:1045–50.

[31] Giovannetti M, Fortuna P, Citernesi AS, Morini S, Nuti MP. The occurrence of anastomosis formation and nuclear exchange in intact arbuscular mycorrhizal networks. New Phytologist 2001;151:717–24.

[32] Rausch C, Daram P, Brunner S, Jansa J, Laloi M, Leggewie G, et al. A phosphate transporter expressed in arbuscule-containing cells in potato. Nature 2001;414:462–70.

[33] Delaux P-M, Séjalon-Delmas N, Bécard G, Ané J-M. Evolution of the plant–microbe symbiotic "toolkit." Trends in Plant Science 2013;18:298–304.

[34] Kucey RMN, Paul EA. Vesicular Arbuscular Mycorrhizal Spore Populations In Various Saskatchewan Soils And The Effect Of Inoculation With Glomus mosseae On Faba Bean Growth In Greenhouse And Field Trials. Canadian Journal of Soil Science 1983;63:87–95.

[35] Simon L, Lalonde M, Bruns TD. Specific amplification of 18S fungal ribosomal genes from vesicular-arbuscular endomycorrhizal fungi colonizing roots. Appl Environ Microbiol 1992;58:291–5.

[36] Helgason T, Fitter AH, Young JPW. Molecular diversity of arbuscular mycorrhizal fungi colonising Hyacinthoides non-scripta (bluebell) in a seminatural woodland. Molecular Ecology 1999;8:659–66.

[37] Stockinger H, Walker C, Schüssler A. "Glomus intraradices DAOM197198", a model fungus in arbuscular mycorrhiza research, is not Glomus intraradices. New Phytol 2009;183:1176–87.

[38] Shokralla S, Spall JL, Gibson JF, Hajibabaei M. Next-generation sequencing technologies for environmental DNA research. Molecular Ecology 2012;21:1794–805.

[39] Öpik M, Zobel M, Cantero JJ, Davison J, Facelli JM, Hiiesalu I, et al. Global sampling of plant roots expands the described molecular diversity of arbuscular mycorrhizal fungi. Mycorrhiza 2013;23:411–30.

[40] Krüger M, Stockinger H, Krüger C, Schüssler A. DNA-based species level detection of Glomeromycota: one PCR primer set for all arbuscular mycorrhizal fungi. New Phytol 2009;183:212–23.

[41] Lumini E, Orgiazzi A, Borriello R, Bonfante P, Bianciotto V. Disclosing arbuscular mycorrhizal fungal biodiversity in soil through a land-use gradient using a pyrosequencing approach. Environmental Microbiology 2010;12:2165–79.

[42] Borriello R, Lumini E, Girlanda M, Bonfante P, Bianciotto V. Effects of different management practices on arbuscular mycorrhizal fungal diversity in maize fields by a molecular approach. Biol Fertil Soils 2012;48:911–22.

[43] Davison J, Öpik M, Zobel M, Vasar M, Metsis M, Moora M. Communities of Arbuscular Mycorrhizal Fungi Detected in Forest Soil Are Spatially Heterogeneous but Do Not Vary throughout the Growing Season. PLoS ONE 2012;7:e41938.

[44] Öpik M, Vanatoa A, Vanatoa E, Moora M, Davison J, Kalwij JM, et al. The online database MaarjAM reveals global and ecosystemic distribution patterns in arbuscular mycorrhizal fungi (Glomeromycota). New Phytologist 2010;188:223–41.

[45] Salvioli A, Bonfante P. Systems biology and "omics" tools: A cooperation for next-generation mycorrhizal studies. Plant Science 2013;203–204:107–14.

[46] Dai M, Bainard LD, Hamel C, Gan Y, Lynch D. Impact of Land Use on Arbuscular Mycorrhizal Fungal Communities in Rural Canada. Appl Environ Microbiol 2013;79:6719–29.

[47] Saks Ü, Davison J, Öpik M, Vasar M, Moora M, Zobel M. Root-colonizing and soil-borne communities of arbuscular mycorrhizal fungi in a temperate forest understory. Botany 2013.

[48] Torrecillas E, Torres P, Alguacil MM, Querejeta JI, Roldán A. A case study of facultative plant epiphytism in semiarid conditions revealed the influence of habitat and climate variables on AM fungi communities distribution. Appl Environ Microbiol 2013:AEM.02466–13.

[49] Ji B, Gehring CA, Wilson GWT, Miller RM, Flores-Rentería L, Johnson NC. Patterns of diversity and adaptation in Glomeromycota from three prairie grasslands. Molecular Ecology 2013;22:2573–87.

[50] Lin X, Feng Y, Zhang H, Chen R, Wang J, Zhang J, et al. Long-Term Balanced Fertilization Decreases Arbuscular Mycorrhizal Fungal Diversity in an Arable Soil in North China Revealed by 454 Pyrosequencing. Environ Sci Technol 2012;46:5764–71.

[51] Dai M, Hamel C, St. Arnaud M, He Y, Grant C, Lupwayi N, et al. Arbuscular mycorrhizal fungi assemblages in Chernozem great groups revealed by massively parallel pyrosequencing. Canadian Journal of Microbiology 2012;58:81–92.

[52] Lekberg Y, Schnoor T, Kjøller R, Gibbons SM, Hansen LH, Al-Soud WA, et al. 454-sequencing reveals stochastic local reassembly and high disturbance tolerance within arbuscular mycorrhizal fungal communities. Journal of Ecology 2012;100:151–60.

[53] Liu Y, Shi G, Mao L, Cheng G, Jiang S, Ma X, et al. Direct and indirect influences of 8 yr of nitrogen and phosphorus fertilization on Glomeromycota in an alpine meadow ecosystem. New Phytologist 2012;194:523–35.

[54] Moora M, Berger S, Davison J, Öpik M, Bommarco R, Bruelheide H, et al. Alien plants associate with widespread generalist arbuscular mycorrhizal fungal taxa: evidence from a continental-scale study using massively parallel 454 sequencing. Journal of Biogeography 2011;38:1305–17.

[55] Dumbrell AJ, Ashton PD, Aziz N, Feng G, Nelson M, Dytham C, et al. Distinct seasonal assemblages of arbuscular mycorrhizal fungi revealed by massively parallel pyrosequencing. New Phytologist 2011;190:794–804.

[56] VAN Diepen LTA, Lilleskov EA, Pregitzer KS. Simulated nitrogen deposition affects community structure of arbuscular mycorrhizal fungi in northern hardwood forests. Mol Ecol 2011;20:799–811.

[57] Balestrini R, Magurno F, Walker C, Lumini E, Bianciotto V. Cohorts of arbuscular mycorrhizal fungi (AMF) in Vitis vinifera, a typical Mediterranean fruit crop. Environmental Microbiology Reports 2010;2:594–604.

[58] Dumbrell AJ, Nelson M, Helgason T, Dytham C, Fitter AH. Relative roles of niche and neutral processes in structuring a soil microbial community. ISME J 2009;4:337–45.

[59] Alguacil MM, Roldán A, Torres MP. Complexity of semiarid gypsophilous shrub communities mediates the AMF biodiversity at the plant species level. Microb Ecol 2009;57:718–27.

[60] Öpik M, Metsis M, Daniell TJ, Zobel M, Moora M. Large-scale parallel 454 sequencing reveals host ecological group specificity of arbuscular mycorrhizal fungi in a boreonemoral forest. New Phytologist 2009;184:424–37.

[61] Cesaro P, van Tuinen D, Copetta A, Chatagnier O, Berta G, Gianinazzi S, et al. Preferential Colonization of Solanum tuberosum L. Roots by the Fungus Glomus intraradices in Arable Soil of a Potato Farming Area. Appl Environ Microbiol 2008;74:5776–83.

[62] Öpik M, Moora M, Zobel M, Saks Ü, Wheatley R, Wright F, et al. High diversity of arbuscular mycorrhizal fungi in a boreal herb-rich coniferous forest. New Phytologist 2008;179:867–76.

[63] Alguacil MM, Lumini E, Roldán A, Salinas-García JR, Bonfante P, Bianciotto V. The impact of tillage practices on arbuscular mycorrhizal fungal diversity in subtropical crops. Ecol Appl 2008;18:527–36.

[64] Schechter SP, Bruns TD. Serpentine and non-serpentine ecotypes of Collinsia sparsiflora associate with distinct arbuscular mycorrhizal fungal assemblages. Molecular Ecology 2008;17:3198–210.

[65] Toljander JF, Santos-González JC, Tehler A, Finlay RD. Community analysis of arbuscular mycorrhizal fungi and bacteria in the maize mycorrhizosphere in a long-term fertilization trial. FEMS Microbiology Ecology 2008;65:323–38.

[66] Hempel S, Renker C, Buscot F. Differences in the species composition of arbuscular mycorrhizal fungi in spore, root and soil communities in a grassland ecosystem. Environ Microbiol 2007;9:1930–8.

[67] Ligrone R, Carafa A, Lumini E, Bianciotto V, Bonfante P, Duckett JG. Glomeromycotean associations in liverworts: a molecular, cellular, and taxonomic analysis. Am J Bot 2007;94:1756–77.

[68] Pivato B, Mazurier S, Lemanceau P, Siblot S, Berta G, Mougel C, et al. Medicago species affect the community composition of arbuscular mycorrhizal fungi associated with roots. New Phytologist 2007;176:197–210.

[69] Santos-González JC, Finlay RD, Tehler A. Seasonal Dynamics of Arbuscular Mycorrhizal Fungal Communities in Roots in a Seminatural Grassland. Appl Environ Microbiol 2007;73:5613–23.

[70] Wu B, Hogetsu T, Isobe K, Ishii R. Community structure of arbuscular mycorrhizal fungi in a primary successional volcanic desert on the southeast slope of Mount Fuji. Mycorrhiza 2007;17:495–506.

[71] Vallino M, Massa N, Lumini E, Bianciotto V, Berta G, Bonfante P. Assessment of arbuscular mycorrhizal fungal diversity in roots of Solidago gigantea growing in a polluted soil in Northern Italy. Environ Microbiol 2006;8:971–83.

[72] Yamato M, Iwase K. Community analysis of arbuscular mycorrhizal fungi in a warm-temperate deciduous broad-leaved forest and introduction of the fungal community into the seedlings of indigenous woody plants. Mycoscience 2005;46:334–42.

[73] Wirsel SGR. Homogenous stands of a wetland grass harbour diverse consortia of arbuscular mycorrhizal fungi. FEMS Microbiol Ecol 2004;48:129–38.

[74] Rosendahl S, Stukenbrock EH. Community structure of arbuscular mycorrhizal fungi in undisturbed vegetation revealed by analyses of LSU rDNA sequences. Molecular Ecology 2004;13:3179–86.

[75] Gollotte A, van Tuinen D, Atkinson D. Diversity of arbuscular mycorrhizal fungi colonising roots of the grass species Agrostis capillaris and Lolium perenne in a field experiment. Mycorrhiza 2004;14:111–7.

[76] Saito K, Suyama Y, Sato S, Sugawara K. Defoliation effects on the community structure of arbuscular mycorrhizal fungi based on 18S rDNA sequences. Mycorrhiza 2004;14:363–73.

[77] Heinemeyer A, Ridgway KP, Edwards EJ, Benham DG, Young JPW, Fitter AH. Impact of soil warming and shading on colonization and community structure of arbuscular mycorrhizal fungi in roots of a native grassland community. Global Change Biology 2004;10:52–64.

[78] Wubet T, Kottke I, Teketay D, Oberwinkler F. Mycorrhizal status of indigenous trees in dry Afromontane forests of Ethiopia. Forest Ecology and Management 2003;179:387–99.

[79] Öpik M, Moora M, Liira J, Kõljalg U, Zobel M, Sen R. Divergent arbuscular mycorrhizal fungal communities colonize roots of Pulsatilla spp. in boreal Scots pine forest and grassland soils. New Phytologist 2003;160:581–93.

[80] Vandenkoornhuyse P, Husband R, Daniell TJ, Watson IJ, Duck JM, Fitter AH, et al. Arbuscular mycorrhizal community composition associated with two plant species in a grassland ecosystem. Molecular Ecology 2002;11:1555–64.

[81] Helgason T, Merryweather J, Cole J, Wilson P, Young J, Fitter A. Selectivity and Functional Diversity in Arbuscular Mycorrhizas of Co-Occurring Fungi and Plants from a Temperate Deciduous Woodland. Journal of Ecology 2002;90:371–84.

[82] Husband R, Herre EA, Turner SL, Gallery R, Young JPW. Molecular diversity of arbuscular mycorrhizal fungi and patterns of host association over time and space in a tropical forest: AM diversity in a tropical forest. Molecular Ecology 2002;11:2669–78.

[83] Husband R, Herre EA, Young JPW. Temporal variation in the arbuscular mycorrhizal communities colonising seedlings in a tropical forest. FEMS Microbiology Ecology 2002;42:131–6.

[84] Daniell TJ, Husband R, Fitter AH, Young JPW. Molecular diversity of arbuscular mycorrhizal fungi colonising arable crops. FEMS Microbiology Ecology 2001;36:203–9.

[85] Helgason T, Daniell T, Husband R, Fitter A, Young J. Ploughing up the wood-wide web? Nature 1998;394:431.

[86] Verbruggen E, Kiers ET. Evolutionary ecology of mycorrhizal functional diversity in agricultural systems. Evolutionary Applications 2010;3:547–60.

[87] Verbruggen E, Röling WFM, Gamper HA, Kowalchuk GA, Verhoef HA, Van Der Heijden MGA. Positive effects of organic farming on below-ground mutualists: large-scale comparison of mycorrhizal fungal communities in agricultural soils. New Phytologist 2010;186:968–79.

[88] Jansa J, Mozafar A, Anken T, Ruh R, Sanders IR, Frossard E. Diversity and structure of AMF communities as affected by tillage in a temperate soil. Mycorrhiza 2002;12:225–34.

[89] Goss M., de Varennes A. Soil disturbance reduces the efficacy of mycorrhizal associations for early soybean growth and N2 fixation. Soil Biology and Biochemistry 2002;34:1167–73.

[90] Kabir Z. Tillage or no-tillage: Impact on mycorrhizae. Canadian Journal of Plant Science 2005;85:23–9.

[91] Sanders IR, Streitwolf-Engel R, Heijden MGA van der, Boller T, Wiemken A. Increased allocation to external hyphae of arbuscular mycorrhizal fungi under CO2 enrichment. Oecologia 1998;117:496–503.

[92] Powell JR, Parrent JL, Hart MM, Klironomos JN, Rillig MC, Maherali H. Phylogenetic trait conservatism and the evolution of functional trade-offs in arbuscular mycorrhizal fungi. Proc Biol Sci 2009;276:4237–45.

[93] Avio L, Castaldini M, Fabiani A, Bedini S, Sbrana C, Turrini A, et al. Impact of nitrogen fertilization and soil tillage on arbuscular mycorrhizal fungal communities in a Mediterranean agroecosystem. Soil Biology and Biochemistry 2013;67:285–94.

[94] De La Providencia IE, De Souza FA, Fernández F, Delmas NS, Declerck S. Arbuscular mycorrhizal fungi reveal distinct patterns of anastomosis formation and hyphal healing mechanisms between different phylogenic groups. New Phytologist 2005;165:261–71.

[95] De La Providencia IE, Fernández F, Declerck S. Hyphal healing mechanism in the arbuscular mycorrhizal fungi Scutellospora reticulata and Glomus clarum differs in response to severe physical stress. FEMS Microbiology Letters 2007;268:120–5.

[96] Derelle D, Declerck S, Genet P, Dajoz I, van Aarle IM. Association of highly and weakly mycorrhizal seedlings can promote the extra- and intraradical development of a common mycorrhizal network. FEMS Microbiology Ecology 2012;79:251–9.

[97] Biermann B, Linderman RG. Use of Vesicular-Arbuscular Mycorrhizal Roots, Intraradical Vesicles and Extraradical Vesicles as Inoculum. New Phytologist 1983;95:97–105.

[98] Hamel C, Hanson K, Selles F, Cruz AF, Lemke R, McConkey B, et al. Seasonal and long-term resource-related variations in soil microbial communities in wheat-based rotations of the Canadian prairie. Soil Biology and Biochemistry 2006;38:2104–16.

[99] Rosendahl S, Mcgee P, Morton JB. Lack of global population genetic differentiation in the arbuscular mycorrhizal fungus Glomus mosseae suggests a recent range expansion which may have coincided with the spread of agriculture. Molecular Ecology 2009;18:4316–29.

[100] Egerton-Warburton LM, Johnson NC, Allen EB. Mycorrhizal Community Dynamics Following Nitrogen Fertilization: A Cross-Site Test In Five Grasslands. Ecological Monographs 2007;77:527–44.

[101] Alguacil MM, Lozano Z, Campoy MJ, Roldán A. Phosphorus fertilisation management modifies the biodiversity of AM fungi in a tropical savanna forage system. Soil Biology and Biochemistry 2010;42:1114–22.

[102] Ryan MH, Herwaarden AF van, Angus JF, Kirkegaard JA. Reduced growth of autumn-sown wheat in a low-P soil is associated with high colonisation by arbuscular mycorrhizal fungi. Plant Soil 2005;270:275–86.

[103] Johnson NC, Rowland DL, Corkidi L, Egerton-Warburton LM, Allen EB. Nitrogen Enrichment Alters Mycorrhizal Allocation At Five Mesic To Semiarid Grasslands. Ecology 2003;84:1895–908.

[104] Johnson NC, Wilson GWT, Bowker MA, Wilson JA, Miller RM. Resource limitation is a driver of local adaptation in mycorrhizal symbioses. PNAS 2010;107:2093–8.

[105] Mäder P, Edenhofer S, Boller T, Wiemken A, Niggli U. Arbuscular mycorrhizae in a long-term field trial comparing low-input (organic, biological) and high-input (conventional) farming systems in a crop rotation. Biol Fertil Soils 2000;31:150–6.

[106] Kiers ET, West SA, Denison RF. Mediating mutualisms: farm management practices and evolutionary changes in symbiont co-operation. Journal of Applied Ecology 2002;39:745–54.

[107] Thrall PH, Hochberg ME, Burdon JJ, Bever JD. Coevolution of symbiotic mutualists and parasites in a community context. Trends in Ecology & Evolution 2007;22:120–6.

[108] Kiers ET, Denison RF. Sanctions, Cooperation, and the Stability of Plant-Rhizosphere Mutualisms. Annual Review of Ecology, Evolution, and Systematics 2008;39:215–36.

[109] Johnson NC, Graham J-H, Smith FA. Functioning of mycorrhizal associations along the mutualism–parasitism continuum. New Phytologist 1997;135:575–85.

[110] Nijjer S, Rogers WE, Siemann E. The Impacts of Fertilization on Mycorrhizal Production and Investment in Western Gulf Coast Grasslands. The American Midland Naturalist 2009;163:124–33.

[111] Hart MM, Forsythe J, Oshowski B, Bücking H, Jansa J, Kiers ET. Hiding in a crowd— does diversity facilitate persistence of a low-quality fungal partner in the mycorrhizal symbiosis? Symbiosis 2013;59:47–56.

[112] Oehl F, Sieverding E, Mäder P, Dubois D, Ineichen K, Boller T, et al. Impact of long-term conventional and organic farming on the diversity of arbuscular mycorrhizal fungi. Oecologia 2004;138:574–83.

[113] Jakobsen I, Read D, Lewis D, Fitter A, Alexander I. Phosphorus transport by external hyphae of vesicular-arbuscular mycorrhizas. Mycorrhizas in Ecosystems Wallingford: CAB International 1992:48–58.

[114] Sieverding E, Leihner DE. Influence of crop rotation and intercropping of cassava with legumes on VA mycorrhizal symbiosis of cassava. Plant Soil 1984;80:143–6.

[115] Hijri I, Sykorovà Z, Oehl F, Ineichen K, Mader P, Wiemken A, et al. Communities of arbuscular mycorrhizal fungi in arable soils are not necessarily low in diversity. Molecular Ecology 2006;15:2277–89.

[116] Jiao H, Chen Y, Lin X, Liu R. Diversity of arbuscular mycorrhizal fungi in greenhouse soils continuously planted to watermelon in North China. Mycorrhiza 2011;21:681–8.

[117] Ryan MH, Angus JF. Arbuscular mycorrhizae in wheat and field pea crops on a low P soil: increased Zn-uptake but no increase in P-uptake or yield. Plant and Soil 2003;250:225–39.

[118] Kirkegaard J, Christen O, Krupinsky J, Layzell D. Break crop benefits in temperate wheat production. Field Crops Research 2008;107:185–95.

[119] Oehl F, Sieverding E, Ineichen K, Mader P, Boller T, Wiemken A. Impact of Land Use Intensity on the Species Diversity of Arbuscular Mycorrhizal Fungi in Agroecosystems of Central Europe. Appl Environ Microbiol 2003;69:2816–24.

[120] Barr J. The Value of Mycorrhizal Fungi for Sustainable and Durable Soils. In: Silva AP, Sol M, editors. Fungi: types, environmental impact, and role in disease, Hauppauge, N.Y: Nova Science Publishers, Inc.; 2011, p. 531.

[121] Baker NJ, Bancroft BA, Garcia TS. A meta-analysis of the effects of pesticides and fertilizers on survival and growth of amphibians. Science of The Total Environment 2013;449:150–6.

[122] Lehman RM, Taheri WI, Osborne SL, Buyer JS, Douds Jr. DD. Fall cover cropping can increase arbuscular mycorrhizae in soils supporting intensive agricultural production. Applied Soil Ecology 2012;61:300–4.

[123] Camargo-Ricalde SL, Dhillion SS. Endemic Mimosa species can serve as mycorrhizal "resource islands" within semiarid communities of the Tehuacán-Cuicatlán Valley, Mexico. Mycorrhiza 2003;13:129–36.

[124] Allen MF. Ecology of vesicular-arbuscular mycorrhizae in an arid ecosystem: use of natural processes promoting dispersal and establishment. Mycorrhizae in the next Decade Practical Applications and Research Priorities 7ª NACOM, IFAS, Gainesville, Florida 1987:133–5.

[125] Gulati A, Cummings JR. Mycorrhiza, a fungal solution for the farm. The Economic Times 2008.

[126] Barr J. Restoration of plant communities in The Netherlands through the application of arbuscular mycorrhizal fungi. Symbiosis 2010;52:87–94.

[127] Dalpé Y, Monreal M. Arbuscular Mycorrhiza Inoculum to Support Sustainable Cropping Systems. Crop Management 2004.

[128] Verbruggen E, van der Heijden MGA, Rillig MC, Kiers ET. Mycorrhizal fungal establishment in agricultural soils: factors determining inoculation success. New Phytologist 2013;197:1104–9.

[129] Berruti A, Borriello R, Della Beffa MT, Scariot V, Bianciotto V. Application of nonspecific commercial AMF inocula results in poor mycorrhization in Camellia japonica L. Symbiosis 2013.

[130] Mohammad A, Khan AG. Monoxenic in vitro production and colonization potential of AM fungus Glomus intraradices. Indian Journal of Experimental Biology 2002;40:1087–91.

[131] Oehl F, Laczko E, Bogenrieder A, Stahr K, Bösch R, van der Heijden M, et al. Soil type and land use intensity determine the composition of arbuscular mycorrhizal fungal communities. Soil Biology and Biochemistry 2010;42:724–38.

[132] Estrada B, Beltrán-Hermoso M, Palenzuela J, Iwase K, Ruiz-Lozano JM, Barea J-M, et al. Diversity of arbuscular mycorrhizal fungi in the rhizosphere of Asteriscus maritimus (L.) Less., a representative plant species in arid and saline Mediterranean ecosystems. Journal of Arid Environments 2013;97:170–5.

[133] Sanders IR. Plant and arbuscular mycorrhizal fungal diversity – are we looking at the relevant levels of diversity and are we using the right techniques? New Phytologist 2004;164:415–8.

[134] Berruti A, Borriello R, Lumini E, Scariot V, Bianciotto V, Balestrini R. Application of laser microdissection to identify the mycorrhizal fungi that establish arbuscules inside root cells. Front Plant Sci 2013;4.

[135] Farmer MJ, Li X, Feng G, Zhao B, Chatagnier O, Gianinazzi S, et al. Molecular monitoring of field-inoculated AMF to evaluate persistence in sweet potato crops in China. Applied Soil Ecology 2007;35:599–609.

[136] Thonar C, Erb A, Jansa J. Real-time PCR to quantify composition of arbuscular mycorrhizal fungal communities—marker design, verification, calibration and field validation. Molecular Ecology Resources 2012;12:219–32.

[137] Tisserant E, Kohler A, Dozolme-Seddas P, Balestrini R, Benabdellah K, Colard A, et
al. The transcriptome of the arbuscular mycorrhizal fungus Glomus intraradices
(DAOM 197198) reveals functional tradeoffs in an obligate symbiont. New Phytologist 2012;193:755–69.

[138] Hildebrandt U, Regvar M, Bothe H. Arbuscular mycorrhiza and heavy metal tolerance. Phytochemistry 2007;68:139–46.

[139] Ruiz-Lozano JM, Azcón R. Symbiotic efficiency and infectivity of an autochthonous
arbuscular mycorrhizal Glomus sp. from saline soils and Glomus deserticola under
salinity. Mycorrhiza 2000;10:137–43.

[140] Vosátka M, Rydlová J, Malcová R. Microbial inoculations of plants for revegetation
of disturbed soils in degraded ecosystems. Nature and Culture in Landscape Ecology
1999:303–17.

[141] Oliveira RS, Vosátka M, Dodd JC, Castro PML. Studies on the diversity of arbuscular
mycorrhizal fungi and the efficacy of two native isolates in a highly alkaline anthropogenic sediment. Mycorrhiza 2005;16:23–31.

[142] Klironomos J, Hart M. Colonization of roots by arbuscular mycorrhizal fungi using
different sources of inoculum. Mycorrhiza 2002;12:181–4.

[143] Tommerup IC. Development of Infection by a Vesicular–Arbuscular Mycorrhizal
Fungus in Brassica Napus L. and Trifolium Subterraneum L. New Phytologist
1984;98:487–95.

[144] Brundrett MC, Abbott LK, Jasper DA. Glomalean mycorrhizal fungi from tropical
Australia. Mycorrhiza 1999;8:305–14.

[145] Pellegrino E, Turrini A, Gamper HA, Cafà G, Bonari E, Young JPW, et al. Establishment, persistence and effectiveness of arbuscular mycorrhizal fungal inoculants in
the field revealed using molecular genetic tracing and measurement of yield components. New Phytologist 2012;194:810–22.

[146] IJdo M, Cranenbrouck S, Declerck S. Methods for large-scale production of AM fungi: past, present, and future. Mycorrhiza 2011;21:1–16.

[147] Mohammad A, Khan AG, Kuek C. Improved aeroponic culture of inocula of arbuscular mycorrhizal fungi. Mycorrhiza 2000;9:337–9.

[148] Abdul-Khaliq, Gupta ML, Alam M. Biotechnological Approaches for Mass Produc-
tion of Arbuscular Mycorrhizal Fungi: Current Scenario and Future Strategies. In:
Mukerji KG, Manoharachary C, Chamola BP, editors. Techniques in Mycorrhizal
Studies, Springer Netherlands; 2002, p. 299–312.

[149] Declerck S, Strullu DG, Plenchette C. In vitro mass-production of the arbuscular mycorrhizal fungus, Glomus versiforme, associated with Ri T-DNA transformed carrot roots. Mycological Research 1996;100:1237–42.

[150] Declerck S, Strullu DG, Plenchette C. Monoxenic Culture of the Intraradical Forms of Glomus sp. Isolated from a Tropical Ecosystem: A Proposed Methodology for Germplasm Collection. Mycologia 1998;90:579–85.

[151] Diop TA, Plenchette C, Strullu DG. Dual axenic culture of sheared-root inocula of vesicular-arbuscular mycorrhizal fungi associated with tomato roots. Mycorrhiza 1994;5:17–22.

[152] Declerck S, Strullu D., Plenchette C, Guillemette T. Entrapment of in vitro produced spores of Glomus versiforme in alginate beads: in vitro and in vivo inoculum potentials. Journal of Biotechnology 1996;48:51–7.

[153] Jolicoeur M, Williams RD, Chavarie C, Fortin JA, Archambault J. Production of Glomus intraradices propagules, an arbuscular mycorrhizal fungus, in an airlift bioreactor. Biotechnology and Bioengineering 1999;63:224–32.

[154] Fernández F, Dell'Amico JM, Angoa MV, de la Providencia IE. Use of a liquid inoculum of the arbuscular mycorrhizal fungi Glomus hoi in rice plants cultivated in a saline Gleysol: A new alternative to inoculate. Journal of Plant Breeding and Crop Science 2011;3:24–33.

[155] Oehl F, Sieverding E, Ineichen K, Mäder P, Wiemken A, Boller T. Distinct sporulation dynamics of arbuscular mycorrhizal fungal communities from different agroecosystems in long-term microcosms. Agriculture, Ecosystems & Environment 2009;134:257–68.

[156] Juge C, Samson J, Bastien C, Vierheilig H, Coughlan A, Piché Y. Breaking dormancy in spores of the arbuscular mycorrhizal fungus Glomus intraradices: a critical cold-storage period. Mycorrhiza 2002;12:37–42.

[157] Plenchette C, Strullu DG. Long-term viability and infectivity of intraradical forms of Glomus intraradices vesicles encapsulated in alginate beads. Mycological Research 2003;107:614–6.

[158] Lalaymia I, Cranenbrouck S, Draye X, Declerck S. Preservation at ultra-low temperature of in vitro cultured arbuscular mycorrhizal fungi via encapsulation–drying. Fungal Biology 2012;116:1032–41.

[159] McGuire KL, Payne SG, Palmer MI, Gillikin CM, Keefe D, Kim SJ, et al. Digging the New York City Skyline: Soil Fungal Communities in Green Roofs and City Parks. PLoS ONE 2013;8:e58020.

[160] Ray DK, Mueller ND, West PC, Foley JA. Yield Trends Are Insufficient to Double Global Crop Production by 2050. PLoS ONE 2013;8:e66428.

[161] Dodd IC, Ruiz-Lozano JM. Microbial enhancement of crop resource use efficiency. Current Opinion in Biotechnology 2012;23:236–42.

Chapter 9

An Overview of the Yeast Biodiversity in the Galápagos Islands and Other Ecuadorian Regions

Enrique Javier Carvajal Barriga,

Patricia Portero Barahona, Carolina Tufiño,

Bernardo Bastidas, Cristina Guamán-Burneo,

Larissa Freitas and Carlos Rosa

Additional information is available at the end of the chapter

http://dx.doi.org/10.5772/58303

1. Introduction

One of the most emblematic natural regions for studies of evolution and biodiversity in the world is the Galápagos Islands, which is the inspiring environment where the naturalist Charles Darwin was moved to propose what eventually became Theory of the Origin of Species launched in the 19th Century.

This Archipelago has been formed by subaquatic volcanic activity around 5 million years ago. The plant and animal populations settled on this group of 21 islands and 107 rocks and islets were introduced mainly by the sea currents and winds that reached the emerging lands in this equatorial region of the sea.

The study of plants and endemic species of animals has fascinated biologists for decades. Giant turtles, finches, marine and terrestrial iguanas and boobies have been the center of studies, as well as other birds and flora of the region. Many adaptations and evolution evidences were found in the macrobiota adapted to the particular environments of each island in the archipelago.

However, not much attention was paid to the microorganisms and, in particular, to yeast biodiversity in the islands. In 2009 in an effort to address this scientific shortfall, a prospective study was started by Ecuadorian-Brazilian-Spanish team that visited four human-inhabited islands (i.e. Floreana, San Cristóbal, Santa Cruz and Isabela).

The substrates chosen by the researchers were mainly flowers from *Datura* and *Ipomoea* genera, as well as *Opuntia* fruits and leaves. Moreover, unique substrates like endemic tree's exudates or even giant turtle's and marine iguana's feces were also taken. Flowers, insect, fungus and rotten vegetal matter was also part of the substrates chosen by the expeditionaries.

The resulting prospection yielded more than 800 yeast isolates. Most of those yeasts have been identified by sequencing of the LSU or the 26S rDNA gene. Among the yeasts recovered, there are several novel yeast species such as *Saccharomycopsis fodiens* and *Kodamaea transpacifica*, and other hitherto non described ones.

About 31% of the yeast biota in the islands is coincident with the species found in Ecuador mainland. Most of the yeast species are hitherto not found in the mainland since 2006 when the Catholic University Yeasts Collection (CLQCA) initiated its identification, characterization and preservation activities, devoted to yeast. Currently this yeasts collection represents the most complete deposit of wild species from Ecuador.

A comparison between the yeast biodiversity in the islands with the yeasts biodiversity in Ecuadorian mainland is done in this chapter in order to draw a first line of understanding of the adaptability, biogeography and interaction of species in an insular territory located about 1000 Km from the nearest South American mainland coasts.

Moreover, an overview of the yeast biodiversity of mainland Ecuador's ecosystems is addressed in this chapter in order to establish the comparisons and the extent in which the closest mainland has had influence in current microbial (yeast) biodiversity in this relatively recently formed archipelago in the Pacific Ocean.

2. The Galápagos Islands features

As Michael H Jackson cites in his book titled Galapagos, a Natural History, "the feature that sets the Galápagos apart from all other archipelagos is its unique geographic position. Situated on the equator under the tropical sun, and yet bathed for much of the year by the cool waters of the Humboldt and Cronwell currents, the islands have a special mix of tropical and temperate environments which is reflected in the ecology of its unusual plants and animals" [1]. It is certainly something extremely unusual finding an archipelago which biota has had influences not only from the nearest American mainland but also from Australasian environments. The remarkable combination of living organisms in this group of oceanic islands and the adaptive radiation that eventually yielded such a particular group of animal and vegetal endemism makes the Galápagos Islands one of the most interesting environments in the world to study evolution, ecology and natural history.

The disharmony as one of the features of oceanic islands in the Galápagos is observed by the over-representation of certain plant groups such as Pteridophyta (ferns), Poaceae (grasses), Asteraceae (sunflower family), Amaranthaceae (pigweed family), Fabaceae (bean family), and Cyperaceae (sedge family), lichens, mosses and liverworts. On the other hand, the under or non-represented families of plants are: Palmaceae, Anacardiaceae, Meliaceae, Labiatae, Scrophulariaceae, Orchidiaceae, Acanthaceae, Melastomataceae and Bromeliaceae[1].

2.1. Climate

In the Galapagos Archipelago the diurnal temperature varies from 5°C in the windward side of the islands, to 10°C in the leeward sides [2]. The annual range is from 11 to 12°C; while during the warm season (from December to June) the temperature reaches 29°C. In August, the upper limit is about 19°C [3].

Because of the equatorial location of the islands, there are only two seasons readily distinguishable, where the rainy weather coincides with warm period (December to June) and the dry, cooler season from July to November, which is a period characterized as mostly foggy and overcasted [3].

In terms of rainfall, the whole archipelago receives less than 750 mm per year; this is due to the geographical situation of the islands which are in the dry zone of the Pacific Ocean.

2.2. Vegetation zones

The islands where the yeast expedition was carried out in October 2009 (during the dry season), are all inhabited by human populations. These four islands: Isabela, Floreana, Santa Cruz and San Cristóbal have littoral zones as well as high zones which exhibit very clear differences in plant species. The zones of the higher elevations such as Sierra Negra volcano are mostly rocky and sterile, where only a few *Opuntia* sp. individuals are seen. In contrast the higher zones of Floreana, Santa Cruz and San Cristóbal all of them exhibit a green cover and even agricultural zones where a number of introduced species are being cultivated.

The list of plants occurring in the littoral zone is small and differs less from a list of plants from a comparable mainland zone [3]. As for the vegetation in this zone, there is a lack of plant species in the sheer cliffs or basalt rocks rising from the sea to 10 or more meters. In Santa Cruz Island there is a predominant and extended area of "palo santo" *Bursera graveolens*, in a rather xerophytic vegetation environment.

There are also arid zones located immediately inland from the littoral zones, where vegetation is xerophytic which remain even up to 80 to 120 m height but sometimes much higher reaching up to 300 m. Predominant genus in this zone are *Opuntia*, *Jasminocereus* and *Brachycereus* which share the spaces with small leaved and spiny shrubs and trees. Most interestingly for yeast communities which "prefer" to degrade vegetal matter is the fact that annual herbs prosper during the wet season, providing a green cover which lasts for several weeks. After that period, the biomass from the green blanket dies and degrades: a micro ecological niche which offers a great opportunity of growth for yeasts and other microorganisms. *Datura* and *Ipomoea* species both of them introduced by human in undetermined times, are frequently found in this zone as well as in the transition zone.

In the transition zone the plant communities are frequently evergreen with ample leaves which provides with a green landscape to this zone. There is possible to find out some epiphytic plants, orchids, lichens and bryophytes. Ferns are also part of the communities: a vast forest composed by arboraceous ferns can be seen in Santa Cruz Island. The transition zone covers a large area in the Galápagos Islands, where the upper and lower limits varies considerably from one island to the other.

There is also a very particular zone inhabited mainly by an endemic species of tree, *Scalesia*. This species occurs in the range of 200 to 400 m.a.s.l. Some exudes of this species were sampled in the search of yeast species along the expedition. In this zone the trunks, branches, twigs and in some species even the leaves bear large numbers of epiphytic liverworts and mosses. Saprophytic fungi are common to abundant during the early part of the rainy season. Amanita and Agaricus-like species with caps as much as 15 cm diameter can be seen as well as several types of puffballs [3]. Additionally it is distinguishable the *Miconia* zone, another endemic plant that is catalogued as shrub which grows between 400 to 550 m.a.s.l. In the community of this zone herbaceous plants are important members: *Lycopodium* species are widely represented in the plant's community.

Beyond the *Miconia* zone and overlapped there is the Fern-Sedge zone that occurs from 500 to 700 m.a.s.l. This zone reaches the top of the peaks in the larger islands of the archipelago. The zone is frequently covered by clouds during the wet season, thus, the vegetation uptakes the water in the form of a fine drizzle or fog [3]. Plants growing in this zone are mostly low growing species with narrow leaves to reduce the scarce water available. This is the habitat of the endemic tree fern *Cyathea weatherbyana*.

2.3. Fauna

The fauna in the Galápagos Islands is disharmonic respect to the mainland: there is a lack of several phyla and groups of animals in the different archipelago's environments. The Class Amphibians is not present in any of the islands, islets or rocks in the Galápagos Archipelago, due to the long distance from the mainland and the scarcity of fresh water; there is no possibility to survive a trip along the 1000 km distance between the mainland and the islands for any amphibian species.

Another scarcely represented group is the mammals, from which it is possible to find three orders: Chiroptera with three endemic species of bats; Pinnipeds, represented by the Galápagos fur seals, a close relative of fur seals from the Southern Hemisphere, and sea lions which are a subspecies of the Californian sea lion. Additionally, there are six species belonging to the Order Rodentia that represent the only terrestrial endemic mammals in the islands. Nevertheless, it is important to mention that a number of mammal species were introduced in the islands namely goats, horses, pigs, dogs, donkeys, cats, rats and cattle and other Guinea pigs. These species are a serious threat to the whole environments of the islands.

Reptiles like marine and terrestrial iguanas, as well as marine and terrestrial tortoises are very representative species of the archipelago. In fact, the giant tortoises, *Geochelone elephantopus* are the animals for which the islands were named. This is a herbivorous species which represented an interesting organism to study from its gut microflora—including the yeasts—point of view. There is only one snake species endemic to the islands and seven species of lizards (lava lizard) from the genus *Tropidurus*. Finally, there are some species of gekos, especially in the inhabited islands, where they've found some empathy with human population.

Birds are fairly well represented in the islands with 108 species, of these, 89 reside and breed in the archipelago and 77 of them are endemic. The commonest and most thoroughly studied birds are the Geospizidae, or Galápagos finches, the flight less cormorant, the Galápagos penguin, the small Galápagos green heron, the endemic dove, the Galápagos hawk, the endemic duck and several subspecies of mocking birds. Additionally there are big colonies of blue footed boobies, red footed boobies and masked boobies.

All the above mentioned animals are interesting from their gut's microflora composition that becomes themselves as vectors for the introduction of yeasts species into the islands environments. The mobility of marine species and birds between the islands constitutes a factor of dispersion for yeasts as well as plant species. But there is another group of animals that are highly involved in the yeasts biodiversity: the invertebrates, and particularly the Arthropoda.

Arachnida and Chilopoda represented by spiders and scorpions are very numerous; nevertheless there are a few species in the islands. Spiders inhabit all the vegetation zones, but are less common in the fern-sedge zone before described. Centipeds are also very common and numerous.

In the case of insects, the fauna is considered relatively poor compared to the closest mainland Ecuador. This is a feature shared with other oceanic islands. "The biological paucity of diversity and abundance was noticed initially by Darwin: "I took great pains in collecting the insects, but, excepting Tierra del Fuego, I never saw in this respect so poor a country" [4].

There are Hymenoptera, especially bees that act as pollinators; butterflies and moths are scarcely represented with only 7 and 12 species respectively [3]. Some bees are thought to be introduced, for instance *Megachile timberlakei* that is an unexpected register in Galápagos since this bee species is common in Hawaii [4].

New studies of lice in the Galápagos report 47 genera and 104 species, 17 of them are endemic species, 79 are native and eight introduced by human agency [5]. On the other hand, the Othopteroid insects are one of the more diverse assemblages found in the archipelago, with 57 species in 37 genera in seven orders. The cricket genus *Grillus* has eight native species, nevertheless some of the crickets species are introduced [6].

One insect species (*Halogates robustus*) from the Hemiptera order occupies a somewhat uncommon habitat in the islands. This is a flightless marine insect which occurs in the surface of the coastal waters, associated to mangrove and lava edges [7].

The Thysanoptera in the Galápagos Islands accounts for 77 species of thrips which belong to 42 genera and four families. This group of insects has been registered in 17 islands. At least nine of the existing species are considered serious pests [8], owing to their herbivorous habits that causes serious damage to crops.

The islands harbor also endemic cockroaches' species which may be of high interest in terms of the gut's microbiota. Eighteen species are reported to occur in the archipelago, where five are endemic [9].

Beetles species are relatively more abundant, occurring in all the vegetation zones. Some of them are found around flowers of cacti, *Ipomoea* sp., *Datura* sp., and a number of other plant species. It has been reported a number of about 200 beetles species [10].

Additionally there is a number of introduced species of insects such as cockroaches, *Drosophila* sp, ants and flies, among the most conspicuous. The changes in the insect biota composition in the islands occurred with the introduction of foreign species by aboriginal peoples and later by colonization activities of Europeans [11]. These insects also play a fairly important role in dispersion and introduction of yeast species into the islands.

3. On the Galápagos biogeography

The Galápagos Islands represent a strategic geographic zone in the planet to carry out studies on biodiversity, speciation, adaptation, ecology and dispersion of species belonging to different kingdoms. In an attempt to explain the current biogeography of the islands, a number of studies have been done since the Darwin's times in the 19th Century [12-14].

It is well known that the islands geological origin is oceanic, while the origin of its biota is continental [15]. The studies on the Galapagos Islands' biogeography and evolution were predominantly (if not exclusively) focused on macro organisms: there is a marked shortfall in the knowledge of microbial populations, communities and biodiversity in the Galapagos Islands.

Standard biogeographic tracks link the archipelago of Galápagos with Central America, western North and South America, the Caribbean, Asia and Australasia [14]. The Galápagos Islands community characteristics share the common features of the biota occurring in oceanic islands such as: disharmony, endemism and relictualism [16].

The archipelago of Galápagos is composed by oceanic islands where the terrestrial colonists such as plants, animals, and part of the microorganisms, should have crossed an oceanic barrier to reach the land. It means that the origin of the biota in this group of islands is entirely explained by the dispersal with no vicariant component, since the islands were never connected to the mainland by any land bridge or island chain [14]. Furthermore, since the Galápagos Islands are separated from the centers of origin of species in the mainland, there are abundant "empty" ecological niche spaces [17].

The colonization of the varied environments within the islands by yeast species must be primarily explained by the occurrence of colonizing terrestrial species of plants and animals that arrived to the islands as a sweepstake along the natural history of the archipelago. The sea water and marine fauna must have been other sources of colonizing yeasts species. Finally, the yeasts flora inhabiting the gut of insects associated to plants as well as birds and terrestrial vertebrates must have completed the cast of yeast diversity in the Islands.

Undoubtedly the strong association of plants and microorganisms must be regarded as a dispersion factor for microorganisms, including yeasts in the Galápagos. In 1976 [18] determined that 236 species of vascular plants are endemic (representing 45% of total species); 155 species are from Neotropical origin (which represents 30% of the total species); 62 different species of vascular plants are from Pantropical origin (12% of total species); 61 species are originated in the Andes (representing 12% of total species); while only 4 and 2 species correspond to Mexico and Central America and non-tropical South America, respectively (2% of total species). This study takes into account the indigenous species of the archipelago, namely, those species that were not introduced by man's activities [18].

4. The expedition to Galápagos in the search of yeasts

In the past, the studies of microorganisms in the Galápagos Islands were focused on topics such as the bacterial dynamics around the islands [19] some studies referring to entomogenous fungi found in the Galápagos Islands [20]; and probably some other works devoted to punctual groups of moulds, etc.

To the best of our knowledge there is not any yeast collection from the Galápagos Islands or any other study involving yeasts in the past. Based upon this, we can say that this is likely the first study focused on the biodiversity of yeasts carried out in this zone of the planet.

The yeast biodiversity in four populated islands is the first approach that we've done in order to try shortening the lack of previous knowledge of the yeasts biodiversity in this archipelago. Nevertheless it is to say that inhabited islands' environments present certain degrees of disturb and where human mediated introduction of certain yeast species is a real issue. The islands herein reported as the localities for our collections were explored for logistical reasons, since those islands that are mostly pristine, represent a major logistic challenge that could be faced in future expeditions.

There's interesting islands to be explored in future works, such as Española, Pinta, Genovesa, Marchena, etc. Additionally there are rock shelters that could be explored in the search of substrates for yeasts isolation. In these environments no fresh water is available, moreover, most of these islets or rocks are inaccessible since their precipitous cliffs, rising directly out of the water present hazards to landing that almost no one has attempted to collect any living organism from those inaccessible and still pristine environments [3].

In this expedition we were able to focus our attention in collecting yeasts from several substrates, so we can draw a general overview of the yeasts biodiversity in the islands. Nevertheless, there is still much work to be done in future expedition in order to refine the current data of the yeasts biodiversity in the archipelago, since it is one of the still unexplored regions in the planet [21].

5. Collection methodologies

In October 2009, from 19th to 23rd, an expedition composed by 5 researchers from Ecuador, Brazil and Spain went to the Galapagos Islands in order to start a pioneer study mostly on yeasts biodiversity within the still natural environments of four inhabited islands. This chapter is the first report of such expedition that explored Santa Cruz, Isabela, Floreana and San Cristóbal Islands, where a number of substrates were sampled by using different culture media and techniques.

Yeast sampling in substrates like flowers, fruits, excrement or fugus was carried out using sterile cotton wool swabs, to inoculate in liquid and solid YM media. In the case of sampling the insect's gut content a technique of catching the living insect in plastic bags for further inoculation by the living insect walking on the surface of Petri dishes containing YM agar medium. Eventually the insects were liberated alive. Additionally a number of substrates were collected in plastic sterile tubes for further culturing in selective culture broths such as YNB-CMC, YNB-D-xylose; YNB-xylan; YNB-L-arabinose, and YNB-raffinose the CLQCA laboratory in Quito, Ecuador. The selective culture media were used especially in the search of yeast strains that exhibit some biotechnological potential use in xylose fermentation as well as cellulose degradation/fermentation.

6. Yeast species identification

Macroscopic and microscopic identification of yeasts is frequently inaccurate due to the high similarity that yeast may show at a glance, either in colony or in a microscopic field. The best way to identify and differentiate yeast species as well as strains is by molecular techniques that are being used since 2000. A variety of techniques have been developed and this fact has greatly boosted the number of new species of yeasts identified. Molecular analyses of the variable D1/D2 regions of the 26S rDNA, 18S, 5.8S and mitochondrial small subunit rDNAs gene, as well as ITS sequencing and RFLP-ITS are very useful ways to identify yeast species and invaluable tools for phylogenetic studies [21, 22].

The D1/D2 domain of the LSU rRNA gene was PCR amplified directly from whole yeast cell suspensions [23], and using the primers NL1 and NL4 [24]. Initial amplification reactions were carried out in Ecuador (The Catholic University Yeast Collection-CLQCA) and Brazil (Collection of Microorganisms and Cells of UFMG). Amplified fragments were checked by agarose gel electrophoresis. The PCR purified products, were further sequenced with the external amplification primers NL1 and NL4. Finally, a sequence similarity search was conducted using NCBI Blast tool. Yeasts with more than 99% identity were considered members of the same species.

6.1. An overview of the yeast species in four Islands of the Galápagos Archipelago

As a result of the collections carried out in Santa Cruz, Isabela, Floreana, and San Cristóbal Islands 881 isolates were recovered from a wide variety of substrates. Currently we can report 614 yeast isolates already identified, while 267 isolates are in process of identification.

The number of yeast isolates collected per island is shown in Table 1.

Santa Cruz Island	Isabela Island	Floreana Island	San Cristóbal Island	Total Isolates
321	269	177	114	**881**

Table 1. Number of yeast isolates collected in four islands of the Galápagos Archipelago during an expedition carried out in October 2009

The difference in the abundance of yeast isolates collected in each island is due to different conditions of time and logistics of the expedition and does not have any relationship with the abundance or diversity of yeasts in each island.

The present chapter reports the isolates and biodiversity recovered in a number of substrates including: flowers, rotten wood, excrement, insects, fruits, exudates, leafs, one sugar cane mill (Santa Cruz Island) and others. Figure 1 shows the shares of yeast isolates by substrate we have preserved in the Catholic University Yeasts Collection (CLQCA).

6.2. Yeast species in Ecuador Mainland and Galápagos Islands

The Catholic University Yeasts Collection in Quito, in its database presents 118 yeast species belonging to the Ecuador mainland and the Galápagos Islands as shown in Table 4. By establishing a comparison between these two regions, there is not a big difference in number of species registered in the mainland and the islands (82 and 78 respectively). It is important to remark that the yeasts registers in this work were taken only from natural environments and substrates (no clinical or industrial samples are taken into account).

In Mainland Ecuador, about 50% of the characterized isolates preserved in the CLQCA (c.a. 250 yeast isolates) belong to four species: *Candida tropicalis, Meyerozyma guilliermondii, Kodamaea ohmeri,* and *Pichia kudriavzevii.* In contrast, the more represented isolates from the Galápagos Islands in the CLQCA (c.a. 615 yeast isolates) correspond to the species: *Candida tropicalis, Hanseniaspora* sp., *Pichia norvegensis, Candida parazyma, Kodamaea transpacifica, Hanseniaspora uvarum, Barnettozyma californica, Candida intermedia,* and *Galactomyces geotrichum.*

Candida tropicalis in both cases is the most abundant yeast species registered in CLQCA: in mainland it is about 21% of the total identified isolates, while in the Galápagos Islands it represents about 18%. *C. tropicalis* is a cosmopolite yeast species that is ubiquitous in a wide range of substrates: from beetles to fermented beverages, but predominantly it is found in rotten vegetal matter, flowers and excrements.

Between the three regions of mainland there are also registers of 10 species that have been collected from different substrates. These species are quite adaptable to a wide range of climatic conditions and substrates. Table 2 shows the species that are ubiquitous in Amazonia, Andes and Pacific Coast.

Matching the coincidences of yeast species between any individual mainland region with the Galápagos Islands yeast isolates it is noticeable an increase in the number of coincident yeast species if compared to the matching between the three regions within the mainland (Table 3).

N°	YEASTS SPECIES
1	*Candida intermedia*
2	*Candida tropicalis*
3	*Galactomyces geotrichum*
4	*Hanseniaspora sp.*
5	*Kodamaea ohmeri*
6	*Pichia kluyveri*
7	*Pichia kudriavzevii*
8	*Pichia manshurica*
9	*Saccharomyces cerevisiae*
10	*Torulaspora delbrueckii*

Table 2. Ubiquitous yeast species within Ecuadorian mainland

This increase represents almost four times the amount of the yeasts species shown in Table 2 (i.e. 38 yeast species) according to updated CLQCA registers.

Despite the fact that it is really difficult to address an accurate center of origin of the yeast species occurring in the natural zones of the Galápagos Islands (the islands are now visited by thousands of tourists along the year), we must assume that the Ecuadorian mainland must be the center of origin of most of the yeast species occurring in the Galápagos based upon the Porter's report [18] where about 30% of the vegetal species in the Galápagos Islands have had a Neotropical origin. From our data, about 31% of yeast species have been found both in Ecuador Mainland and the Galápagos Islands. These figures reveal the disharmony of taxa between mainland and oceanic islands [16].

Table 3 shows the species that matched at least once between any single regions of Ecuador's mainland with yeast species found in anyone of the four islands explored.

N°	SPECIES FOUND IN MAINLAND AND THE ISLANDS
1	*Aureobasidium pullulans*
2	*Barnettozyma californica*
3	*Candida carvajalis*
4	*Candida humilis*
5	*Candida intermedia*
6	*Candida oleophila*
7	*Candida orthopsilosis*
8	*Candida parapsilosis*

N°	SPECIES FOUND IN MAINLAND AND THE ISLANDS
9	*Candida pseudointermedia*
10	*Candida quercitrusa*
11	*Candida saopaulonensis*
12	*Candida sinolaborantium*
13	*Candida theae*
14	*Candida tropicalis*
15	*Cryptococcus humícola*
16	*Cryptococcus laurentii*
17	*Debaryomyces hansenii*
18	*Debaryomyces nepalensis*
19	*Galactomyces geotrichum*
20	*Hanseniaspora sp.*
21	*Kazachstania exigua*
22	*Kodamaea ohmeri*
23	*Meyerozyma guilliermondii*
24	*Pichia fermentans*
25	*Pichia kluyveri*
26	*Pichia kudriavzevii*
27	*Pichia manshurica*
28	*Pichia occidentalis*
29	*Pichia terricola*
30	*Rhodosporidium paludigenum*
31	*Rhodotorula glutinis*
32	*Rhodotorula mucilaginosa*
33	*Saccharomyces cerevisiae*
34	*Torulaspora delbrueckii*
35	*Trichosporon asahii*
36	*Trichosporon coremiiforme*
37	*Wickerhamomyces anomalus*
38	*Yamadazyma mexicana*

Table 3. Yeast species represented both in mainland and Galápagos Islands

On the other hand, mainland Ecuador is one of the richest biodiversity zones in the world [21], due to the varied environments as consequence of the topography that provokes the occurrence of uncountable micro-ecosystems as a consequence of a number of biotic and abiotic factors such as the altitude variation (from 0 to more than 6000 m.a.s.l), geographic location, marine currents, as well as the natural history involving the Neotropic which provides biogeographic unique features to this region of the planet [21].

The total yeast biodiversity currently registered both in Ecuadorian mainland and the Galápagos Islands (i.e. 118 yeast species identified) is shown in Table 4. The shaded cases represent the register of occurrence of the species either in the mainland or in the Galápagos Islands.

N°	YEAST SPECIES	ECUADOR MAINLAND	GALAPAGOS ISLANDS
1	*Aureobasidium pullulans*	X	
2	*Barnettozyma californica*	X	
3	*Candida albicans*		X
4	*Candida apicola*	X	
5	*Candida asparagi*		X
6	*Candida boidinii*	X	
7	*Candida boleticola*	X	
8	*Candida bombi*	X	
9	*Candida carpophila*	X	
10	*Candida carvajalis*	X	
11	*Candida cylindracea*		X
12	*Candida dendronema*		X
13	*Candida ecuadorensis*	X	
14	*Candida gigantensis*	X	
15	*Candida glabrata*	X	
16	*Candida humilis*	X	
17	*Candida intermedia*	X	
18	*Candida leandrae*		X
19	*Candida naeodendra*	X	
20	*Candida natalensis*		X
21	*Candida oleophila*	X	

N°	YEAST SPECIES	ECUADOR MAINLAND	GALAPAGOS ISLANDS
22	*Candida orthopsilosis*		
23	*Candida parapsilosis*		
24	*Candida parazyma*		
25	*Candida pomicola*		
26	*Candida pseudointermedia*		
27	*Candida pseudolambica*		
28	*Candida quercitrusa*		
29	*Candida rugosa*		
30	*Candida saopaulonensis*		
31	*Candida silvae*		
32	*Candida sinolaborantium*		
33	*Candida sonorensis*		
34	*Candida sorbosivorans*		
35	*Candida sorboxylosa*		
36	*Candida stellimalicola*		
37	*Candida tallmaniae*		
38	*Candida theae*		
39	*Candida tropicalis*		
40	*Candida trypodendroni*		
41	*Candida xylopsoci*		
42	*Candida zeylanoides*		
43	*Clavispora lusitaniae*		
44	*Clavispora opuntiae*		
45	*Cryptococcus albidus*		
46	*Cryptococcus flavescens*		
47	*Cryptococcus flavus*		
48	*Cryptococcus humicola*		
49	*Cryptococcus laurentii*		
50	*Cryptococcus rajastharensis*		
51	*Cryptococcus saitoi*		

N°	YEAST SPECIES	ECUADOR MAINLAND	GALAPAGOS ISLANDS
52	*Debaryomyces hansenii*		
53	*Debaryomyces nepalensis*		
54	*Dekkera anomala*		
55	*Dekkera bruxellensis*		
56	*Filobasidium uniguttulatum*		
57	*Galactomyces geotrichum*		
58	*Geotrichum silvicola*		
59	*Hanseniaspora guilliermondii*		
60	*Hanseniaspora meyeri*		
61	*Hanseniaspora opuntiae*		
62	*Hanseniaspora sp.*		
63	*Hanseniaspora uvarum*		
64	*Hanseniaspora valbyensis*		
65	*Kazachstania exigua*		
66	*Kazachstania unispora*		
67	*Kluyveromyces lactis*		
68	*Kluyveromyces marxianus*		
69	*Kodamaea ohmeri*		
70	*Kodamaea transpacifica*		
71	*Kurtzmaniella zeylanoides*		
72	*Kwoniella mangrovensis*		
73	*Lindnera sp.*		
74	*Lindnera fabianii*		
75	*Lindnera jadinii*		
76	*Lindnera saturnus*		
77	*Metschnikowia kipukae*		
78	*Metschnikowia koreensis*		
79	*Metschnikowia reukaufii*		
80	*Meyerozyma guilliermondii*		
81	*Pichia fabianii*		

N°	YEAST SPECIES	ECUADOR MAINLAND	GALAPAGOS ISLANDS
82	*Pichia fermentans*		
83	*Pichia kluyveri*		
84	*Pichia kudriavzevii*		
85	*Pichia manshurica*		
86	*Pichia nakasei*		
87	*Pichia norvegensis*		
88	*Pichia occidentalis*		
89	*Pichia terricola*		
90	*Rhodosporidium babjevae*		
91	*Rhodosporidium paludigenum*		
92	*Rhodotorula glutinis*		
93	*Rhodotorula minuta*		
94	*Rhodotorula mucilaginosa*		
95	*Rhodotorula slooffiae*		
96	*Rhodotorula* sp.		
97	*Saccharomyces cerevisiae*		
98	*Saccharomycodes ludwigii*		
99	*Saccharomycopsis fodiens*		
100	*Saccharomycopsis vini*		
101	*Saturnispora quitensis*		
102	*Scheffersomyces stipitis*		
103	*Sporidiobolus ruineniae*		
104	*Sporopachydermia* sp.		
105	*Torulaspora delbrueckii*		
106	*Trichosporon asahii*		
107	*Trichosporon coremiiforme*		
108	*Trichosporon dermatis*		
109	*Trichosporon insectorum*		
110	*Trichosporon jirovecii*		
111	*Trichosporon laibachii*		

N°	YEAST SPECIES	ECUADOR MAINLAND	GALAPAGOS ISLANDS
112	*Trichosporon multisporon*		
113	*Wickerhamiella occidentalis*		
114	*Wickerhamomyces anomalus*		
115	*Wickerhamomyces onychis*		
116	*Yamadazyma mexicana*		
117	*Yarrowia lipolytica*		
118	*Zygotorulaspora florentina*		

Table 4. Species occurrence in Mainland Ecuador and/or the Galápagos Islands

It is noticeable that in terms of biodiversity registered in Ecuador's mainland and the Galápagos Islands, the figures are very similar. Nevertheless, it should be understood that the total sample that is herein analyzed, represents only a fraction of the total yeasts species which may be found in this biodiverse region of the planet. At this point of the research the registers of yeast species in Ecuador are still in constant updating as well as the registers in the rest of the world. We believe that less than 5% of the world's yeast biodiversity has been described [21, 25]. A big effort and a long time of research are needed to try fulfilling the CLQCA database. Despite the data analyzed in this first overview of the biodiversity of yeasts in the four regions of Ecuador are still in process, we can certainly draw some features of the yeast biota of the Galápagos and its closest mainland territory.

6.3. A brief ecological approach to the most remarkable yeast species in Galápagos and Mainland

The yeast species that are part of the communities found both in the Galápagos Islands and the mainland are: *Aureobasidium pullulans, Barnettozyma californica, C. carvajalis, C. humilis, C. intermedia, C. oleophila, C. orthopsilosis, C. parapsilosis, C. pseudointermedia, C. saopaulensis, C. sinolaborantium, C. tropicalis, Debaryomyces nepalensis, Galactomyces geotrichum, Hanseniaspora* sp., *Pichia terricola, Rhodotorula glutinis,* and *Rh. mucilaginosa.*

Strains recovered of *B. californica* in other studies includes substrates such as soil, ox dung, insect frass, pond water, sewage water, river water, and stream water [25]. In the Ecuadorian collection this species was found in flowers, beetles captured in flowers of *Ipomoea alba,* giant turtle's dung, sugar cane bagasse, fungus, and rotten wood. Insects may have played a role in the dispersion of this species from mainland into the Galápagos.

Candida humilis in the Galápagos Islands has been found in rotten wood and flowers, but other registers of this species such as the strains CBS 6312 and CBS 6099 reported in The Yeasts a Taxonomic Studies [25] which were found in frass and gut of auger beetles. In this case we have maybe another evidence of insects as vectors of this yeast species and the occupation of the habitats in the Islands may not be so different than in mainland for this yeast species. In

Ecuadorian mainland *C. humilis* was isolated from mango fruit, chicha de jora (corn fermented beverage), spider network, rotten wood and some flowers from the Asteraceae family. As for *C. intermedia* and *C. parapsilosis* are considered widespread yeasts that can be found in clinical samples, caterpillar frass, beer contaminants and other substrates. In the case of Ecuadorian isolates of these both species, the origin is predominantly found in vegetal substrates and in archaeological fermentation pots from ancient cultures.

In the case of *C. saopaulonensis* we can find that in the mainland Ecuador this species was isolated from a Heliconiaceae as well as the type strain collected in Brazil [26]; in contrast, this yeast species was found in nine different flowers from Asteraceae family in Santa Cruz Island, which is an interesting case of adaptation to new ecological niches in oceanic islands [16].

C. sinolaborantium [27] is a yeast species that in the Ecuadorian mainland as well as in the Galápagos Islands was found chiefly in vegetal substrates except for one strain that was collected in *Drosophila* sp. Nevertheless the type strain was isolated from the gut of handsome fungus beetle in Panamá, and two other strains were isolated from a cerambycid larva. The original study was carried out in insect's guts. The results obtained by [27] suggest that this yeast species is part of the microbial community that can be found in the intestine of insects (beetles) whose excretions are dropped on flowers and other vegetal matter.

Another remarkable yeast species found in this survey both in mainland and the Galápagos Islands is *C. tropicalis*. This ubiquitous species is by far the most abundant in isolates number that has been collected since 2006 in our surveys in Ecuador; *C. tropicalis* is practically distributed in every kind of substrate, which is a sign of the high adaptability of this species. As part of a study on traditional fermented beverages we were able to find this yeast species even in unexpected substrates such as cassava fermented beverages which are produced in Yasuní National Park, located in the deep Ecuadorian Amazonia. This millenary beverage is still being produced in the same traditional way by autochthonous tribes of Waorani people [28]. In the literature *C. tropicalis* is reported to be collected from clinical samples. This species belongs to the *Lodderomyces* clade, where *C. albicans*, *C. theae* [29] and other potential pathogens have been accommodated. Nevertheless its presence by itself does not mean a hazard of infection in human, at least to those who possess a strong immune response due to permanent exposition to *Candida* sp. cell wall antigens [30].

A very few understood yeast species in terms of its ecology is *Debaryomyces nepalensis*, since it was isolated from a number of different substrates like soil, fermenting tobacco, spoiled sake and others [25]. The isolates from the Galápagos were found in rotten wood, but also in flowers, orchids, and leafs among others. This yeast species was also collected in a range of altitudes in Ecuadorian territory: from 150 to 1820 m.a.s.l.

Galactomyces geotrichum [31] is another yeast species which is widespread along the four natural regions of Ecuador. Its distribution has been registered from 75 to 3500 m.a.s.l. This fact demonstrates its high adaptability to a number of ecosystems and growth temperatures. It is a vigorous yeast species characterized by its fast growth in laboratory conditions. The strength of this yeast is a feature that has been useful in the prevention of microbial infections in greenhouse crops [25] because of its highly competitive way of occupying micro substrates and ecological niches. As for the substrates where this yeast species was found in this survey, these includes rotten wood, flowers, turtle's feces, insects, vegetal residues, fruits, etc. Remarkably, by 1970 just a few strains of this species were collected from soil samples in Puerto Rico [25].

The genus *Hanseniaspora* was also found in flower samples from the Galápagos and Mainland Ecuador. This genus is characterized by its bipolar budding, presenting apiculate and ovoid cells that can even be long ovoidal or elongate [32]. This genus is composed by a number of species: *H. clermontiae, H. guilliermondii, H. lachancei, H. meyeri, H. occidentalis, H. opuntiae, H osmophila, H. pseudoguilliermondii, H. uvarum, H. valbyensis, H. vinae, H. singularis and H. thailandica.* In Ecuador mainland it was found *H. meyeri, H. guilliermondii, H. valvyensis, H. opuntiae,* and *H. uvarum.* Nevertheless, we couldn't yet identify the Galápagos Islands isolates to the species level.

In terms of the substrates where the yeast species have been found both in mainland and the islands there is a wide range of sources: insects, flowers, leaves, feces, sugar cane mills, fungus, fruit, moss, and a number of samples taken from endemic plants in the Galápagos Islands, including *Miconia robinsoniana, Scalesia* sp., *Opuntia* sp. *Castela galapageia,* etc. This genus is widely represented in the CLQCA, where about 100 isolates were collected from all the regions in mainland and Galápagos. The isolates from Galápagos represent about 60% of the total isolates of this genus in the CLQCA.

In the case of *Pichia terricola* [33] has been scarcely represented in the mainland with only one isolate in the Andean province of Pichincha. The range of altitude in which this species is distributed in the Ecuadorian territory, according to the registers of the CLQCA, is from 100 to 1120 m.a.s.l., and it's found in a variety of habitats, from dry forests in the Galápagos Islands to cloudy forests in the mainland. The description of this species was based on soil samples from South Africa, but it was also found in cherry juice, from pressed grapes, spoiled figs, and wine [25]. This is an ethanol assimilating yeast, since it can use this compound as an alternative carbon source. This yeast can also use other carbon compounds such as glycerol and glucose, as well as succinate.

This yeast species was collected in flowers, rotten wood, fruit, insects, and exude of trees. In Galápagos the microhabitats for this species are quite different to those of the original description in 1957, where *P. terricola* was named based on the soil samples where it was found in South Africa by [33]. The appellative given to this species illustrates a frequent issue found in taxonomic accommodation; the species names derived from the substrates where the isolates were found, may not provide an accurate idea of its ecological niche. Frequently, yeasts isolates may represent allochthonous members of microbial communities [34].

Rhodotorula glutinis [35] and *Rhodotorula mucilaginosa* [36] are both basidiomycetous yeast species characterized by the production of carotenoid pigments which protect them from the UV irradiation that provokes damage in DNA [37-39]; especially in zones where the solar rays are particularly severe, like in the equatorial latitude, the need for a protective strategy can make the difference for yeast cells that are exposed to aggressive light stress [40].

Both yeast species are ubiquitous being found in a wide range of latitudes, including in Antarctica glaciers (data not published). Some registers of *R. glutinis* are in atmosphere, trees, leaves, grapes, soil, spoiled leather, sea water, water supply of a brewery, sputum of pneumonia patient, exudates, limph nodes, feces, pasteurized beer and a number of other substrates [41]. These two yeast species are not fermentative, which is a trait of basidiomycetous yeast species.

From Ecuador we were able to collect 46 strains belonging to these two species. Astonishingly, 18 out of 46 were isolated from ancient-dormant yeast communities found in fermentation vessels and other utensils used by ancient cultures from the Andes used in their daily life. The studies of Microbial Archaeology [42] yielded other yeast species which will be taken later on in this chapter.

Other substrates where *R. glutinis* and *R. mucilaginosa* were isolated from are fermented beverages, sugar cane juice, insects (Orthoptera, Coleoptera, Hemiptera), as well as in flowers, and moss.

C. carvajalis [43] was the first yeast species described from Ecuador. This yeast was found in the course of a yeast biodiversity survey in the Amazonia. The substrate sampled was rotten wood and fallen leaf debris, collected around crude oil wells, close to Dayuma town. One Isolate of this species was collected in Santa Cruz Island from *Psidium guajava* mucilage. The closest relatives of *C. carvajalis* are *C. asparagi*, *C. fructus*, and *C. musae*. This group of yeasts belongs to the *Clavispora* clade. In Mainland Ecuador it has not collected any *C. asparagi* [44] isolate, but in the Galápagos Islands we have one register from Santa Cruz Island where this species was collected from a nitidulid beetle.

6.4. Ecology of the yeasts species in the Galápagos Islands and Mainland Ecuador

The yeast isolates that were identified, characterized and preserved in the CLQCA since 2006 represent an invaluable platform for studies in ecology of yeasts form Ecuador. Our data permit a better understanding of the situation of the yeast species and communities in natural environments of Ecuador, which is a contribution for the knowledge of the yeast ecology and biology. In this collaborative work other laboratories and centers have been involved, namely the Collection of Microorganisms and Cells at the Universidade Federal de Minas Gerais in Belo Horizonte, Brazil; the National Collection of Yeast Cultures in Norwich, England; the Department of Biology at the University of Western Ontario, In Ontario, Canada, among the more active collaborators in this survey.

In this part of the chapter we will analyze not only the yeast species occurrence in Ecuador, but also the relationships we have found respect to the substrates where the yeasts were collected; the ecological roles of the species as reported by other authors and contrasting information we got in our own work. Finally, we will present some ecological similitudes and differences in the roles and behavior of yeasts depending on their biogeographic zones where they occur.

6.5. Yeast species by substrate

We intended to compare the communities of yeasts that are present in the different substrates we have sampled in mainland and the islands. The most common substrates represented in this survey are: flowers, rotten wood, beetles, excrement, rotten vegetal matter, fruits, insects (*Drosophila* sp. and Nitidulid beetles), exudates, leaves, fungus, and human related substrates such as artisanal sugar cane mills. Some substrates like bodies of water, moss, wood, etc. are not analyzed due to the fact that in mainland we do not have the correspondent substrates to compare with.

The data were homogenized in order to get comparable ecological niches in Galápagos and Ecuador mainland. Consequently, a comparison of the yeast communities found in analogous substrates sampled both in the islands and the mainland is made in this section.

As a consequence of this grouping of substrates, the number of species having correspondence between mainland and the islands decreased from 38 to 22, since the variety of substrates in mainland is much larger than those sampled in the Galapagos, furthermore, it was not possible to match all the samples for a comparison.

Nevertheless, an independent analysis of the species found only in mainland and the Galápagos was also performed in order to get some biodiversity plots of the yeast communities. Based on that, we were able to analyze the situation of yeast biodiversity in Ecuador, taking into account the general characteristics of oceanic islands.

The substrates from which more samples were taken are flowers, insects and rotten wood. This fact responds to the specific objectives of the expeditionaries who centered their interest in the surveys of these substrates. Nonetheless, other substrates were also sampled with different intensity. The share of the samples taken in the expedition is shown in Figure 1.

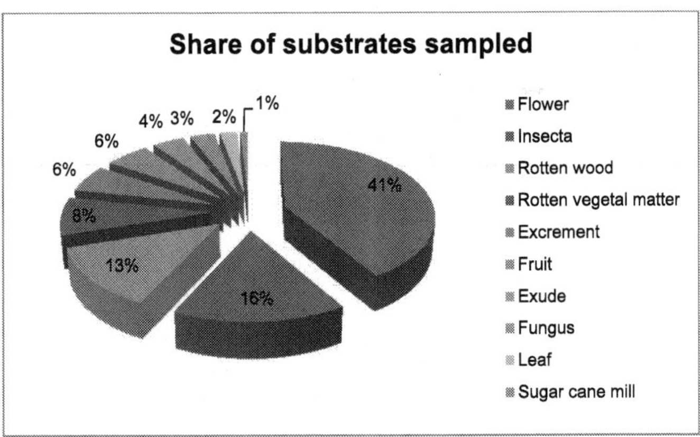

Figure 1. Substrates sampled in the basis of the number of samples taken in the Galápagos Islands.

Another issue is the number of isolates that were collected by substrate. Given the fact that the number of collections in the field made by each substrate varies substantially from one substrate to the other, a calculation was employed to obtain the average yeast isolates obtained by substrate. This calculation was done using the equation (1):

$$\bar{A} = \frac{N}{S} \tag{1}$$

Where $\bar{A}$ represents the average number of isolates per substrate, N represents the number of isolates with morphological differences gotten from each substrate; and, S corresponds to the number of samples taken from each substrate. In that way it was possible to determine the number of potentially different species based upon the macroscopic morphology. This approach provides an idea about the composition of the communities in terms of the complexity, based on the morphological differences from its members in a variety of substrates.

From a general point of view, we can say that the potential number of different species isolated is predominant in those substrates where the conditions favor the yeast reproduction and activity. Thus, in the sugar cane mill we could recover an average of 5 yeast isolates per sampling (regardless its biodiversity), while in flowers we could recover an average of 1.9 yeast isolates per sample. The abundance of yeasts also seems to be higher in insect's guts as well as in rotten wood: from the guts of beetles staying within *Datura* sp. and *Ipomoea* sp. flowers we could recover 3.1 and 4 isolates in average, which is slightly lower than the number of isolates from the sugar cane mill. In rotten wood the number of isolates was relatively high since we could recover an average of 3.5 yeast colonies. Table 5 resumes the average of isolates recovered by substrate in the Galápagos Islands. Nevertheless, these data does not provide enough information about the biodiversity of yeasts living in the substrates.

A further analysis of these figures reveals that the occurrence of yeast isolates in the samples has not a direct relationship with the biodiversity that can be found in each substrate, considering that macroscopic differences of yeast colonies does not necessarily represent different yeast species, but strains of a same species. A clear example of that is the biodiversity of yeasts found in the sugar cane mill (5 possible different species) which reports the highest average of isolates per sample (5 isolates), compared to flowers which present a rather low average of isolates per sample (1.9 isolates) but the highest biodiversity (14.2 species per sample). In Figure 2 the correlation between the number of samples, the number of isolates per sample, and the number of species by sample is drawn.

The insects were the substrates that yielded the highest biodiversity of yeasts in the Galápagos Islands (38 species), followed by rotten wood (35 species) samples and flowers (28 species). The substrates like trees exudates, rotten plant matter (leaves and fruits), fruits, fungus, leaves, and excrement, yielded from 7 to 11 different yeast species, while the sugar mill and fermented sugar cane juice sampled presented 5 isolates. Nonetheless, the highest average number of species per sample was registered in flowers, followed by insects and rotten wood.

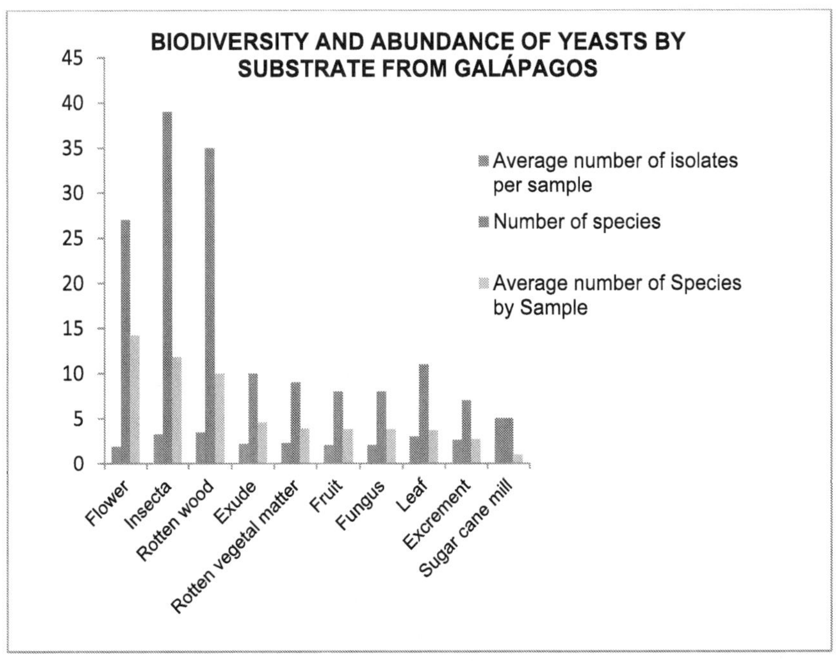

Figure 2. Correlation between number of samples from substrates, number of isolates per sample, and biodiversity of yeasts in substrates

On the other hand, the fermentation vessels and utensils where sugar cane juice is extracted and then transformed into fermented liquor containing high ethanol degree, constitutes a restrictive environment for yeast, where only a few species such as *Saccharomyces cerevisiae* can be found, owing to its high tolerance to ethanol provided by the fatty acids contained in the cells [45]. The rather poor biodiversity of this kind of yeast community is inversely proportional with the population abundance of the community. This inversely proportional relationship confirms that the highest biodiversity of yeasts can be found in those substrates where yeast communities play a fundamental ecological role. Most yeast species isolated from flowers are supposedly nectar-inhabiting yeasts. Dense yeast communities often occur in the floral nectar of animal-pollinated plants, where they can behave as parasites of plant-pollinator mutualisms [46-50].

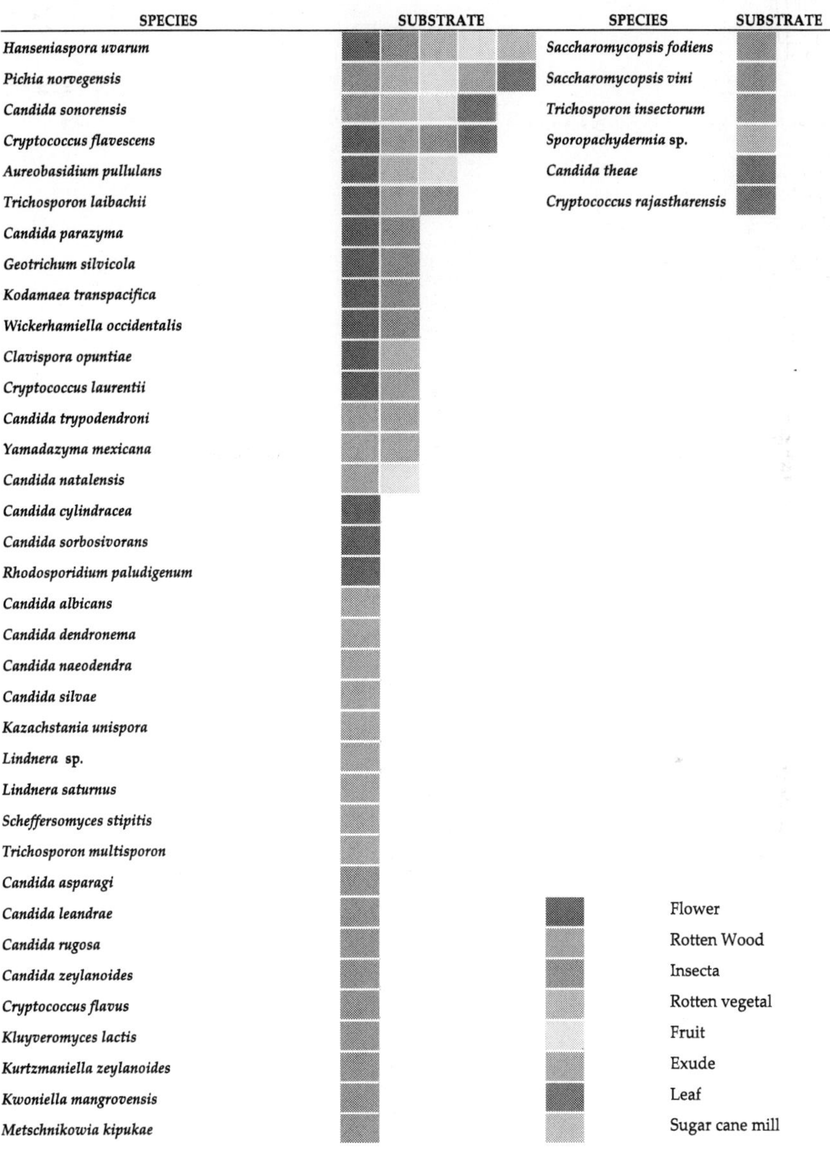

(b)

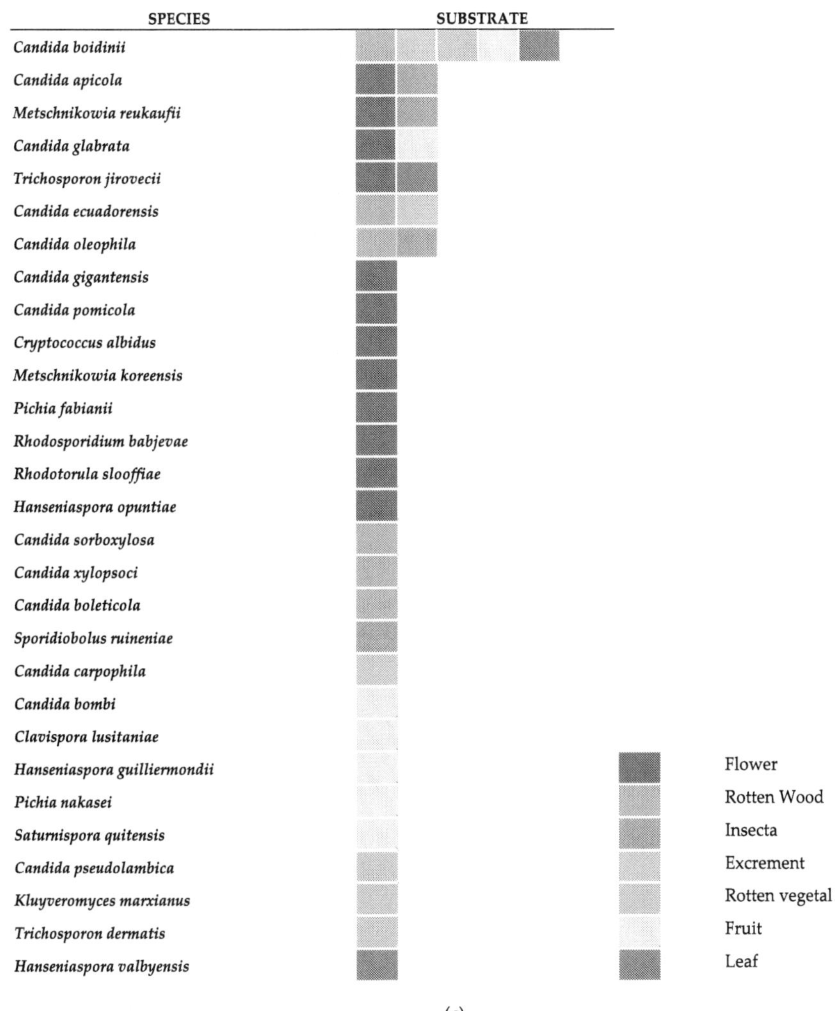

Figure 3. (a) Shared species between Galápagos Islands and Mainland Ecuador and its occurrence in different substrates; (b) Species found in Galápagos and the substrates these yeast colonize; (c) Species found in Ecuador Mainland and the substrates these yeasts colonize

7. Ecological parameters and models developed

The degree of specialization of a yeast species is given by its metabolic abilities and tolerance to environmental factors such as UV radiation [40], inhibition substances and even microbial predators [34], etc. Leaves are exposed to fluctuations of temperature and relative humidity values, which may have an impact on the yeast communities. Large fluxes of UV radiation are also one of the most prominent features of the leaves, fruits and other substrates in the environment to which microorganisms have presumably had to adapt [21, 51]. Many plants contain a number of compounds whose adaptive significance may be a defense against invertebrates and microorganisms; for instance, all parts of *Datura* sp. contain toxic belladonna alkaloids, the concentration of which is highest in the petioles of the flowers [52]. These compounds also act, in some cases, as selective agents which shape the yeast community composition [53].

The yeasts which are highly adaptable to different ecological niches—like *Candida tropicalis*—are the ones that display a wide range of responses and developed defense strategies [45]. This particularly abundant and cosmopolitan yeast species is the most frequently isolated in mainland and in the islands according to CLQCA database. This yeast species has been found in flowers, rotten wood, insects, excrement, rotten vegetal matter, and fungus (Figure 3a). *Candida tropicalis* has been reported to grow even at pH so high as 10 [41], which is a remarkable feature in terms of tolerance facing unusually hard environmental conditions.

On the other hand there are yeasts that show a narrower repertoire of metabolism and even metabolic deficiencies such as *Saccharomycopsis fodiens*, a predacious yeast species that is deficient in sulfate uptake and require supplementation of organic sulfur sources [54]. This yeast species is highly specialized in predation of other yeasts and fungi. Nevertheless it appears to be quite rare in the environment and very few is known about its natural history and ecology. Only three isolates from Costa Rica, Australia and the Galápagos islands were found so far. A further expedition in Taiwan registered other isolates of this species associated to *Drosophila* flies [34].

7.1. How the ecological niches are occupied by yeasts in mainland and the islands?

It is certainly ventured to establish in an accurate way the ecological relationships of yeasts in the islands and the mainland with general data and relatively few samples. For this report we have addressed this issue by means of a global comparison of yeast species number that has been found in different substrates both in mainland and the islands. Despite the kind of substrate, we marked the repetitions of registers in substrates from each one of the species. This analysis gave us a general scheme of the degree of adaptation that is currently facing the yeasts.

In Figure 4 it is seen the general state of the yeasts based on the number of substrates they occur. This analysis was performed on those species found in mainland and the islands and those which are represented in both ecosystems.

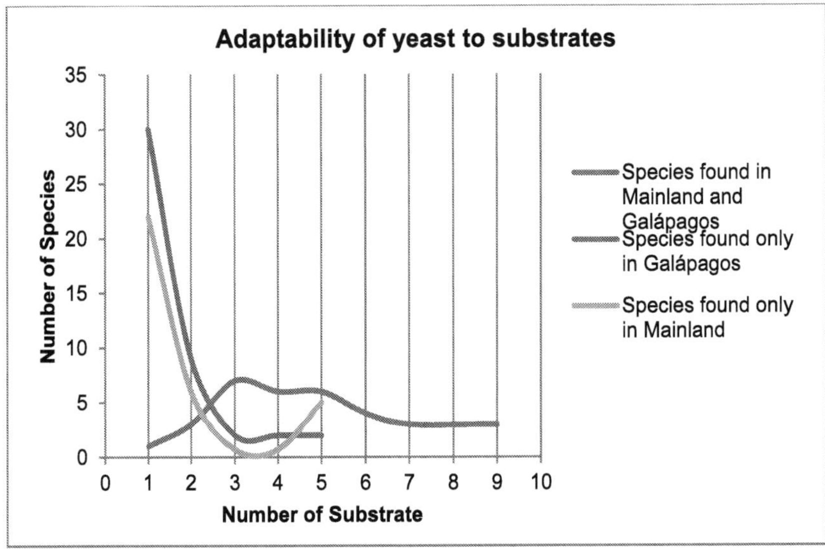

Figure 4. Global comparison in terms of adaptability to different substrates of yeast species.

Noteworthy, those yeast species which are common to Galápagos and mainland are represented by various yeasts that are able to colonize up to 9 different substrates. Only 1 yeast species looks confined to one kind of substrate. In the case of the species found in mainland exclusively, the figures are different, since the adaptability to various substrates looks lower, while the colonization of a single substrate is higher than the previous case, namely 22 yeasts occupy only one kind of substrate. As for the Galápagos yeasts the degree of adaptability is seen in more species but the specialization is much higher than in the before mentioned cases, this is 30 yeast species are confined to one specific substrate.

The yeast species shared between Galápagos and the mainland are mostly generalistic and show a wide range of substrates as ecological niches occupied. In contrast, the majority of the yeast species which have been found exclusively, both in the islands and the mainland, is less generalistic and occupy a narrow range of substrates. In other words, we can see that those yeast species which are more adapted to the mainland and the island ecosystems will probably find some ecological barriers that may impede the colonization of those ecosystems which they do not belong to. This can be regarded as a disharmony example.

No endemism has been studied or detected in this survey. There is only one yeast isolate representing a hitherto non described species, this is *Lindnera* sp. that was collected from *Scalesia* sp. rotten wood. This genus of tree is endemic to the Galápagos Islands.

The fact that in mainland the competition for substrates may be much more intensive than in the islands [16] could lead to shape more generalist yeast communities, while in the islands the yeasts species show a different trend that may be explained by a lower competence level and lower diversity in terms of ecological niches as compared to mainland.

7.2. Yeast specialization in Ecuadorian environments

From the data obtained in this work, it has been developed a simple model of calculation which allows us to establish the Index of Abundance (I_a) which is understood as the product of the total ecosystems where the species has been registered and the total different substrates that the species is able to colonize. The index maximum value is 1 and minimum is 0. The equation (2) to calculate such index is as follows:

$$I_a = \frac{(S_c {}^* O)}{(S_m {}^* O_m)}$$

(2)

Where I_a is the Index of Abundance; S_c is defined as the substrates colonized by the species; O is the occurrence in the different ecosystems that are being analyzed; S_m is the maximum number of substrates analyzed in the survey; and, O_m represents the maximum occurrence and is equal to the total number of ecosystems analyzed (seven ecosystems in the case of this study).

The I_a provides figures to compare the relative abundance of each yeast species in the ecosystems. Moreover it gives an idea about the generalist species, where the higher the I_a, the lower the specialization (more generalist species).

On the other hand, it is also necessary to define the extent of specialization of each yeast species. The Index of Specialization (S_i) is calculated from the following equation (3):

$$S_i = \frac{-\log_{10} I_a}{2}$$

(3)

Where S_i is defined as the Index of Specialization; and I_a is the Index of Abundance.

The values of S_i are inversely proportional to those of I_a given that the higher the abundance, the lower the specialization.

In Table 6 the I_a and S_i calculations can be seen based on the data from CLQCA and the analysis of seven ecosystems and 10 substrates (no yeast have been collected in all the substrates). It can be seen that *C. tropicalis* has been found in nine of 10 substrates compared in this survey, being the species with the higher I_a and lower S_i. The number 1 represents the positive occurrence and 0 means the lack of register of the species in the particular substrate. The number of substrates colonized by the species was taken from Figure 3a, 3b, and 3c.

SPECIES	AMAZON REGION	COASTAL REGION	HIGHLAND REGION	FLOREANA ISLAND	ISABELA ISLAND	SAN CRISTOBAL ISLAND	SANTA CRUZ ISLAND	OCCURRENCE IN ECOSYSTEMS	SUBSTRATES COLONIZED	INDEX OF ABUNDANCE (I_a)	SPECIALIZATION INDEX (S_i)
Candida tropicalis	1	1	1	1	1	1	1	7	9	0.90	0.02
Candida intermedia	1	1	1	1	1	0	1	6	9	0.77	0.06
Hanseniaspora sp.	1	1	1	0	1	1	1	6	9	0.77	0.06
Galactomyces geotrichum	1	1	1	1	1	0	1	6	7	0.60	0.11
Barnettozyma californica	0	0	1	1	1	1	1	5	6	0.43	0.18
Candida quercitrusa	0	1	1	1	1	1	1	6	5	0.43	0.18
Kodamaea ohmeri	1	1	1	0	1	1	0	5	6	0.43	0.18
Meyerozyma guilliermondii	0	1	1	0	1	1	1	5	6	0.43	0.18
Candida parapsilosis	0	1	1	0	1	0	1	4	7	0.40	0.20
Saccharomyces cerevisiae	1	1	1	0	0	0	1	4	7	0.40	0.20
Pichia kluyveri	1	1	1	1	0	0	1	5	5	0.36	0.22
Pichia kudriavzevii	1	1	1	0	1	1	0	5	5	0.36	0.22
Wickerhamomyces anomalus	0	1	1	1	0	0	1	4	6	0.34	0.23
Debaryomyces nepalensis	1	0	1	1	0	1	1	5	4	0.29	0.27
Hanseniaspora uvarum	0	0	0	1	1	1	1	4	5	0.29	0.27
Pichia terricola	0	0	1	1	1	0	1	4	5	0.29	0.27
Rhodotorula mucilaginosa	0	1	1	1	1	0	1	5	4	0.29	0.27
Candida pseudointermedia	0	0	1	1	1	0	1	4	4	0.23	0.32
Pichia manshurica	1	1	1	1	0	0	0	4	4	0.23	0.32
Candida sinolaborantium	0	0	1	1	0	0	1	3	5	0.21	0.33
Kazachstania exigua	0	0	1	0	1	0	1	3	5	0.21	0.33
Pichia norvegensis	0	0	0	1	0	1	1	3	5	0.21	0.33
Torulaspora delbrueckii	1	1	1	1	0	0	1	5	3	0.21	0.33
Candida sonorensis	0	0	0	1	0	1	1	3	4	0.17	0.38
Pichia fermentans	0	1	1	0	0	0	1	3	4	0.17	0.38
Trichosporon coremiiforme	0	0	1	1	0	0	1	3	4	0.17	0.38
Candida boidinii	0	1	1	0	0	0	0	2	5	0.14	0.42

SPECIES	AMAZON REGION	COASTAL REGION	HIGHLAND REGION	FLOREANA ISLAND	ISABELA ISLAND	SAN CRISTOBAL ISLAND	SANTA CRUZ ISLAND	OCCURRENCE IN ECOSYSTEMS	SUBSTRATES COLONIZED	INDEX OF ABUNDANCE (I_a)	SPECIALIZATION INDEX (S_i)
Candida humilis	0	1	1	0	0	0	1	3	3	0.13	0.45
Cryptococcus humicola	1	0	1	0	0	0	1	3	3	0.13	0.45
Debaryomyces hansenii	0	0	1	0	1	0	1	3	3	0.13	0.45
Pichia occidentalis	0	1	1	0	1	0	0	3	3	0.13	0.45
Trichosporon asahii	0	1	1	0	0	0	1	3	3	0.13	0.45
Cryptococcus flavescens	0	0	0	1	1	0	0	2	4	0.11	0.47
Yamadazyma mexicana	0	1	0	1	1	0	1	4	2	0.11	0.47
Aureobasidium pullulans	0	0	0	1	0	0	1	2	3	0.09	0.53
Candida carvajalis	1	0	0	0	0	0	1	2	3	0.09	0.53
Candida orthopsilosis	0	1	1	0	1	0	0	3	2	0.09	0.53
Candida parazyma	0	0	0	1	1	1	0	3	2	0.09	0.53
Cryptococcus laurentii	0	0	1	0	1	0	1	3	2	0.09	0.53
Kodamaea transpacifica	0	0	0	1	1	1	0	3	2	0.09	0.53
Trichosporon laibachii	0	0	0	1	1	0	0	2	3	0.09	0.53
Candida ecuadorensis	1	0	1	0	0	0	0	2	2	0.06	0.62
Candida natalensis	0	0	0	1	0	0	1	2	2	0.06	0.62
Candida oleophila	0	0	1	0	0	0	1	2	2	0.06	0.62
Geotrichum silvicola	0	0	0	0	1	1	0	2	2	0.06	0.62
Hanseniaspora meyeri	0	0	1	1	0	0	0	2	2	0.06	0.62
Rhodotorula glutinis	0	0	1	0	1	0	0	2	2	0.06	0.62
Wickerhamiella occidentalis	0	0	0	0	1	1	0	2	2	0.06	0.62
Candida apicola	0	0	1	0	0	0	0	1	2	0.03	0.77
Candida glabrata	0	0	1	0	0	0	0	1	2	0.03	0.77
Candida saopaulonensis	0	0	1	0	0	0	1	2	1	0.03	0.77
Candida theae	0	0	1	0	0	1	0	2	1	0.03	0.77
Candida trypodendroni	0	0	0	0	0	0	1	1	2	0.03	0.77
Clavispora opuntiae	0	0	0	1	0	0	1	2	1	0.03	0.77

SPECIES	AMAZON REGION	COASTAL REGION	HIGHLAND REGION	FLOREANA ISLAND	ISABELA ISLAND	SAN CRISTOBAL ISLAND	SANTA CRUZ ISLAND	OCCURRENCE IN ECOSYSTEMS	SUBSTRATES COLONIZED	INDEX OF ABUNDANCE (I_a)	SPECIALIZATION INDEX (S_j)
Kazachstania unispora	0	0	0	1	0	0	1	2	1	0.03	0.77
Kluyveromyces lactis	0	0	0	0	1	1	0	2	1	0.03	0.77
Metschnikowia reukaufii	0	0	1	0	0	0	0	1	2	0.03	0.77
Rhodosporidium paludigenum	0	0	1	0	0	0	1	2	1	0.03	0.77
Trichosporon dermatis	1	0	1	0	0	0	0	2	1	0.03	0.77
Trichosporon jirovecii	0	0	1	0	0	0	0	1	2	0.03	0.77
Candida albicans	0	0	0	1	0	0	0	1	1	0.01	0.92
Candida asparagi	0	0	0	0	0	1	0	1	1	0.01	0.92
Candida boleticola	0	0	1	0	0	0	0	1	1	0.01	0.92
Candida bombi	0	0	1	0	0	0	0	1	1	0.01	0.92
Candida carpophila	0	0	1	0	0	0	0	1	1	0.01	0.92
Candida cylindracea	0	0	0	0	0	0	1	1	1	0.01	0.92
Candida dendronema	0	0	0	1	0	0	0	1	1	0.01	0.92
Candida gigantensis	0	0	1	0	0	0	0	1	1	0.01	0.92
Candida leandrae	0	0	0	0	0	1	0	1	1	0.01	0.92
Candida naeodendra	0	0	0	1	0	0	0	1	1	0.01	0.92
Candida pomicola	0	0	1	0	0	0	0	1	1	0.01	0.92
Candida pseudolambica	0	0	1	0	0	0	0	1	1	0.01	0.92
Candida rugosa	0	0	0	0	1	0	0	1	1	0.01	0.92
Candida silvae	0	0	0	1	0	0	0	1	1	0.01	0.92
Candida sorbosivorans	0	0	0	0	0	0	1	1	1	0.01	0.92
Candida sorboxylosa	1	0	0	0	0	0	0	1	1	0.01	0.92
Candida xylopsoci	0	0	1	0	0	0	0	1	1	0.01	0.92
Candida zeylanoides	0	0	0	0	1	0	0	1	1	0.01	0.92
Clavispora lusitaniae	0	0	1	0	0	0	0	1	1	0.01	0.92
Cryptococcus albidus	0	0	1	0	0	0	0	1	1	0.01	0.92
Cryptococcus flavus	0	0	0	0	1	0	0	1	1	0.01	0.92

SPECIES	AMAZON REGION	COASTAL REGION	HIGHLAND REGION	FLOREANA ISLAND	ISABELA ISLAND	SAN CRISTOBAL ISLAND	SANTA CRUZ ISLAND	OCCURRENCE IN ECOSYSTEMS	SUBSTRATES COLONIZED	INDEX OF ABUNDANCE (I_a)	SPECIALIZATION INDEX (S_i)
Cryptococcus rajastharensis	0	0	0	1	0	0	0	1	1	0.01	0.92
Hanseniaspora guilliermondii	0	0	1	0	0	0	0	1	1	0.01	0.92
Hanseniaspora opuntiae	0	0	1	0	0	0	0	1	1	0.01	0.92
Hanseniaspora valbyensis	0	0	1	0	0	0	0	1	1	0.01	0.92
Kluyveromyces marxianus	0	0	1	0	0	0	0	1	1	0.01	0.92
Kurtzmaniella zeylanoides	0	0	0	0	1	0	0	1	1	0.01	0.92
Kwoniella mangrovensis	0	0	0	0	1	0	0	1	1	0.01	0.92
Lindnera sp.	0	0	0	0	0	0	1	1	1	0.01	0.92
Lindnera saturnus	0	0	0	0	0	0	1	1	1	0.01	0.92
Metschnikowia kipukae	0	0	0	0	1	0	0	1	1	0.01	0.92
Metschnikowia koreensis	0	0	1	0	0	0	0	1	1	0.01	0.92
Pichia fabianii	0	0	1	0	0	0	0	1	1	0.01	0.92
Pichia nakasei	0	0	1	0	0	0	0	1	1	0.01	0.92
Rhodosporidium babjevae	0	0	1	0	0	0	0	1	1	0.01	0.92
Rhodotorula slooffiae	0	0	1	0	0	0	0	1	1	0.01	0.92
Saccharomycopsis fodiens	0	0	0	0	0	1	0	1	1	0.01	0.92
Saccharomycopsis vini	0	0	0	0	0	0	1	1	1	0.01	0.92
Saturnispora quitensis	0	0	1	0	0	0	0	1	1	0.01	0.92
Scheffersomyces stipitis	0	0	0	0	0	0	1	1	1	0.01	0.92
Sporidiobolus ruineniae	0	0	1	0	0	0	0	1	1	0.01	0.92
Sporopachydermia sp.	0	0	0	1	0	0	0	1	1	0.01	0.92
Trichosporon insectorum	0	0	0	0	1	0	0	1	1	0.01	0.92
Trichosporon multisporon	0	0	0	1	0	0	0	1	1	0.01	0.92

Table 6. Occurrence in different ecosystems and different substrates of yeast species in Galápagos and Mainland and calculation of I_a and S_i.

With these data it is possible to obtain a plot S_i versus I_a which provides the curve of Adaptability of yeast species in the ecosystems herein analyzed. Figure 5 shows the curve:

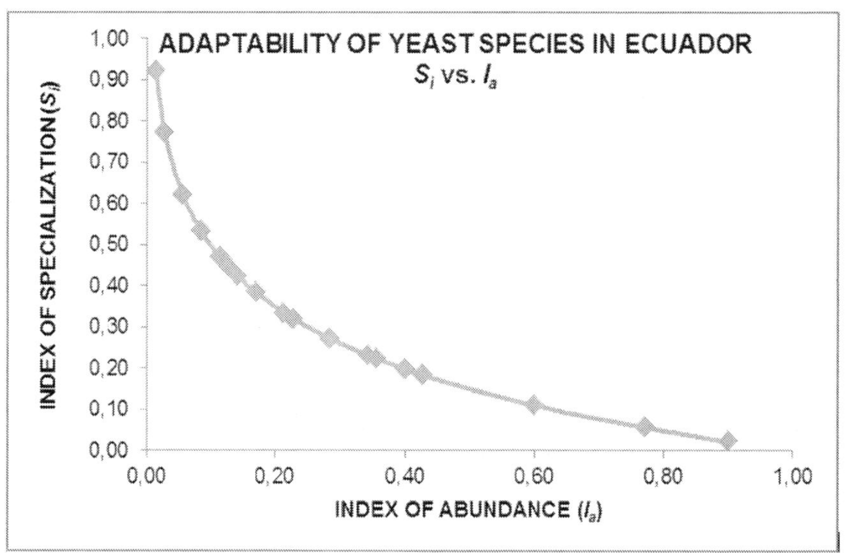

Figure 5. Curve of adaptability of the yeasts species studied in Mainland Ecuador and four Islands in the Galápagos Archipelago.

Analyzing this curve it is possible to infer that the higher the slope, the more specialized is the yeast species and vice versa. This kind of modeling is useful to have a general understanding of the behavior of yeast species in the ecosystems. As the information of new collects and characterization of yeasts is updated in the data base of the CLQCA this model will become more robust and accurate.

8. Concluding remarks

Despite the current appearance of human-disturbing habitats, especially in those islands that are inhabited, where this survey took place, the Galápagos Archipelago are the world's most pristine, best preserved and protected, tropical oceanic island ecosystem where more than the 95% of the land area in the islands is part of the Galápagos National Park [55].

The present work is a first approach to the study of the yeasts biodiversity and occurrence in the Galápagos Islands and Mainland Ecuador. The data herein presented show, nevertheless, some interesting relationships in both environments. Adaptation to the new conditions as well as the disharmony in term of biodiversity between both ecosystems appears to be aligned to the situation of some macro organisms such as plants. We have found a coincidence of 31% of the yeast species between the explored islands and the mainland; remarkably, 30% of plant species represented in the Galápagos Islands has also been registered in mainland. We could not proof endemism in the present work.

Endangered plant species may also put in danger some insects associated to those endemic plants, since the insect-plant associations generates different degrees of dependence. Studies carried out between 2001 and 2002 showed that 19 endangered plant species exhibit a range of interactions (with different degrees) with 108 different insect species that use the plants as refugee or food. The study carried out by [56] shows that 77% of the insects are endemic. From our studies, insects have shown to be the second source of yeasts biodiversity after flowers. Moreover, studies focused on insect-yeast interactions, have shown a remarkable role of insects acting as "wet nurses" for yeasts during certain periods during the year [57]. Consequently, some yeast species could be endangered along with their hosts who have a tight mutualism relationship with the endangered plant species.

To date, two novel yeast species collected in this survey were already described (i.e. *Saccharomycopsis fodiens* and *Kodamaea transpacifica*). Both yeasts species were isolated from nitidulid beetles in ephemeral flowers of *Ipomoea* sp. and *Datura* sp. These species are regarded as biomarkers for ancient migrations of Polinesyan sailors, who took the sweet potato plants (*Ipomoea* sp.) from the Andes (currently Ecuador and Peru) and introduced them into Polynesia and beyond [58].

In this chapter we have developed a new ecological approach by means of a mathematical model which is useful for a better understanding of the adaptability of yeasts as well as the specialization degree of these microorganisms in Ecuadorian ecosystems. The data herein processed will be completed in future expeditions, but constitutes a base for the upcoming ecological studies of the yeasts in the Galápagos Islands and Ecuadorian Mainland. The mathematical model shows an inverse correlation between the "Index of Specialization" (S_i) and the "Index of Abundance" (I_a). Moreover, it can be seen that the trend is towards the specialization since 70 out of 104 yeasts species analyzed (c.a. 67%) showed a S_i between 0.92 to 0.53, which means that they were isolated from a maximum of three out of seven ecosystems and a maximum of three out of nine substrates; 30 yeast species (c.a. 29%) showed an intermediate S_i, between 0.18 and 0.47, meaning that these yeasts species were found in a maximum of six different ecosystems and six different substrates; finally, only four yeast species (c.a. 4%) showed a very low S_i, between 0.02 and 0.11, which means that these species were found in up to seven ecosystems and nine substrates analyzed. These four yeast species are considered the more generalist and exhibit the highest adaptability, but represents a minority in the complete pool of yeast species studied. The total number of ecosystems analyzed was seven and the total number of substrates studied were 10. No yeast species were found in all the 10 substrates.

Moreover, we have found that in the Galápagos Islands the percentage of yeast species that are colonizing a single substrate is about a 30% higher than the correspondent figure in mainland (Figure 4). We can hypothesize that in mainland the ecosystems have had much more exchanges along the natural history, furthermore, the yeasts species were able to adapt to a wider range of substrates. In the Galápagos archipelago, a rather young group of volcanic islands, the exchange as well as the opportunity to colonize new substrates and—in evolutionary terms—to adapt to new micro environments, has been certainly lower than in the interconnected mainland ecosystems. In other words, we can see that adaptability of yeasts in mainland and the archipelago is clearly different.

Undoubtedly there is much work to do in order to attain a better understanding on ecology of yeasts in the archipelago, since the sampling was done in 4 of the 21 islands. Future expeditions may focus on substrates such as flowers, insects and rotten vegetal matter, in order to fill the shortfalls from the previous collections, such as more novel yeast species as well as endemic species from the Galápagos Islands.

Acknowledgements

The yeast research carried out by the CLQCA was funded by Pontificia Universidad Católica del Ecuador and The National Secretariat of Science and Technology, SENESCYT. This work was also funded by the Conselho Nacional de Desenvolvimento Cientifico e Tecnológico (CNPq – Brazil, awarded to CAR) and Fundação do Amparo a Pesquisa do Estado de Minas Gerais (FAPEMIG, awarded to CAR).

Author details

Enrique Javier Carvajal Barriga[1*], Patricia Portero Barahona[1], Carolina Tufiño[1], Bernardo Bastidas[1], Cristina Guamán-Burneo[1,2], Larissa Freitas[2] and Carlos Rosa[2]

*Address all correspondence to: ejcarvajal@puce.edu.ec

1 Centro Neotropical para la Investigación de la Biomasa, Pontificia Universidad Católica del Ecuador, Quito, Ecuador

2 Departamento de Microbiologia, Universidade Federal de Minas Gerais, Belo Horizonte, Brazil

References

[1] Jackson, M. H. Galapagos, a Natural History. Calgary, Alta: University of Calgary Press; 1993

[2] Bowman, R. The Galápagos. Berkeley: Univ. Calif. Press; 1966

[3] Wiggins, I., Porter, D. Flora of the Galápagos Islands. Stanford, California: Stanford University Press;1971

[4] Rasmussen, C. *Megachile timberlakei* Cockerell (Hymenoptera: Megachilidae): Yet another adventive bee species to the Galápagos Archipelago. Pan-Pac Entomol. 2012; 88 (1), 98-102.

[5] Palma, R., Peck, S. An annotated checklist of parasitic lice (Insecta: Phthiraptera) from the Galápagos Islands. Zootaxa. 2013; 3627 (1): 001–087.

[6] Otte, D., Peck, S. Crickets of the Galapagos Islands, Ecuador (Orthoptera: Gryllidae; Nemobiinae and Trigonidiinae). J. Orthoptera Res. 1998 Dec; 231-240. Foster, W., Treherne, J. Feeding, predation and aggregation behaviour in a marine insect, *Halobates robustus* Barber (Hemiptera:Gerridae). Proc. R. Soc. Lond. 1980; B 209: 539-553.

[7] Hoddle, M., Mound, L. Thysanoptera of the Galápagos Islands. Pac Sci. 2011; 65 (4), 507-513.

[8] Peck, S., Roth, L. Cockroaches of the Galápagos Islands, Ecuador, with descriptions of three new species (Insecta: Blattodea). Can J Zool.1992; 70, 2202-2217.

[9] Terraquest [Internet]. [Place unknown]: Mountain Travel-Sobek and WorldTravel Partners; 1996 [date unknown]. Available from: http://www.doc.ic.ac.uk/~kpt/terraquest/galapagos/

[10] Carlquist, S. Island Life. Natural History. NY: Natural History Press; 1965

[11] Baur, G. On the Origin of the Galapagos Islands. Am Nat. 1891; 25 (291), 217-229.

[12] Bowman, R., Berson, M., Leviton, A. Patterns of evolution in Galapagos organisms. San Francisco, California: Pacific Division, AAAS; 1983.

[13] Grehan, J. Biogeography and evolution of the Galapagos: integration of the biological and geological evidence. Biol J Linn Soc Lond. 2001;74, 267-287.

[14] Craw, R., Grehan, J., Heads, M. Panbiogeography: Tracking the history of life. New York: Oxford University Press; 1999

[15] Gillespie, R. Oceanic islands: Models of diversity. S.A. Levin (ed.) Encyclopedia of Biodiversity, Elsevier Ltd, Oxford. 2007; Pp. 1-13. Available from: http://nature.berkeley.edu/~gillespie/Publications_files/EncycloBiodiversity07517.pdf

[16] Gillespie, R., Roderick, G. Arthropods on Islands: Colonization, Speciation and Conservation. Annu Rev Entomol. 2002; 47, 595-632.

[17] Porter, D. Geography and dispersal of Galapagos Islands vascular plants. Nature. 1976 Dec; 745 – 746.

[18] LeFevre, A. Bacterial growth rates and bacterivory around the Galápagos Islands. Senior Thesis. University of Washington, School of Oceanography; 2006.

[19] Evans, H., Samson, R. Entomogenous fungi from the Galápagos Islands. Can J Bot. 1982; 60 (11), 2325-2333.

[20] Carvajal Barriga, E., Libkind, D., Briones, A., Úbeda, J., Portero, P., Roberts, I., James, S., Morais, P., Rosa C. Yeasts Biodiversity and Its Significance: Case Studies in Natural and Human-Related Environments, Ex Situ Preservation, Applications and Challenges. In: Grillo O. Venora G.(Eds.) Changing Diversity in Changing Environments (pp. 55-86). Croatia: INTECH; 2011. p55-86.

[21] Kurtzman, C., Robnett, C. Identification and phylogeny of ascomycetous yeasts from analysis of nuclear large subunit (26S) ribosomal DNA partial sequences. Antonie Van Leeuwenhoek. 1998; 73, 331-371.

[22] James, S., Collins, M., Roberts, I. Use of an rRNA internal transcribed spacer region to distiguish phylogenetically closely related species of the genera Zygosaccharomyces and Torulaspora. Int J Syst Bacteriol. 1996; 46, 189-194.

[23] O'Donnell, K. Fusarium and its near relatives. In: D. R. Reynolds and J. W. Taylor (ed.), The fungal holomorph: mitotic, meiotic and pleomorphic speciation in fungal systematics (pp. 225-233). CAB International, Wallingford, Oxon, United Kingdom; 1993

[24] Kurtzman, C., Fell, J., Boekhout, T. The Yeasts, a Taxonomic Study 2. London: Elsevier; 2011 pp. 335.

[25] Ruivo, C., Lachance, M., Rosa, C., Bacci, M., Pagnocca, F. *Candida heliconiae* sp. nov., *Candida picinguabensis* sp. nov. and *Candida saopaulonensis* sp. nov., three ascomycetousyeasts from *Heliconia velloziana* (Heliconiaceae). Int J Syst Evol Microbiol. 2006; 56, 1147-1151.

[26] Suh, S.O., Nguyen, N., Blackwell, M. Nine new Candida species near C. *membranifaciens* isolated from insects. Microbiol Res. 2005;109 (9), 1045-1056.

[27] Steinkraus, K. Handbook of indigenous fermented foods. New York: M. Dekker; 1996

[28] Chang, C., Lin, Y., Chen, S., Carvajal Barriga, E., Portero, P., James, S., Bond, J., Ian, N., Lee, C. *Candida theae* sp. nov., a new anamorphic beverage-associated member of the Lodderomyces clade. Int J Food Micorbiol. 2012; 153, 10-14.

[29] Alam, F., Mustafa, A., Khan, Z. Comparative evaluation of (1-3)-b-D-glucan, mannan and anti-mannan antibodies, and Candida species-specific snPCR in patients with candidemia. BMC Infect Dis. 2007; 7 (103), 1-9.

[30] Redhead, S., Malloch, D. The Endomycetaceae: new concepts, new taxa. Can J Bot. 1977; 55 (13), 1701-1711.

[31] Meyer, S., Smith, M., Simione, F. Systematics of Hanseniaspora Zikes and Kloeckera Janke. Antonie van Leeuwenhoek. 1978; 79-96.

[32] van der Walt, J. Three new sporogenous yeasts from soil. Antonie van Leeuwenhoek. 1957; 23 (1), 23-29.

[33] Lachance, M., Rosa C., Carvajal Barriga, E., Freitas, L., Bowles, J. *Saccharomycopsis fodiens* sp. nov., a rare predacious yeast from three distant localities. Int. jour. Syst and Evol Microb. 2012; 62, 2793-2798.

[34] Harrison, F. *Rhodotorula glutinis*. Trans R Soc Can. 1927; 3rd series 21.

[35] Jörgensen, A., Harrison, F. *Rhodotorula mucilaginosa*. Trans R Soc Can. 1929; 22 (191).

[36] Brash, D., Haseltine, W. UV-induded mutation hotspots occur at DNA damage hotspots. Nature. 1982; 298, 189-192.

[37] Sakaki, H., Nakanishi, T., Tada, A., Miki, W., Komemushi, S. Activation of torularhodin production by *Rhodotorula glutinis* using weak white light irradiation. J Biosci Bioeng. 2001; 92 (3), 294-297.

[38] Moliné, M. F. Photoprotection by carotenoid pigments in the yeast *Rhodotorula mucilaginosa*: the role of torularhodin. Photochem Photobiol Sci. 2010; 9, 1145-1151.

[39] Libkind, D., Brizzio, S., van Brook, M. *Rhodotorula mucilaginosa*, a carotenoid producing yeast strain from a Patagonian high-altitude lake. Folia Microbiol (Praha). 2004; 49 (1), 19-25.

[40] Barnett, J., Payne, R., Yarrow, D. Yeasts: Characteristics and identification (3th ed.). Cambridge: Cambridge University Press; 2007.

[41] Gomes, F., Lacerda, I., Libkind, D., Lopes, C., Carvajal Barriga, E., Rosa, C. Traditional Foods and Beverages from South America: Microbial Communities and Production Strategies. In: J. F. Krause, Industrial Fermentation. Food processes, Nutrient Sources and Production Strategies (pp. 79-114). New York, USA: Nova Science Publishers, Inc; 2010

[42] James, S. C. *Candida carvajalis* sp. nov., an ascomycetous yeast species from the Ecuadorian Amazon jungle. FEMS Yeast Res. 2009; 9 (5), 784-788.

[43] Lu, H., Jia, J., Wang, Q., Bai, F. *Candida asparagi* sp. nov., *Candida diospyri* sp. nov. and *Candida qinlingensis* sp., nov., novel anamorphic, ascomycetous yeast species. Int J Syst Evol Microbiol. 2004; 54, 1409-1414.

[44] You, K., Rosenfield, C., Knipple, D. Ethanol tolerance in the yeast *Saccharomyces cerevisiae* is dependent on cellular oleic acid content. Appl Environ Microbiol. 2003; 69 (3), 1499-1503.

[45] Brysch-Herzberg, M. *Candida bombiphila* sp. nov., a new asexual yeast species in the Wickerhamiella clade. Int J Syst Evol Microbiol. 2004; 28, 369-377.

[46] Canto, A., Herrera, C., Medrano, M., Perez, R., García, I. Pollinator foraging modifies nectar sugar composition in Helleborus foetidus L.(Ranunculaceae): an experimental test. Am J Bot. 2008; 95, 315-320.

[47] Herrera, C., De Vega, C., Canto, A., Pozo, M. Yeasts in floral nectar: a quantitative survey. Ann Bot. 2009; 103, 1415-1423.

[48] Herrera, C., García, I., Pérez, R. Invisible floral larcenies: microbial communities degrade floral nectar of buble bee-pollinated plants. Ecology. 2008; 89, 2369-2376.

[49] de Vega, C., Herrera, C., Johson, S. Yeasts in floral nectar of some South African plants: Quantification and associations with pollinaro type and sugar concentration. S Afr J Bot. 2009; 75, 798-806.

[50] Lindow, S. B. Microbiology of the phyllosphere. Appl Enfiron Microbiol. 2003; 69, 1875-1883.

[51] Krenzelok, E. Aspects of Datura poisoning and treatment. Clin Toxicol (Phila). 2010; 48, 104-110.

[52] Starmer, W., Lachance, M. Yeast Ecology. In: The Yeasts A Taxonomic Study. C. F. Kurtzman (Ed.) Amsterdam: Elsevier; 2011

[53] Lachance, M., Pupovac Velikonja, A., Natarajan, S., Schlag Elder, B. Nutrition and phylogeny of predacious yeast. Can J Microbiol. 2000; 46, 495-505.

[54] Peck, S., Heraty, J., Landry, B., Sinclair, B. Introduced insect fauna in an Oceanic Archipelago: The Galápagos Islands, Ecuador. American Entomologist. 1998; 44 (4) 218-237(20)

[55] Boada, R. Insects associated with endangered plants in the Galápagos Islands, Ecuador. Boletín de Entomología Venezolana. 2005; 20(2), pp. 77-88. http://www.bioline.org.br/request?em05014 (accessed 20 October 2013).

[56] Brysch-Herzberg M. Lachance, M. Ecology and taxonomy of yeast associated with the plant-bumblebee mutualism in Central Europe. FEMS Microbiol Ecol. 2004; 50(2), 87-100.

[57] Freitas, L., Carvajal-Barriga, E., Portero-Barahona, P., Lachance, M., Rosa, C. *Kodamaea transpacifica* f.a., sp. nov., a yeast species isolated from ephemeral flowers and insects in the Galápagos Islands and Malaysia: further evidence for ancient human transpacific contacts. Int J Syst Evol Microb. 2013; 63, 4324-4329.

Chapter 10

Ecosystem Biodiversity of India

Vivek Khandekar and Anita Srivastava

Additional information is available at the end of the chapter

http://dx.doi.org/10.5772/58431

1. Introduction

Forests are amongst the most biologically-rich terrestrial systems. Tropical, temperate and boreal forests together offer diverse sets of habitats for plants, animals and micro-organisms, and harbour the vast majority of the world's terrestrial species. In the past, timber production was regarded as the dominant function of forests. However, in recent years these perception has changed towards recognizing and acknowledging the diverse ecological services and functions offered by forests. Today, it is understood that forest biodiversity underpins a wide range of goods and services for over all human well-being. Ecologically intact forests store and purify drinking water, mitigate natural disasters such as droughts and floods, help store carbon and regulate the climate, provide food and produce rainfall and provide a vast array of goods and services for medicinal, cultural and spiritual purposes. The health of forests and the provision of forest ecosystem services depend on the diversity between species, the genetic diversity within species, and the diversity of forest types.

Vegetative biodiversity, heretofore referred to forest biodiversity of the country is under severe threat due to various factors such as increasing population, environmental degradation, indiscriminate resource utilization etc. Social, economic and spatial constraints have made the value of biodiversity irredeemable. The alarming rate of loss of biodiversity particularly in terms of ecological, genetic, economic and evolutionary consequences became a matter of universal concern when the eventful Earth Summit work place at Rio de Janiero in 1992. This later culminated in 1993 in the ratification of a global agenda on biodiversity, now referred as the UN Convention on Biodiversity. India being signatory to the Convention on Biological Diversity (CBD), is committed not only to the conservation of its biodiversity but also to sustainable and equitable utilization of its genetic resources. With long history of conservation in India, the conservation of forest biodiversity becomes an integral part of the development process. Over the years, India has developed a strong legal and policy framework along with a number of programmes promoting biodiversity

conservation in the country. In this paper the intrinsic nature of biodiversity and multiple nature of its stakeholder are explored in order to explore the inextricable link between human welfare and conservation biodiversity. The paper is a compendium of practice, a synthesis of insights into biodiversity conservation related research and technology in the country, and a source of ideas for way forward.

2. Literature review

Biological diversity refers to variety within the living world, and is commonly used to describe the number, variety and variability of living organisms. Thus, biodiversity is the variation of taxonomic life forms within a given ecosystem, biome or for the entire Earth. It is often used as a measure of the health of biological systems.

The term 'Biodiversity', a contraction of the term 'biological diversity' was first coined by Walter Rosen in the 1986 Forum on Biodiversity (Wilson 1988). The term biodiversity entails more than just the accumulation of species. The 1992 United Nations Earth Summit in Rio de Janeiro defined 'biodiversity' as "the variability among living organisms from all sources, including, 'inter alia', terrestrial, marine, and other aquatic ecosystems, and the ecological complexes of which they are part: this includes diversity within species, between species and of ecosystems" (UNEP 1992). This comes closest thing to a single legally accepted definition of biodiversity and also the definition adopted by the United Nations Convention on Biological Diversity (CBD). The concept of biodiversity involves an "understanding that all organisms interact, like a web of life, with every other element in their local environment" (SCBD 2010).

An estimated 1.7 million species have been described to date although estimates for the total number of species existing on earth at present vary from five million to nearly 100 million. However, biodiversity is not distributed evenly on Earth. It is consistently richer in the tropics and in other localized regions. Forests are more biologically diverse than any other land-based ecosystem, and contain more than two-thirds of the world's terrestrial species (ibid).

2.1. India – A megadiverse country with diverse landscape

India is situated north of the equator between 66°E to 98°E and 8°N to 36°N. The varied edaphic, climatic and topographic conditions have resulted in a wide range of ecosystems and habitats such as forests, grasslands, wetlands, coastal and marine ecosystems, and deserts. The mountainous region covers an area close to 100 mha, arid and semi-arid zones are spread over 30 mha and the coastline is about 8000 km long (MoEF 2009). India represents: (i) Two 'Realms'- the Himalayan region represented by Palearctic Realm and the rest of the sub-continent represented by Malayan Realm; (ii) Five Biomes e.g. Tropical Humid Forests; Tropical Dry Deciduous Forests (including Monsoon Forests); Warm Deserts and Semi-deserts; Coniferous Forests; Alpine Meadows; and (iii) Ten biogeographic zones and Twenty-seven biogeographic provinces(ibid).

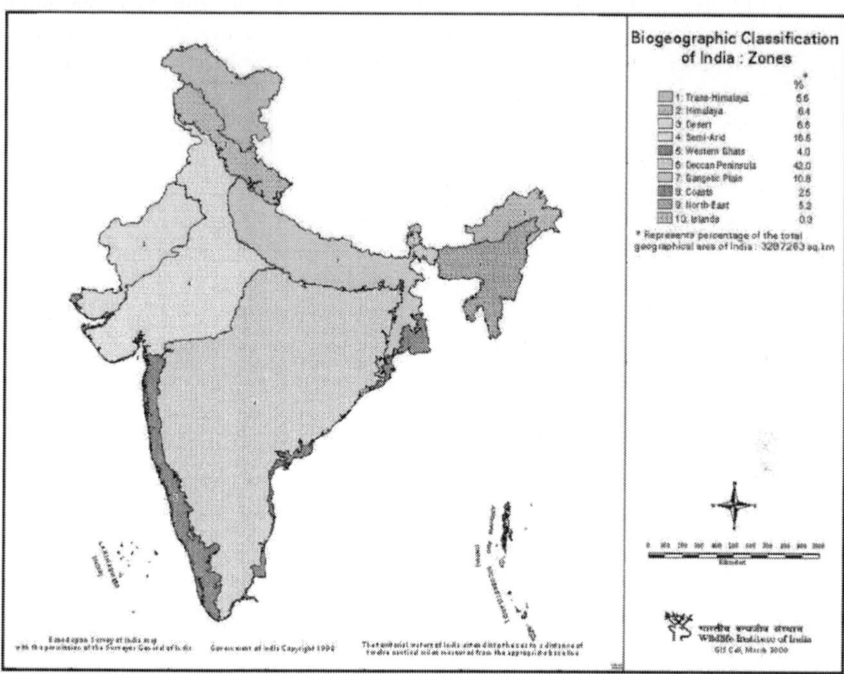

(Source: Rodgers and Panwar, 1988)

Figure 1. Biogeographic zones in India

Indian region has over 130,000 species of plants and animals which have been scientifically documented. The country has been referred to as one of the top mega diversity region of the globe with only 2.5% of the global land area. Of the 34 globally identified biodiversity hotspots, India harbours four hotspots, i.e., Himalaya, Indo-Burma, Western Ghats and Sri Lanka, Sundaland. The richness of the biodiversity of the region is largely due to the occurrence of rich diversity of species, genetic and ecological variabilities in different biogeographically and bioclimatically defined zones.

In terms of plant diversity, India ranks tenth in the world and fourth in Asia. India represents nearly 11% of the world's known floral diversity with over 45,500 plant species. The richness of Indian plant species as compared to the world is shown in Table 1.

Endemism pertains to the restricted distribution of the flora and fauna. The probable causes for same are geographical isolation, land degradation, close and distinct ecosystem like mountain and oceanic systems etc. About 11,058 species are endemic to Indian region, 6,200 of which belong to flowering plants alone. Eastern Himalaya and north-eastern region (about 2,500 species), peninsular India including western and Eastern Ghats (about 2,600 species), north-western Himalaya (about 800 species) and Andaman & Nicobar Islands (about 250 species) are the areas rich in endemic plants. Endemism in different plant groups of India is given in Table 2.

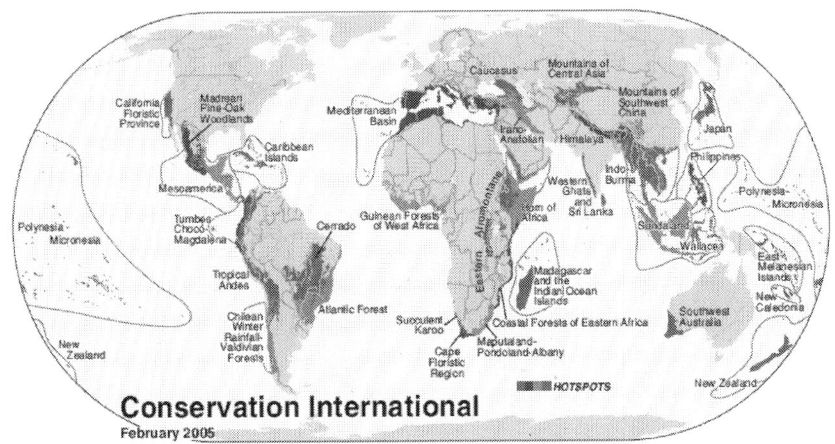

(Source: http://www.conservation.org)

Figure 2. Global biodiversity hotspots

Plant groups	No. of species		% of India to the world
	India	World	
Virus/Bacteria	850	8,050	10.6
Algae	7175	40,000	17.9
Fungi	14,500	72,000	20.1
Lichens	2223	13,500	16.4
Bryophytes	2500	14,500	17.2
Pteridophytes	1,200	10,000	12.0
Gymnosperms	67	650	10.3
Angiosperms	17,527	2,50,000	7.0

(Source: BSI, 2009)

Table 1. Number of species in major groups of plants and microorganisms in relation to same at international level showing extent of diversity

S. N.	Plant group	Total number of species in India	Nos. of Endemic Species	%
1	Angiosperms	17527	6200	35.3%
2	Gymnosperms	67	7	14.9%
3	Pteridophytes	1200	193	16.0%
4	Bryophytes	2500	629	25.1%
5	Lichens	2223	527	23.7%
6	Fungi	14500	3500	24.0%
7	Algae	7175	1925	26.8%

(Source: Botanical Survey of India, 2009)

Table 2. Endemism in different plant groups of India showing high % of endemism to emphasize need of conservation

As per the IUCN Red List (2008), India has 246 globally threatened floral species, which constitute approximately 2.9% of the world's total number of threatened floral species (8457). Distribution of various IUCN threat categories of Indian plants as compared to global trends is given in Figure 3 and 4.

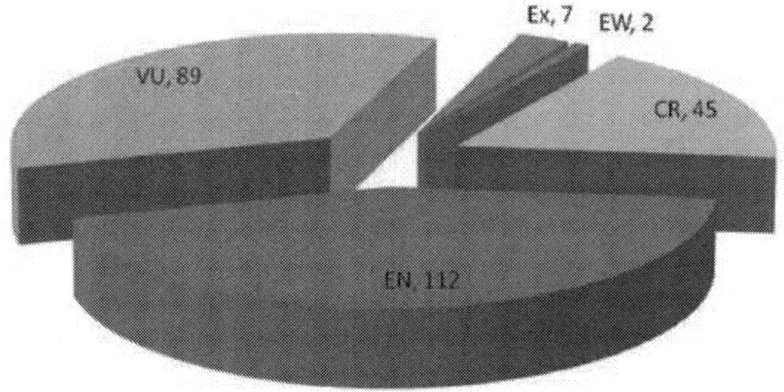

Figure 3. Indian plants – representation in IUCN threat categories

2.2. Forest biodiversity

India is endowed with vast forest resources. Forests play a vital role in social, cultural, economic and industrial development of the country and in maintaining its ecological balance. The forest resources are storehouse of biodiversity. Other land use practices are benefitted by forests. Realizing the crucial role of forests in maintaining the ecological balance and socio-economic development, the National Forest Policy, 1988 aims at maintaining a minimum of

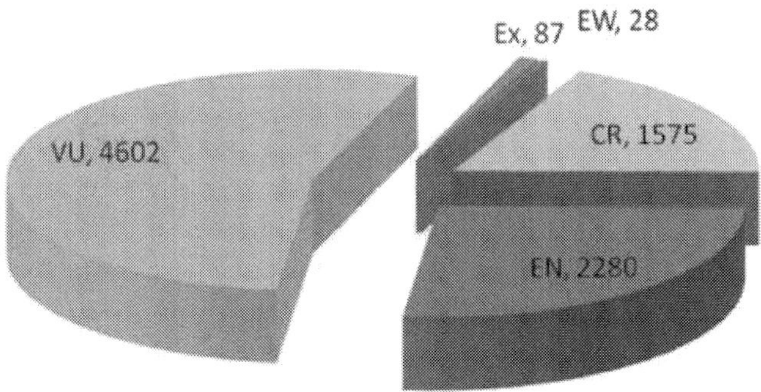

Figure 4. Plants – global representation in IUCN threat categories

33% of country's geographical area under forest and tree cover. Currently, total forest cover of India is 692,027 km² which forms 21.05% of the geographical area of the country (FSI 2011). The state of Madhya Pradesh has the largest forest cover (77,700 km²) in the country followed by Arunachal Pradesh (67,410 km²), Chhattisgarh (55,674 km²), Maharashtra (50,646 km²) and Orissa (48,903 km²) in terms of percentage of forest cover with respect to total geographical area.

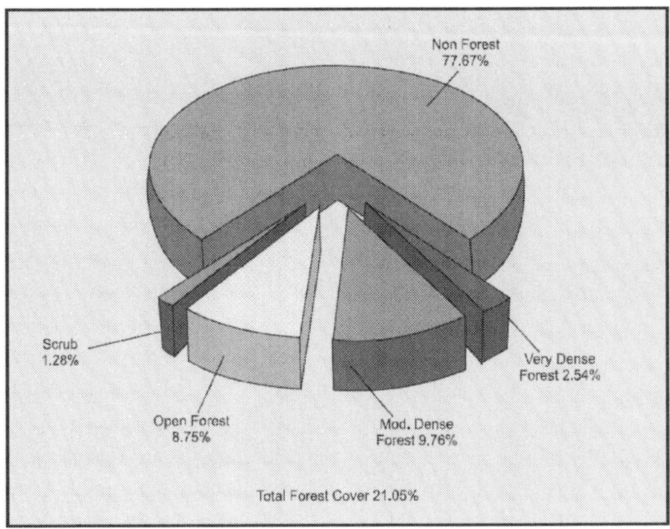

Figure 5. Forest cover of India

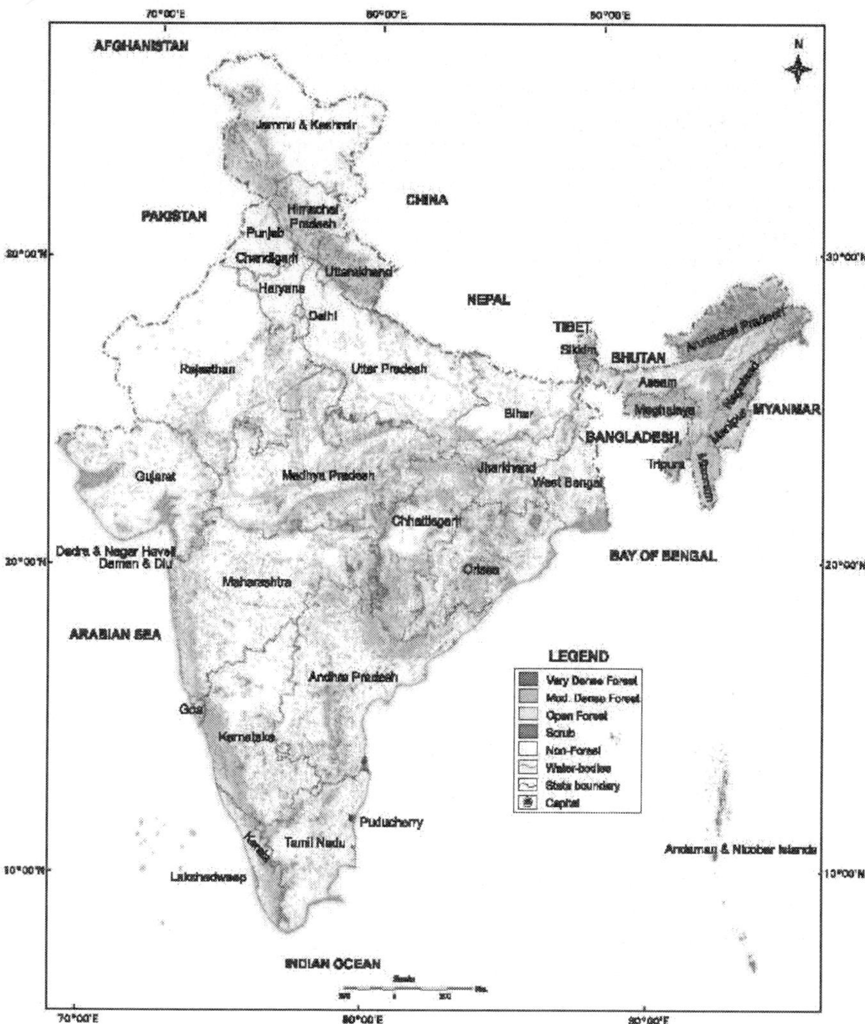

(Source: FSI 2001)

Figure 6. Forest cover map of India

The forests in the country have been classified into 16 major types and 251 subtypes by Champion and Seth (1968) on the basis of climatic and edaphic features. Forest Survey of India has mapped forest types of India, based on Champion and Seth classification on 1:50,000 scale. Distribution of diverse forest types across the country is presented in Table 3.

Sl.No	Group	% of Forest Cover
1.	Group1- Tropical Wet Evergreen Forest	2.92
2.	Group2- Tropical Semi –Evergreen Forest	13.79
3.	Group3- Tropical Moist Deciduous Forest	19.73
4.	Group 4-Littoral and Swamp Forest	0.69
5.	Group5- Tropical Dry Deciduous	41.87
6.	Group 6-Tropical Thron Forest	2.25
7.	Group 7- Tropical Dry Evergreen Forest	0.13
8.	Group 8- Subtropical Broadleaved Hill Forest	2.69
9.	Group 9 Subtropical Pine Forest	2.63
10	Group 10- Subtropical Dry Evergreen Forest	0.03
11	Group 11- Montane Wet Temperate Forest	0.69
12	Group 12- Himalayan Moist Temperate Forest	4.12
13	Group13- Himalayan Dry Temperate Forest	0.84
14	Group 14,15 ,16- Sub Alpine and Alpine Forest	2.55
15	Plantation/TOF	5.07

(Source: FSI 2011)

Table 3. Percentage of total forest cover in different forest type groups supporting harbouring diverse life forms

Forests are one of the most biologically rich terrestrial systems with diversity in their forest types. These types provide habitats for plants, animals and micro-organisms, and harbour the vast majority of the world's terrestrial species. Furthermore, forest biodiversity is interlinked to a web of other socio-economic factors, providing an array of goods and services that range from timber and non-timber forest resources to mitigating climate change and genetic resources. At the same time, forests provide livelihoods for people worldwide and play important economic, social, and cultural roles in the lives of many indigenous communities. Therefore, forests and forest biological diversity are innately linked to ecosystem and human well-being.

2.3. Salient features of India's forest phytodiversity

i.   The Himalayas

Himalayas-one of the largest and youngest mountain chains in the world-cover roughly 10% of India total land surface. Variations in terms of its size, climate and altitudinal ranges have created environments unique and characteristic to this region only. The diverse climate and the varied environmental conditions of Himalayas support diverse habitat and ecosystems with equally diverse life forms.

The Indian Himalayan flora represents 71 endemic genera and 32% endemic species. The families such as Tetracentraceae, Hamamelidaceae, Circaeasteraceae, Butomaceae, and Stachyuraceae are endemic families of flowering plants. Over 90% of the species in the family Berberidaceae and Saxifragaceae are endemic to the Himalaya, several of the species are disjunct elements of phytogeographical significance. A large number of orchids, many representing neoendemic taxa occur in Sikkim and Arunachal Pradesh.

Except perhaps in the higher elevations of Himalayas and in the more arid parts of Tahr Desert, the natural vegetation of India is essentially arboreal. It is conifers in the Himalayas, which generally resemble that from the north. The most conspicuous trees here are *Shorea robusta*, *Dipterocarpus tuberculatus*, *Dalbergia sissoo*, *Acacia catechu* and *Acacia nilotica*, ssp Indigenous palms like *Corypha*, *Phoenix sylvestris* and *Borassus flabellifer* as well as Bamboos are common here. Tree ferns are abundant in the forests of Eastern Himalays. The occurrence of *Rhododendron*, an Ericaceae member of high altitude, both in the Himalayas and in the Western Ghats is phytogeographically interesting. The tropical humid elements in the Western Himalays is represented by the Malesian-Deccanian-Pre Himalayan forms such as *Terminalia bellirica*, *Bombax ceiba*, *Toona cilitata*, *Syzygium cumini*, *Lagerstroemia* spp., *Daphniphyllum* spp. and *Shorea robusta*. The tropical semi-arid elements are represented by *Woodfordia* spp., *Dalbergia sissoo* and *Lannea coromandelica*.

The evergreen forests in the Eastern Himalays consist of tree species such as *Aesculus panduana*, *Artocarpus chamba* (*A. chaplasha*), *Michelia chmapaca*, *Cinnamomum* spp., *Schima wallichii* and *Dillenia indica*. The common Bamboo seen here is *Dendrocalamus hamiltonii*. The Savannas in this part of the Himalays are usually moist and consist of trees like *Albizia procera*, *Bischofia javanica* and *Bombax*. The vegetation in the lower region is dominated by broad leaved species of Oaks such as *Quercus lamellosa*, *Q. lineate*, *Michelia daltsopa* (*M. excelsa*), *Pyrus* spp., *Symplocos* spp., *Eurya* spp., *Meliosma* spp., and *Castanopsis* spp.

ii.   Desert region

In Indo-Pacific region Thar desert ecoregion is considered the world's seventh largest desert. The eco-region lies to the west of Aravalli Range and characterized by extreme climate with temperature touching subzero in winters and nearly 50°C during the summer. Rainfall in such region ranges between 100-150 mm.

Flora of Indian desert consists of approximately 700 species (352 genera and 87 families including solitary gymnosperm family). 37 genera and 63 species are exotic to the region. A large number of plants species found in the Thar Desert are known to be endemic (Khan and Frost 2001). High endemism and the occurrence of a large number of sub-species provide evidence of high phylogenetic plasticity and intense speculation on account of extreme dynamics of the ecological conditions, and the recent geological, physiographic, topographical those of the Sahara (Quezel 1965). The plant taxa endemic to the Thar Desert include 23 species and 14 sub-species, forming 6.4 percent of the recorded total. High endemism and the occurrence of a large number of subspecies provide evidence of high phylogenetic plasticity. A complex of extreme environmental factors, the dynamics of the ecological conditions, recent geological, physiographic and topographical changes, can induce speciation. The proportion of endemic plant species in the Indian desert is higher (6.4 percent) than that of the Sahara which has only 3 to 5 percent of endemic species (Gupta 1998). The Indian desert endemic flora includes: *Calligonum polygonoides* (Polygonaceae), *Prosopis cineraria* (Mimosaceae), *Acacia nilotica* (Mimosaceae), *Lasiurus sindicus* (Poaceae), *Cenchrus biflorus* (Poaceae), *Tecomella undulata* (Bignoniaceae), *Citrullus colocynthis* (Cucurbitaceae), *Anogeissuspendula* (Combretaceae), *Tamarix aphylla* (Tamaricaceae), *Salvadora oleoides* (Salvadoraceae),*Commiphora weightii* (Burseraceae), *Haloxylon salicornicum* (Chenopodiaceae), *Capparis decidua* (Capparaceae), *Sueda fructicosa* (Chenopodiaceae), *Aerva pseudotomentosa* (Amaranthaceae) *Crotolaria burhia* (Fabaceae), *Leptadenia pyrotechnica* (Ascelpiadaceae).

The once glorious forests in this region stand, however degraded because of the biotic interference and comprise mostly of bushy, thorny plants. The common trees found are *Acacia nilotica*, ssp. *indica*, *Prosopis cineraria*, *Acacia senegal*, *Anogeissus latifolia*, *Syzygium cumini*, *Dalbergia sissoo*, *Albizia lebbeck*, *Boswellia serrata*, *Balanites aegyptiaca* (*B. roxburghii*), *Sterculia urens*, *Commiphora mukul* and *Acacia leucophloea*. *Prosopis chilensis* (*P. juliflora*) is extensively planted all over the plains.

Pandey *et al.* (1983) reported 41 and Sharma (1983) reported 106 rare, endangered and threatened (RET) taxa from Rajasthan. Pandey and Shetty (1985) listed RET taxa, whereas Singh (1985) dealt with threatened taxa and their scope of conservation. The present flora of Rajasthan has 2090 species belonging to 819 genera under 159 families of vascular plants (Shetty and Singh, 1987-1993). This included 63 RET taxa, reasons for their depletion and strategies for conservation as well as 45 species of crop and other cultivated plants having 66 species of their wild relatives. However, various researchers are working on under explored areas in search of new species and re-defining the status of RET species. According to Khan et al. (2003), Thar Desert has its own importance and specific characteristics with respect to endemic and medicinal plants. In this study forty-five plant species were considered to be rare and/or endangered and a large number of plants have been categorized as of economic importance and medicinal use. The high population of the region exerts pressure on the biological resources of the Thar Desert causing lack of sustainability and necessitates conservation of biodiversity actions.

iii.   Gangetic Plain

The forests of Gangetic Plain have largely been destroyed and in some tracts Sal forests are left. The Sal forests have species like *Terminalia elliptica* (*T. alata*), *Terminalia bellirica*, *Bauhinia* spp., *Syzygium cumini*, *Phyllanthus emblica* and *Woodfordia* sp. The mangrove forests of Sunderbans have species like *Aechmanthera gossypina* (*Aegiceras majus*), *Cynometra iripa* (*C. mimosoides*) and *Ceriops tagal* (*C. candolleana*).

iv.   North-east India

In the north-east region, the tree flora is luxuriant, consisting of species like *Artocarpus chamba*, *Michelia champaca*, *Ficus elastica*, *Mesua ferrea*, *Alstonia scholaris*, *Pterygota alata*, *Morus macroura* (*M. laevigata*) and *Stereospermum chelonoides*. The common Bamboo available here is *Dendrocalamus hamiltonii*. The hill forests are dominated by *Alnus nepalensis*, *Rhododendron arboreum*, *Michelia champaca* and *Prunus* sp. The Pine forests are composed of *Pinus kesiya*.

Samati and Gogoi (2007) carried out a study for the documentation of the ethnobotanical wealth of Jaintia hills. Population expulsion and unemployment also compel people to exploit these SGs, leading to a rapid dwindling of many rare and threatened taxa of both plants and animals from the region. In this context he mentions that an extensive awareness programme is needed to educate the locals about SGs. The State Forest Department and MoEF can join hands with the local NGOs to create a network of all the SGs and bring them under state-sponsored conservation programmes. The Tourism Department also should come forward to focus on SGs as a destination for tourists. The local community should be provided with adequate funds and the responsibility to manage the SGs. This will help in the protection of the SGs. Eco-restoration and afforestation programmes of the government conservation agencies should also include these degraded SGs.

Deb et al (2008) stated that the areca-nut based traditional agroforests and the natural tropical rainforests have multi-layered vegetal structures with comparable tree density, but showed significant differences in soil nutrients and microbial biomass that recorded lower values in traditional agroforests as compared to the tropical forests. Nonetheless, the percentage contribution of soil microbial biomass, litter and fine roots to soil C was similar and competitive in traditional agroforests, but substantially lower than that in the tropical forests. Litter had a major role to play in soil nutrient turnover in tropical forests followed by soil microbes and fine roots. The traditional agroforesty systems studied are, however not well-managed, but when subjected to scientific management might prove to be a sustainable food production land use system in the hills and flood plains and consequently can potentially promote conservation and sustainability of the tropical forests.

Deb and Sundriyal (2007) stated that small and medium size gaps had limited impacts on the species composition. Such gaps however are crucial for regeneration of top canopy as well as pioneer species and hence the important for maintaining species diversity in Namdapha National Park. As the gaps showed difference in species composition, it clearly illustrated that the plant species behavior in low-land tropical forest is independent of gap size and is mainly governed by availability of seeds at the time of creation of gaps. The information on the tree species gap performance has implications for the management of the forest stand.

Mishra *et al* 2005 conducted a study in a subtropical humid forest (sacred grove) at Mawnai, West Khasi hills district of Meghalaya. *Citrus medica* (Rutaceae), the dominant species, was the only species exhibiting random distribution. All other species showed contagious distribution. *Cryptocarya amygdalina*, family Lauraceae, was the co-dominant species. Lauraceae (17 species) was the species rich family in the grove and exhibited maximum tree density and basal area. However, generic composition was highest (9 genera) in the case of Euphorbiaceae, which is the co-dominant family in the grove. Majority of the families were represented by single genus and single species. Log-normal dominance-distribution curves at the levels of species and family, and wide girth structure signify the complexity and stability of the community.

v.   The Western Ghats

The Western Ghats are the main peninsular hill range extending over 1400 Km, starting in the north from near the Tapti River and ending in the south near Kanyakumari. Undoubtedly it is the most important topographic feature of peninsular India. It is more or less homogenous but biologically distinctive geographical zone with total area of about 1,60,000 sq. Km of which about one third is forests. It is known that 90% of the country's medicinal flora occurs in these forests.

The Western Ghats region is considered as one of the most important biogeographic zones of India as well as one of the 34 'Hot Spots' of biodiversity recognized in the world. About 1500 endemic species of dicotyledonous plants have been reported from the Western Ghats. It is one of the richest centers of endemism. Due to varied topography and micro-climatic regimes, some areas within the region are considered to be active zones of speciation. The region has 490 arborescent taxa, of which as many as 308 are endemics. 56 genera and 1500 species of flowering plants and 63% of India's evergreen forest plants are endemic to the Western Ghats.

The Western Ghats forests are characterized by conditions of high humidity and temperature, favouring vigorous growth of trees that often attains a height of 60 m or more. The tropical wet evergreen and semi-evergreen forests are typically rain forests. Luxuriant vegetation in more or less virgin condition is a characteristic feature of the Malabar region of the Western Ghats. The region has tropical evergreen rain forests, mixed deciduous or monsoon forests and subtropical or temperate forests. Majority of the tree species in this region belong to the families like Dipterocarpaceae, Guttiferae, Myristicaceae, Tiliaceae, Euphorbiaceae, Annona-ceae, Anacardiaceae, Fabacae, Caesalpiniaceae, Mimosaceae, Meliaceae, Myrtaceae, Rutaceae, Rubiaceae, Sapotaceae, Urticaceae and Palmaceae. The tropical evergreen forests consists of tree species such as *Toona ciliata, Dysoxylum malabaricum, Vateria indica, Dipterocarpus indicus, Hopea parviflora, Hopea ponga, Cullenia exarillata, Artocarpus hirsuta, Vitex altissima, Hydnocarpus laurifolia, Humboldtia* spp., *Haldinia cordifolia* and *Garcicina* spp.

The mixed deciduous or monsoon forests consist of dominant tree species like *Terminalia elliptica* (*T. alata*), *T. paniculata, T. bellirica, T. chebula, Lagerstroemia* spp., *Dalbergia latifolia, Xylia xylocarpa, Pterospermum* spp., *Sterculia urens, S. guttata* and *Stereospermum* spp. The bamboos occurring commonly in this forests are *Bambusa bambos* and *Dendrocalamus strictus*.

The subtropical or temperate evergreen forests are commonly known as "Sholas" and they usually exist above an altitude of 1800 m. They are composed mainly of *Gordonia obtusa, Michelia nilagirica, Ternstroemia japonica, Syzgium mudagam, Eugenia* spp., *Meliosma simplicifolia, Symplocos cochinchinensis, Litsea coriacea, Litsea floribunda* and *Actinodaphne* spp. The flora of the Nilgiri hills are interesting, as it shows affinities with Assam flora and with that of the southern slopes of Himalayas. Shola forests in the mountain slopes with trees of *Rhododendron arbor-eum*, ssp. *nilagiricum, Turpinia nepalensis, Elaeocarpus serrtus, E. recurvatus* and *Viburnum* spp. are unique in composition.

Fresh water swamps with characteristic vegetation occur in the Malabar region of the Western Ghats, dominated by different species of *Myristica*. Therefore, they are also called as "Myristica Swamps". The Myristica species in the swampy areas produce "Knee roots" which are very unique. The Myristica swamps consist of species such as *Myristica fatua* var. *magnifica* (*M. magnifica*), *M. malabarica, M. dactyloides, M. beddomei, M. contorta* and *Knema attenuata*

The vegetation of the Western Ghats of the Malabar region is very rich and phytogeographically interesting. The species occurring here show affinities with that of Sri Lanka and Malaysia. The region is also rich in endemic species. The conifer, *Nageia wallichiana (Podocarpus wallichianus)*, confined to the hills of Tirunelveli and Southern Kerala is known only from Burma and South East Asia.

vi.   Deccan Plataeu

The deciduous forests of Deccan have species like *Sterculia urens, Boswellia serrata* and *Cochlospermum religiosum*. The dry slopes have *Anogeissus latifolia, Ougenia oojeinensis, Lannea coromadelica, Cleistanthus collinus, Zizyphus xylopyrus, Buchanania* spp., *Terminalia* spp., *Bauhinia* spp., *Shorea* spp., Dalbergia spp., *Maduca longifolia*, var. *latifolia (Bassia latifolia), Diospyros* spp., *Pterocarpus marsupium, Pterocarpus santalinus, Eugenia* spp. and *Wendlandia thyrsoidea*. The areas of having black cotton soil are covered by *Capparis divaricata, Acacia nilotica*, ssp. *indica, Prosopis cineraria, Parkinsonia aculeata* and *Zizyphus mauritiana*.

Except the more arid margins, the whole of Peninsular India was formerly densely forested, but at present only *Acacia* shrubs occur here. The thorn forests in the west, the closed monsoon forests of *Shorea* in Chota-Nagpur and the open deciduous forests in between are only relics.

vii.   Andaman Islands

The vegetation of Andaman Islands consist of Mangrove forests, Littoral forests, Evergreen forests, Deciduous forests and the hill forests on shallow soiled slopes of hills. The major tree species of the island are *Pterocarpus dalbergioides* (Andaman Padauk), *Thespesia populnea, Pongamia pinnata, Barringtonia* sp., *Erythrina indica, Calophyllum inophyllum, Gyrocarpus americanus, Terminalia catappa, Rhizophora mucronata, Brugueira gymnorrhiza* and *Ceiops tagal* (Parkinson, 1923). The Andamans and the Nicobar Islands possess the best quality of mangrove forests. The total area estimated under mangrove vegetation in India is 4827 km^2 and out of this, 966 km^2 area of mangroves occurs in Andaman and Nicobar Islands (i.e. 20 per cent of the total mangrove area of the Indian territory) (Ramakrishana *et al.* 2010). There are 45 species of mangroves, coming under 27 genera, represented in the island zone.

viii.   Coastal zone and Lakshadweep

The distribution of different mangrove genera shows that the greatest number of genera and species occur along the shores of the Indian and Western Pacific oceans. The Indian mangorves comprise approximately 59 species of 41 genera, belonging to 29 families. Of these, 32 species belonging to 24 genera and 20 families are present along the west coast. There are about 21 mangrove species reported from Gujarat coast, 28 from Maharashtra, 20 from Goa, 21 from Karnataka, 14 from Kerala and 4 from Lakshadweep. The Arabian Sea coast is characterised by the typical funnel shaped estuaries of the major rivers like Narmada and Tapti and numerous small rivers. The entire west coast is thus dominated by estuarine backwater type of mangroves, unlike the deltaic mangroves of the east coast. Mangrove forests of the west coast are evergreen or deciduous, characterized by the presence of *Avicennia marina, Rhizophora mucronata, Kandelia candel, Brugueira gymnorrhiza* and *Carallia brachiata*. *Sonneratia caseolaris* reported by Blatter in 1905 is fast disappearing from the west coast, while *S. apetala* is found only along the Maharashra coast. Species like *Ceriops decandra, Xylocarpus* spp., *Lumnitzera littoralis, Nypa fruticans, Phoenix paludosa* and *Cerbera manghas* are limited along the east coast. The species, which commonly occur and uniformly distributed along the east and west coast of India are *Rhizophora mucronata, R. apiculata, Ceriops tagal, Brugueira gymnorrhiza, Lumnitzera racemosa, Sonneratia apetala, Acanthus ilicifolius, Avicennia marina, A. officinalis, Excoecaria agallocha* and *Acrostichum aureum*. The east coast of India has deltaic type of mangroves and it covers about 70 percent of the total mangrove forested area in India (Deshmukh, 1991).

2.4. Value of forest biodiversity

Life has value and meaning beyond monetary measure, and so does biodiversity. While it is not really possible to put a monetary figure on the value of forests, it is now widely recognized that we need to improve the way our societies and economies account for ecosystem services. We often take these services for granted, such as the ability of the forest to filter water or produce oxygen. Human well-being depends on the goods and services provided by nature — Earth's "natural capital". Recent initiatives, such as the global study on 'The Economics of Ecosystems and Biodiversity' (TEEB) have resulted in a better understanding of the economic value of forests and other ecosystems for societies.

Sl.No	Ecosystem services	Value of Ecosystem Services (US$/ha/year-2007 values)	
		Average	Maximum
Provisioning services			
1.	Food	75	552
2.	Water	143	411
3.	Genetic resources	483	1756
4.	Medicinal resources	181	562
Regulating services			
1.	Influence on air quality	230	449
2.	Climate regulation	1965	3281
3.	Water flow regulation	1360	5235
4.	Waste treatment/ water purification	177	506
5.	Erosion prevention	694	1084
Cultural services			
1.	Recreation and tourism opportunities	381	1171
Total		**6120**	**16362**

(Source: TEEB Climate Issues Update 2009)

Table 4. Values of ecosystem services in tropical forests for influencing economic decisions about the future of forests for Societies

The TEEB study estimates that, on average, one hectare of tropical forest provides US$ 6,120 per year in ecosystem services, such as watershed protection, climate regulation; soil stabilization, coastal protection, nutrient cycling, and carbon storage (Table 4). This also includes the numerous products from tropical forests, such as timber, wild food and non-timber forest products—rubber, oils and fibres that are economically important both locally and nationally in many tropical forest countries. Yet, only a fraction of this value is currently accounted for when we make economic decisions about the future of forests. The Green Economy Initiative of the United Nations Environment Programme (UNEP) and other efforts are now underway to improve the way we value and account for nature in our economic decision-making.

2.5. Support for people's livelihoods

Forests, with their rich biodiversity are essential for human livelihoods and for sustainable development. For example, fuel wood is the primary source of energy for heating and cooking for an estimated 2.6 billion people. The World Bank estimates that forests directly contribute to the livelihoods of some 90 per cent of the 1.2 billion people living in extreme poverty. The Millennium Ecosystem Assessment also found that as many as 300 million people, many of them very poor, depend substantially on forest ecosystem services for their subsistence and survival. Many non timber forest products derived from forest biodiversity, such as wild cocoa, honey, gums, nuts, fruits, flowers, seeds, rattan, fungi and wild meat and berries are essential for the food, medicine and confectionary building material used by indigenous and local communities to sustain their way of life, including their cultural and religious traditions.

India has a huge population living close to the forest with their livelihoods critically linked to the forest ecosystem. There are around 1.73 lakh villages located in and around forests (MoEF, 2006). Though there is no official census figures for the forest dependent population in the country, different estimates put the figures from 275 million (World Bank, 2006) to 350-400 million (MoEF, 2009). People living in these forest fringe villages depend upon forest for a variety of goods and services which includes collection of edible fruits, flowers, tubers, roots and leaves for food and medicines; firewood for cooking (some also sale in the market); materials for agricultural implements, house construction and fencing; fodder (grass and leave) for livestock and grazing of livestock in forest; and collection of a range of marketable non-timber forest products. Therefore, with such a huge population and extensive dependence pattern, any over exploitation and unsustainable harvest practice can potentially degrade forest. Moreover, a significant percentage of the country's underprivileged population happened to be living in its forested regions (Saha and Guru, 2003). It has been estimated that more than 40 per cent of the poor of the country are living in these forest fringe villages (MoEF, 2006). Apart from this, a significant percentage of India's tribal population lives in these regions. Several field based studies have documented the adverse impact of such dependence pattern on the forest quality. The forest fringe communities not just collect these forest products for their own consumption but also for commercial sale, which fetch them some income. The income from sale of the forest products for households living in and around forest constitutes 40 to 60 per cent of their total income (Bharath Kumar *et al*, 2011; Sadashivappa *et al*, 2006; Mahapatra and Kant, 2005; Sills *et al*, 2003; Bahuguna, 2000). A study (Saha and Sundriyal,

2012) on the extent of NTFP use in north east India suggest that the tribal communities use 343 NTFPs for diverse purposes like medicinal (163 species), edible fruits (75 species) and vegetables (65 species). The dependence for firewood and house construction material is 100% and NTFPs contributed 19–32% of total household income for the communities under study (Saha and Sundriyal, 2012). Forests are not only a source of subsistence income for millions of poor households but also provide employment to poor in these hinterlands. This makes forests an important contributor to the rural economy in the forested landscapes in the country. The widespread poverty and lack of other income generating opportunities often make these people resort to over-exploitation of forest resources. The collection of firewood for sale in the market, though it is illegal, is also extensive in many parts of the forested regions in the country and constitutes the source of livelihood for 11 per cent of the population (IPCC, 2007). However, many other forest products have been sustainably harvested by local communities for many years, and are a constant source of household income.

2.6. Threats to the biodiversity

Threats to species are principally due to decline in the extent of their habitat, fragmentation of habitat, decline in habitats quality, shrinking genetic diversity; invasive alien species; declining forest resource base; climate change and desertification; over exploitation of resources; impact of development projects and impact of pollution. For terrestrial species decline habitat quality and quantity arise from conversion of forest and grasslands to agriculture, of natural forests into monoculture plantations and from grazing and firewood collection. In some areas, invasion of exotic species also results in habitat degradation. For aquatic and semi aquatic species, decline in habitat quality are due to diversion of ground and surface water resulting in drying of streams and other water bodies apart from siltation and pollution from pesticides. In this century, the Indian cheetah, Lesser Indian rhino, Pink-headed duck, and the Himalayan mountain quail are reported to have become extinct and several other species (39 mammals, 72 birds and 1,336 plants) are identified vulnerable or endangered.

The constraints and challenges to biodiversity conservation which flow inter alia from these threats include: biodiversity information base; implementation of Biological Diversity Act and safeguarding traditional knowledge; new and emerging biotechnologies; economic valuation and natural resource accounting; policy, legal and administrative measures; and institutional support. Taking clue from the preceding lines, the Cold Desert areas lie in the Trans-Himalayan zone and some species in the region those are endemic to Tibetan plateau and also include oasitic elements that comprises a variety of exotic as well as indigenous species. The area represents common herbaceous, shrubby and woody elements of temperate vegetation and alpine species also dot the region. The region is also a house of species growing in glacial moraines and also harbours threatened medicinal plants. Talking of faunal diversity this area harbours rare and endangered fauna pointed out earlier and avifauna endemic to the region or migrating adds to its uniqueness. Livestock rearing, agricultural & horticultural practices and mode of agro-forestry are entirely different and the people living here have succeeded in developing their own distinguish culture.

The above explanation reveals that the area is unique to the region and hence requires special attention especially with respect of ecology of India. Hence laxity on our part in conserving the cultural and natural resources will put the area under severe thereat thereby exposing the region to all sorts of ecological disasters.

2.7. Habitat fragmentation, degradation and loss, and shrinking genetic diversity

Habitat destruction is identified as the main threat to biodiversity. Under diverse natural conditions, over a billion people in rural and urban areas live in harmony under a democratic system in India. Their pressing needs for food, fiber, shelter, fuel and fodder combined with compelling need for economic development exert enormous pressure on natural resources. With half the total land under agriculture, and approximately 23 or 20% per cent under forests, the protection of diverse habitats poses a formidable challenge. The loss and fragmentation of natural habitats affects all animal and plant species. We need to not only stop any further habitat loss immediately but also to restore a substantial fraction of the wilderness that has been depleted in the past. Various species of plants and animals are on the decline due to habitat fragmentation and over-exploitation, e.g. habitats of Great Indian Bustard in Madhya Pradesh, Gujarat and Rajasthan and of the Lion-tailed Macaque in Western Ghats. The major impact of developmental activities involves diversion of forest land. Since the enactment of Forest (Conservation) Act in 1980, about 14,997 development projects involving diversion of 11.40 lakh hectare forest area for non forestry uses, have been granted clearance. Against this diversion, Compensatory Afforestation has been stipulated over 12.10 lakh hectare.

Habitat fragmentation is also one of the primary reasons leading to cases of man animal conflict. Common property resources like pastures and village forests, which served as a buffer between wildlife habitat and agriculture, have been gradually encroached upon and converted into agricultural fields and habitation. Due to this, the villagers are brought into a direct conflict with wild animals.

Sacred groves (India has over 19,000 sacred groves) are also getting eroded or getting converted to plantations. Because there are several medicinal plants and wild relatives of crop plants occurring naturally in these areas, the sacred groves need to be conserved. Traditional norms and practices for conservation of neighborhood forest and common land are also diminishing, although certain rural and tribal communities continue to safeguard their biological resource base even at the cost of their livelihood and sustenance. Himalayan Forest Research Institute (HFRI), Shimla is in the process of documenting the sacred groves of Kullu Valley and it is proposed that such efforts are required to be replicated elsewhere in this part of the country. It is pertinent to add here that strengthening the database of sacred groves will usher in development of strong bonds between ecological and social ethos relevant to the society at last ultimately reflecting upon the conservation of biodiversity. The point gains significance here, because the areas under sacred groves, otherwise the property of local deity, are being encroached upon by the local population thereby creating loss of biodiversity (Horticulture is now being tried in these areas) on one hand and threatening the social ethos of the area on the other.

2.7.1. Invasive alien species

Among the major threats faced by native plant and animal species (and their habitats), the one posed by the invasive alien species is truly scaring since it is considered second only to that of the habitat loss. The major plant Forest Invasive Species (FIS) include *Lantana camara, Eupatorium glandulosum, Parthenium species, Mimosa species, Eichhornia crassipes, Mikania micrantha, Ulex europaeus, Prosopis juliflora, Cytisus scoparius, Euphorbia royleana* etc. Alien aquatic weeds like water hyacinth and water lettuce are increasingly choking waterways and degrading freshwater ecosystems. Lantana and carrot grass cause major economic losses in many parts of India. Highly invasive climbers like *Chromolaena* and *Mikania* species have over-run the native vegetation in North-East Himalayan region and Western Ghats. Numerous pests and pathogens such as coffee berry borer, turnip stripe virus, banana bunchy top virus, potato wart and golden nematode have invaded agro-ecosystems becoming serious menace. HFRI, Shimla has identified some of the plant and insect species which though invasive have naturalized itself in the region thereby, posing a serious threat to the ecology. No particular attention to these invasive species have been paid over a period of time which had then resulted into the present alarming situation and all the countries have now converged over a single platform for fighting the menace caused by the same. Accordingly special efforts towards assessment of these invasive species are required to be made and if we fail in this direction it will lead to loss of endemic biodiversity on one hand and will expose the area to the exotic species on the other.

2.7.2. Impact of development projects

India, with its large population, is poised for rapid economic growth. Large infrastructural and industrial projects, including highways, rural road network and the special economic zones (SEZs), are coming up. With cities and townships expanding often at the cost of agriculture, and agriculture expanding at the cost of tree cover, fresh threats to biodiversity are emerging. In addition, changing lifestyles of the people with rising incomes, in both rural and urban areas, are placing increasing demands on biodiversity.

No doubt that infrastructural development is essential for the welfare of human beings inhabiting the planet since; it brings more comfort to the society. However, it may specifically be mentioned over here that over utilization of the resources for substantial increase in the comforts is directly impinging upon the environmental health. Developmental activities no doubt, are essentially required for the larger interest of the human kind yet, over utilization of natural resources in the process certainly require a relook into some of the criterion otherwise required to be fitted in the process for sustainable development. The government machinery should therefore, be not averse to the development but should ask the implementers and managers to devise suitable strategies for paying required attention towards the development of issue based parameters for ultimate protection of the environment.

Many river valley projects are being implemented for the last 100 years or so all across the world. These projects besides providing safeguards against floods also provide electricity for increasing overall productivity of the region or of the country. Though efforts in the direction are still continuing, yet it is estimated that approximately 99 per cent of precipitation in the form of rain, snow etc. directly merges with the oceans without being utilized properly on their way. This certainly requires sincere thoughts/ efforts to harness it for the ultimate benefit of the mankind.

In order to harmonize developmental activities with protection of environment, environmental impact assessment (EIA) was made mandatory by the EIA Notification issued in 1994 for notified categories of developmental projects in the sectors of industry, thermal and nuclear power, mining, river valley and infrastructure projects. This Notification has been revised and notified on 14 September 2006 to make the EIA process more efficient, decentralized and transparent. What is required now is the effective implementation of these legislations by making it site and species specific. So that such plans become more relevant to the Environmentalists, implementers and managers. The issue gains significance in the state of Himachal Pradesh. Since large reservoirs are repeatedly coming over on the same river within a short distance and hence, are damaging the fragile ecology of Himalayas and directly impinging upon the loss of biodiversity. Large tunnel projects are also affecting the aquatic fauna-due to diversion of water through these tunnels especially during the winters when the water flow in the rivers gets reduced considerably.

2.7.3. Pollution

Biodiversity in India is facing threat from various sources of pollution, both point and non-point, sources. The major threats are from improper disposal of municipal solid waste, inadequate sewerage, excessive use of chemical pesticides and continuous use of hazardous chemicals even where non-hazardous alternatives are available. New industrial processes are generating a variety of toxic wastes, which cannot be dealt with by currently available technology in the country. Besides, economic constraints and problems related to the indigenization makes the substitution of these technologies difficult.

3. Major findings and management status

India has a long history of conservation and sustainable use of natural resources and over a period of time has developed a stable organizational structure for environment protection. Conservation and sustainable use of biodiversity has been integrated into national decision making through:

1. Policy statements (e.g. National Forest Policy, National Conservation Strategy, National Wildlife Action Plan, Draft National Environment Policy.)

2. Legislative measures (e.g. Environment (Protection) Act, Wildlife (Protection) Act, Biological Diversity Act, Environment Impact Assessment Notification, Coastal Regulation Zone Notification, Notifications on ecologically fragile areas)

India's strategies for conservation and sustainable utilization of biodiversity in the past have comprised providing special status and protection to biodiversity – rich areas by declaring them as National Parks, Wildlife Sanctuaries, Biosphere Reserves, ecologically fragile and sensitive areas, off loading pressure from reserve forests adopting by alternative measures for fuel wood and fodder by afforestation of degraded areas and wastelands, creation of *ex-situ* conservation facilities such as gene banks etc.

Although there were several Acts existed in India for ensuring conservation of biodiversity, the International Convention on Biological Diversity that came in to effect in 1993 and the Biodiversity Act, 2002 enacted by the Indian Parliament subsequently gave impetus to the conservation efforts in the country.

3.1. The Convention on Biological Diversity

The Convention on Biological Diversity (CBD) is a landmark in the environment and development field, as it envisages for the first time a comprehensive rather than a sectoral approach to the conservation of Earth's biodiversity and sustainable use of biological resources. It was in the year 1984 that the need to have in place a global convention on biological diversity started gaining momentum. In response to it, the United Nations Environment Programme (UNEP) in the year 1987 recognized the need to streamline international efforts to protect biodiversity. The Convention on Biological Diversity (CBD) was negotiated and signed by nations at the UNCED Earth Summit at Rio de Janeiro in Brazil in June 1992. The Convention came into force on December 29, 1993. India became a Party to the Convention in 1994. At present, there are 175 Parties to this Convention (NBA, 2004).

The main objectives of the Convention are:

- Conservation of biological diversity

- Sustainable use of the components of biodiversity and

- Fair and equitable sharing of benefits arising out of the utilisation of genetic resources.

3.2. Biological Diversity Act, 2002

The Central Government has brought Biological Diversity Act, 2002 with the following salient features:-

1. To regulate access to biological resources of the country with the purpose of securing equitable share in benefits arising out of the use of biological resources and associated knowledge relating to biological resources.

2. Conservation and sustainable use of biological diversity.

3. To respect and protect knowledge of local communities related to biodiversity;

4. To secure sharing of benefits with local people as conservers of biological resources and holders of knowledge and information relating to the use of biological resources.

5. Conservation and development of areas of importance from the standpoint of biological diversity by declaring them as biological diversity heritage sites.

6. Protection and rehabilitation of threatened species.

7.   Involvement of institutions of State governments in the broad scheme of the implemen-
     tation of the Biological Diversity Act through constitution of committees (NBA, 2004).

3.3. Conservation of forest biodiversity in protected areas

In order to conserve variability within and among different species in its authority Protected
Areas (PAs) have been established for coordinated conservation of ecological units and
corridors with multilateral cooperation of the neighbouring nations. There are different types
of PAs like Biosphere Reserve (BR), National Parks (NPs), Wildlife Sanctuaries (WLS),
Conservation Reserves (CR) and Community Reserves (Com.R).

Different workers have carried out isolated work on evaluation of genetic conservation of
forest trees and woody species in the country. Nageswara Rao *et al.* (2001) assessed the genetic
diversity parameters of Sandal (*Santalum album*) populations of peninsular India and suggested that *in-situ* conservation of Sandal genetic resources to be focused at populations and sites
in the Deccan plateau. Similarly, Ravikanth *et al.* (2001) mapped the genetic diversity of rattans
in central Western Ghats and suggested to have conservation stands at three sites in southern
Western Ghats. Anandarao (2003) and Tikader *et al.* (2001) studied the germplassm of different
species of *Morus* and identified diverse populations in different locations in Andamans, North-
East India. Padmini *et al.* (2001) analysed Genetic diversity of *Phyllanthus emblica* in forests of
South India and identified different locations with high diversity for *in-situ* conservation.
However, these findings couldn't be utilized fully as certain identified sites fall outside the
already established PAs. Vasudeva *et al* (2002) studied the available population of *Semecarpus
kathalekanensis*, an Endangered tree and its diversity in Myristica swamp in Karnataka and
suggested the requirement of special *in-situ* conservation measures.

3.4. Biosphere Reserves, National Parks and Wildlife Sanctuaries

Biosphere Reserve is an international conservation designation given by UNESCO under its
Programme on Man and the Biosphere (MAB). According to "The Statutory Framework of the
World Network of Biosphere Reserves," biosphere reserves are created to promote and
demonstrate a balanced relationship between humans and the biosphere. Under article 4,
biosphere reserves must "encompass a mosaic of ecological systems," and thus consist of
combinations of terrestrial, coastal, or marine ecosystems. In India there are 15 biosphere
reserves with total area of 58,645 sq.km (MoEF, 2009) (Table 5). They encompass one or more
protected areas like National Parks, sanctuaries or conservation reserves.

A network of 667 PAs has been established, extending over 157826.773 sq. kms. (4.80 % of total
geographic area of the country), comprising 102 National Parks, 514 Wildlife Sanctuaries, 47
Conservation Reserves and 4 Community Reserves. The State / Union Territory wise details
of PAs in the country with extent are given in (Table 5.). The protected areas in India are mainly
meant for large mammals, birds and some specific conservation dependant flagship species.
However, when the whole habitat or ecosystems are protected, whole forest biodiversity also
enjoys the protection. Some protected areas have also been designated recently for the
conservation of certain plant species, considering their importance. They are Kurinjimala

S.No.	Name	Area of Biosphere (sq km)	Date of establishment	State
1	Achanakmar-Amarkantak	3835.51	2005	Madhya Pradesh & Chhattishgarh
2	Agasthyamalai	3500.36	2001	Tamilnadu & Kerala
3	Dehang-Debang	5111.5	1998	Arunachal Pradesh
4	Dibru-Saikhowa	765.00	1997	Assam
5	Great Nicobar	885.00	1989	Andaman and Nicobar
6	Gulf of Mannar	10500.00	1989	Tamil Nadu
7	Khangchenjunga	2619.92	2000	Sikkim
8	Manas	2837.00	1989	Assam
9	Nanda Devi	5860.69	1988	Uttaranchal
10	Nilgiri	5520.4	1986	Tamil Nadu, Kerala, and Karnataka
11	Nokrek	820.00	1988	Meghalaya
12	Pachmarhi	4926.00	1999	Madhya Pradesh
13	Simlipal	4374.00	1994	Orissa
14	Sunderbans	9630.00	1989	West Bengal
15	Kachcha	12454		Gujarat
	Total			

(Source: MoEF, 2009)

Table 5. Biosphere Reserves in India encompass a mosaic of ecological systems consist of combinations of terrestrial, coastal, or marine ecosystems

National Park, in Idukki district, Kerala for *Strobilanthes*; the *Rhododendron* Sanctuary at Singba in Sikkim, the *Nepenthes* Sanctuary at Jarain in Meghalaya and the Orchid Sanctuary at Sessa in Arunachal Pradesh.

3.5. Conservation of forest biodiversity outside protected areas

In addition to the PAs, there are several other means of *in-situ* conservation like Sacred Groves (SG), Gene Pool Conservation Areas (GPCA), Seed Production Areas (SPAs) Medicinal Plant Conservation Areas (MPCA) and Permanent Preservation Plots (PPP).

3.6. Sacred groves

Conservation of habitats and individual species has been practiced in India since time immemorial. Concern for nature conservation is deeply embedded in the multiracial Indian society. The Sacred groves are patches of natural vegetation, which are protected through some religious faiths and they exist throughout India. They shelter many economically important, medicinal, endemic, rare and endangered species. The extent of sacred groves varies from $1m^2$ to $1000000\ m^2$. Although, there has been no comprehensive study on the Sacred groves of the entire country, experts estimate the total number of sacred groves in India could be in the range of 100,000 – 150,000. As per some reports, India is having 13270 sacred groves (Malhotra *et al.* 2001; Kunhikannan and Gurudev Singh, 2005; Warrier *et al.* 2008) (Table 6).

State/Union Territories	National Parks (NP)	Wildlife Sanctuaries (WS)	Conser. Reserves (CR)	Community reserve (Com.R)	Area of state (sq km)	Total Area covered (sq km)	% of protected area in State
Andhra Pradesh	6	21	0	0	275068	13006.514	4.73
Arunachal Pradesh	2	10	0	0	83743	9778.57	11.68
Assam	5	18	0	0	78438	3909.80	4.98
Bihar	1	12	0	0	94163	3187.33	3.39
Chhattisgarh	3	11	0	0	135194	6382.27	4.79
Goa	1	6	0	0	3702	754.91	20.39
Gujarat	4	23	1	0	196024	17323.48	8.83
Haryana	2	8	2	0	44212	348.84	0.75
Himachal Pradesh	5	32	0	0	55673	10016.85	17.99
Jammu & Kashmir	4	15	34	0	222235	11688.36	5.26
Jharkhand	1	11	0	0	79714	2182.14	2.74
Karnataka	5	22	2	1	191791	6482.52	3.38
Kerala	6	16	0	1	38863	2382.52	6.13
Madhya Pradesh	9	25	0	0	308252	10814.76	3.51
Maharashtra	6	35	1	0	307690	15429.75	5.02
Manipur	1	1	0	0	22327	224.4	1.01
Meghalaya	2	3	0	0	22429	301.68	1.35
Mizoram	2	8	0	0	21081	1240.75	5.89
Nagaland	1	3	0	0	16579	222.35	1.34
Orissa	2	18	0	0	155707	7959.85	5.11
Punjab	0	12	1	2	50362	344.72	0.68
Rajasthan	5	25	3	0	342239	9548.60	2.79
Sikkim	1	7	0	0	7096	2183.10	30.76
Tamil Nadu	5	21	1	0	130058	3829.82	2.95

State/Union Territories	National Parks (NP)	Wildlife Sanctuaries (WS)	Conser. Reserves (CR)	Community reserve (Com.R)	Area of state (sq km)	Total Area covered (sq km)	% of protected area in State
Tripura	2	4	0	0	10486	603.62	5.76
Uttar Pradesh	1	23	0	0	240926	5711.00	2.37
Uttaranchal	6	6	2	0	53485	7376.33	13.79
West Bengal	5	15	0	0	88752	2896.53	3.26
Andaman & Nicobar	9	96	0	0	8249	1543.33	18.71
Chandigarh	0	2	0	0	114	26.009	22.81
Dadra & Nagar Haveli	0	1	0	0	491	92.16	18.77
Daman & Diu	0	1	0	0	112	2.18	1.95
Delhi	0	1	0	0	1483	27.82	1.88
Lakshadweep	0	1	0	0	32	0.01	0.031
Pondicherry	0	1	0	0	493	3.90	0.79
Total			47	4			4.80

(Source: MoEF, 2012)

Table 6. State-wise details of the Protected Area Network of the country for conservation of certain plant species, mammals and birds

A recent study conducted by (Sambandan and Dhatchanamoorthy, 2012) on the floristic composition of angiosperms occurring in a sacred grove of 0.2 ha area, located in Karaikal area of the UT of Puducherry brought out presence of 59 plants species of flowering plants coming under 55 genera and 30 families. They also found that, many rural people in the district were using the plants from the sacred groves to cure many common diseases. They suggested that, this kind of degraded sacred groves should be immediately restored or regenerated using appropriate technologies and by creating awareness among the rural people regarding the importance of sacred grove and its conservation.

3.7. Gene Pool Conservation Areas

Gene pool conservation is necessary for human welfare. Several species have become extinct and some others are already threatened and may become extinct if appropriate measures for their conservation are not taken. Some State Forest Departments like Kerala, Tamilnadu and West Bengal have initiated establishment of Gene Pool Conservation Areas (GPCA) for providing specific protection to certain areas through participatory approaches, involving local people. The Govt. of Kerala has identified some of the areas to be protected as GPCAs and issued guidelines for identification and mapping of GPCAs in the State (KFD, 2005) (Table 7).

S. No.	State	No. of Sacred Groves
1	Andhra Pradesh	800
2	Arunachal Pradesh	58
3	Assam	40
4	Chhattisgarh	600
5	Gujarat	29
6	Haryana	248
7	Himachal Pradesh	5000
8	Jharkhand	21
9	Jammu & Kashmir	150
10	Karnataka	1424
11	Kerala	3500
12	Maharashtra	1600
13	Manipur	365
14	Meghalaya	79
15	Orissa	322
16	Pondicherry	108
17	Rajasthan	9
18	Sikkim	56
19	Tamil Nadu	499
20	Uttarakhand	1
21	Utter Pradesh	6
22	West Bengal	670
Total		**15585**

Table 7. Status of Sacred Groves in India harbouring original and pristine vegetation and biodiversity of area

3.8. Seed Production Areas (SPAs)

With an aim to improve the productivity and profitability of planting forest species and offering an attractive land use option, many SFDs have established Seed Production Areas, in collaboration with various research organizations, for different forest species. Such SPAs act as means of conservation of Forest Genetic Resources, especially of forest plantation species of high commercial value. Species wise list and total area under SPAs are given in Table 8.

Species (scientific name)	Purpose for establishing conservation unit	Number of populations or stands conserved	Total Area (Ha)
Abies pindrow	Seed production	1	13.25
Acacia catechu	Seed production	14	230.00
Acacia nilotica	Seed production	7	87.00
Acrocarpus fraxinifolius	Seed production	1	2.00
Haldinia cordifolia	Seed production	1	255.00
Aegle marmelos	Seed production	1	0.5
Ailanthus excelsa	Seed production	1	10.00
Ailanthus triphysa	Seed production	1	7.00
Albizia amara	Seed production	1	2.00
Amoora wallichii	Seed production	1	11.00
Anogeissus latifolia	Seed production	5	57.00
Artocarpus chaplasha	Seed production	1	2.00
Artocarpus heterophyllus	Seed production	2	5.00
Bombax cieba	Seed production	7	51.50
Borassus flabellifer	Seed production	1	30.00
Buchnania lanzan	Seed production	1	20.00
Calophyllum inophyllum	Seed production	1	315.00
Cedrus deodara	Seed production	6	86.8
Chloroxylon swietenia	Seed production	1	10.00
Chukrasia tabularis	Seed production	4	29.00
Cupressus torulosa	Seed production	1	5.00
Dalbergia latifolia	Seed production	5	37.30
Dalbergia sissoo	Seed production	19	197.00
Diospyros melanoxylon	Seed production	1	5.00
Dipterocarpus macrocarpus	Seed production	5	39.00
Dipterocarpus turbinatus	Seed production	1	2.00
Ficus spp.	Seed production	2	8.00
Garcinia indica	Seed production	1	78.00
Gmelina arborea	Seed production	7	59.50
Hardwickia binata	Seed production	7	80.40
Hopea parviflora	Seed production	4	50.70
Lagerstroemia lanceolata	Seed production	2	8.30
Limonia acidissima	Seed production	2	3.50
Madhuca longifolia var. latifolia	Seed production	1	10.00
Michelia champaca	Seed production	1	1.00
Mitragyna parvifolia	Seed production	1	5.00
Morinda tinctoria	Seed production	1	10.00
Morus laevigata	Seed production	1	1.00

Species (scientific name)	Purpose for establishing conservation unit	Number of populations or stands conserved	Total Area (Ha)
Pinus caribaea	Seed production	2	6.00
Pinus kesiya	Seed production	1	15.00
Pinus patula	Seed production	1	1.50
Pinus roxburghii	Seed production	17	215.00
Pinus wallichiana	Seed production	5	87.00
Prosopis cineraria	Seed production	1	10.00
Pterocarpus dalbergioides	Seed production	1	29.11
Pterocarpus marsupium	Seed production	5	57.00
Pterocarpus santalinus	Seed production	2	32.40
Rhododendron arboreum	Seed production	1	0.50
Santalum album	Seed production	6	32.60
Schleichera oleosa	Seed production	1	5.00
Semicarpus anacardium	Seed production	1	186.00
Shorea robusta	Seed production	9	501.80
Sterculia villosa	Seed production	1	4.00
Swietenia mahogany	Seed production	1	10.00
Tamarindus indica	Seed production	1	5.00
Tectona grandis	Seed production	223	6014.34
Terminalia alata	Seed production	6	51.74
Terminalia bellirica	Seed production	1	67.00
Terminalia chebula	Seed production	1	5.00
Vateria indica	Seed production	1	4.00
Xylia xylocarpa	Seed production	1	33.00
Ziziphus mauritiana	Seed production	2	14.50
Taxus baccata	Conservation of threatened species	16	89
Tachycarpus takil	Conservation of rare palm species endemic to Kumaon hills	1	10
Shorea robusta	To preserve high quality sal crop.	6	75
Haldinia cordifolia, Albizzia procera, Shorea robusta, Diospyros embroptria, Terminalia belerica	To preserve an area of primeval fresh water swamp forest.	4	57
Dalbergia sissoo	To study natural succession.	2	2
Pterospermum acerifolium	To maintain this interesting group of kanakchampa trees in perpetuity and study the natural succession in this type of forest.	1	3.7

Table 8. Target forest species as Forest Genetic Resources in *in situ* conservation programmes/ units.

3.9. Preservation plots

The Silvicultural conference held in 1929 recommended laying out of preservation plots in India, by selecting representative areas of major forest types and exceptional trees to be selected for permanent protection. Subsequently, 309 preservation plots were laid out throughout the country, 187 in natural forests and 122 in plantations covering a total area of about 8, 500 ha (Table 9.) In addition to this, 537 trees in various States were protected (Khullar, 1992).

State	Number of preservation plots		
	Natural Forests	Plantations	Total
Andhra Pradesh	11	-	11
Arunachal Pradesh	1	-	1
Assam	9	1	10
Bihar	9	-	9
Gujarat	18	-	18
Haryana	-	-	0
Himachal Pradesh	6	6	12
Jammu & Kashmir	3	-	3
Karnataka	11	-	11
Kerala	8	43	51
Madhya Pradesh	28	-	28
Maharashtra	11	-	11
Manipur	-	-	0
Meghalaya	2	-	2
Nagaland	-	-	0
Orissa	6	-	6
Punjab	-	-	0
Rajasthan	3	-	3
Tamil Nadu	2	67	69
Union Territories	-	-	0
Uttar Pradesh	32	-	32
West Bengal	27	5	32
Total	**187**	**122**	**309**

Table 9. *In-situ* conservation and Status of preservation plots in India (1991) for observing and preserving the original vegetation and natural animals

Rodgers (1991) opined that, a network of forest preservation plots within a larger extent of natural forests, which covers the entire range of forest types in India could play a major role in national efforts to protect biodiversity. He pointed out the virtual lack of co-ordination of management efforts in maintaining the network of plots. He therefore advocated that there is need for extending the network of preservation plots to ensure that, it covers (a) the "Vegetation Series" of vegetation mapping studies carried out by the French Institute, Puducherry (b) biogeographic classification of India prepared for conservation and planning purposes and (c) Project Tiger areas, established in different parts of India. He concluded that such a network would complement the conservation zone of Biosphere Reserves and their conservation capability into a much wider variety of forest types.

3.10. Plus trees

Plus tree selection is one of the methods to conserve diversity at species level. Plus tree is a phenotypically superior tree. Most of the tropical forest trees are out crossers and therefore, there is wide variability among individual trees of a particular species in terms of growth, form and wood characters. In some cases much of the variations may be genetic and in others environmental. It is the utilization part of gene conservation where individual selection and breeding within locally adapted provenance will provide additional improvement in selected characteristics. A list of superior phenotypes of different species (Emmanual *et al*. 1990) selected in different States is given in Table 10.

State	Species with number of trees
Andhra Pradesh	*Tectona grandis-75.*
Arunachal pradesh	*Acrocarpus fraxinifolius-16, Ailanthus grandis-18, Altangia excelsa 27, Neolamarkia cadamba-08, Bombax ceiba-21, Canarium resiniferum-7, Cinnamomum cecidodaphne-4, Chukrasia tabularis-5, Dipterocarpus macrocarpus-39, Mesua ferrea-8, Michelia champaca-30, Phoebe goalparensis-21, Shorea assamica-27, Terminalia myriocarpa-28, Tectona grandis-2, Pinus roxburghii-4, Gmelina arborea-20.*
Himachal Pradesh	*Pinus roxburghii-47* high resin yielder and 59 for timber production.
Karnataka	*Tectona grandis-50, Artocarpus heterophyllus-31, Phyllanthus emblica -11, Limonia acidissima-40, Ziziphus mauritiana-3, Syzygium cumini-1*
Kerala	*Tectona grandis-29, Bombax ceiba-11, Santalum album-3, Azadirachta indica-300*
Madhya Pradesh	*Tectona grandis-360, Albizia procera-55, Azadirachta indica-200*
Maharashtra	*Tectona grandis-33, Dalbergia sissoo-12, Bombax ceiba-1, Acacia catechu-23.*
Manipur	*Tectona grandis-25.*
Tamil Nadu	*Tectona grandis-24.*
Tripura	*Gmelina arborea-50, Shorea robusta-50, Tectona grandis-50.*
Uttar Pradesh & Uttarakhand	*Pinus roxburghii- 54* for tree form and *39* for high resin yield, *D. sissoo 302,*
Rajasthan	*Dalbergia sissoo-50, Azadirachta indica –350*
Gujrat	*Tectona grandis -63*
North eastern states	*Gmelina arborea-119,Tectona grandis-46,Dipterocarps-93*

Table 10. Plus trees selected in different States for *in situ* conservation

3.11. Conservation of Bamboo Genetic Resources

India is the 2nd richest country in bamboo genetic resources. Large forest areas have been declared as National Bamboo Reserve and maintained. Considering the limitation in seed supply, vegetative methods for *ex situ* conservation and tissue culture work have been started in Asian countries. National Bamboo Mission has been launched by the Ministry of Agriculture for bringing more areas under bamboos. National Mission on Bamboo Applications (NMBA), focuses on Wood substitutes and composites, construction & structural applications, agro-processing, machinery & process technologies, propagation & cultivation, industrial products and product applications in bamboos. A National Mission on Bamboo Technology and Trade Development was established, considering its role in rural economy and poverty alleviation and potential use in handicrafts and industrial development. A Bamboo Information Centre established at KFRI, Peechi disseminates information on 137 species of Indian bamboos (Table 10). To deal with the gregarious flowering of Muli bamboos (*Melocanna baccifera*) in N. E. States, 3 task forces on 'regeneration', 'harvesting and marketing' and 'rodent control' have been constituted.

3.12. Mangrove Conservation Program

Mangrove forests in India covers an area of 6,000 km^2 and they shelter 59 plant species coming under 41 genera and 29 families. Taking into consideration the ecological and economic significance of Mangroves, the Ministry of Environment and Forests had launched a Scheme for Conservation and Management of Mangroves and Coral Reefs in 1986 with following objectives:-

- Conservation and protection from further degradation of the Mangrove Ecosystem;

- Afforestation of degraded Mangrove areas

- Restoration of degraded Coral Reef areas

- Maintenance of genetic diversity especially of the threatened and endemic species

- Creation of awareness among the people on importance of Mangrove/Coral Reef Ecosystem and the need for their conservation.

This was launched in 1987 and 35 mangrove areas were identified for intensive conservation and management (Table 11). Financial support is given under Management Action Plans for raising mangrove plantations, protection, catchment area treatment, siltation control, pollution abatement, biodiversity conservation, sustainable resource utilization and creating awareness. A National Mangrove Genetic Resource Centre was established in Orissa, in the east coast of India, for conservation, afforestation and regeneration of mangrove species.

3.13. Medicinal Plants Conservation Program

India has probably the oldest, richest and most diverse cultural traditions in the use of medicinal plants. The total number of medicinal plant species recorded from India is about 7500 and still the health care system based on herbal medicine is very much prevalent in the

SN	State/ UT	Mangrove Area
1.	West Bengal	Sunderbans
2.	Orissa	Bhitarkanika, Mahanadi, Subernarekha, Devi and Dhamra
3.	Andhra Pradesh	Coringa, East Godavari and Krishna
4.	Tamil Nadu	Pichavaram, Muthupet and Ramnad
5.	Andaman & Nicobar	North Andamans and Nicobar
6.	Kerala	Vembanad
7.	Karnataka	Coondapur and Dakshin Kannada
8.	Goa	Goa
9.	Maharashtra	Achra-Ratnagiri, Devgarh-Vijay Durg, Veldur, Kundalika- Revdanda, Mumbra-Diva, Vikroli, Shreevardhan, Vaitarna, Malvan and Vasai- Manori
10.	Gujarat	Gulf of Kutchh and Gulf of Khambat

Table 11. State-wise list of Mangrove areas identified by MoEF, Govt. of India for Conservation and Management for *in situ* conservation of costal vegetation

country. A National Medicinal Plants Board (NMPB) was established for co-ordination and implementation of policies relating to conservation, harvesting, cultivation, research and marketing of medicinal plants through 32 State Medicinal Plant Boards. At the national level 32 medicinal plant species have been selected for research and development. A network of 54 Medicinal Plant Conservation Areas (MPCAs)-"as forest gene bank sites" have been established by the State Forest Departments of Andhra Pradesh, Karnataka, Kerala, Tamil Nadu and Maharashtra (Ravikumar, 2010) in consultation with the Foundation for Revitalization of Local Health Traditions (FRLHT) and with the support of DANIDA, which harbour 45 percent of recorded populations of flowering medicinal plants of Peninsular India, including 70 percent of the red-listed species. To conserve wild germplasm, revitalize the indigenous health care and livelihood security, a UNDP-CCF-II project has been implemented in 9 states. A list of MPCAs in Karnataka, Kerala and Tamilnadu is provided below in Table 12.

3.14. Biodiversity Hotspots in India

Certain tropical forest areas rich in diverse endemic species, which are on the verge of destruction, have been designated as 'Hot Spots'. The conservation of these areas is indispensable for the survival of these species. About 34 'Hot Spots' of biodiversity have been identified around the world. Among them four are in India-Eastern Himalaya, Indo-Burma, Western Ghats and Sri Lanka, and Sundaland located in Nicobar Islands. These 'Hot Spots' together have about 5330 endemic species including flowering plants, mammals, reptiles, amphibians and butterflies (Ramakrishna, 2010).

1	Agumbe	Karnataka
2	BRT hills	Karnataka
3	Charmadi	Karnataka
4	Devimane	Karnataka
5	Devrayandurga	Karnataka
6	Karpakapalli	Karnataka
7	Kemmanagundi	Karnataka
8	Kollur	Karnataka
9	Kudremukh	Karnataka
10	Sandur	Karnataka
11	Savanadurga	Karnataka
12	Subramanya	Karnataka
13	Talacauvery	Karnataka
14	Athirapally	Kerala
15	Eravikulam	Kerala
16	Kulamavu	Kerala
17	Peechi	Kerala
18	Silent Valley	Kerala
19	Triveni	Kerala
20	Waynad	Kerala
21	Agasthiarmalai	Tamilnadu
22	Alagarkovil	Tamilnadu
23	Kodaikanal	Tamilnadu
24	Kodikaria	Tamilnadu
25	Kollihills	Tamilnadu
26	Kurumburam	Tamilnadu
27	Kutrallum	Tamilnadu
28	Mundanthurai	Tamilnadu
29	Petchiparai	Tamilnadu
30	Thaniparia	Tamilnadu
31	Thenmalai	Tamilnadu
32	Topslip	Tamilnadu

Table 12. Medicinal Plants Conservation Programme for *in situ* conservation of medicinal plants in Karnataka, Kerala and Tamilnadu

The following are some of general criteria for identification of *in-situ* genetic conservation units:-

- Hot spot areas of biodiversity and endemism

- Representative and unique forest types

- Representative forest plantations of valuable timber species

- Areas with high concentration of medicinal plants of conservation concern

Generally, it is true that *in-situ* conservation measures are preferred over *ex-situ* means, because of habitat specificity of constituent species; maintenance of diversity and providing opportunities for evolutionary process to continue; endemic and threatened nature of some of the species; high cost factor and technological need involved in *ex-situ* conservation measures.

Constraints and problems for in-situ conservation

- The greatest constraints to improving *in situ* conservation in the country are lack of scientific know-how, anthropogenic and biotic pressures like fire, grazing, encroachment and illicit felling, to mention a few.

- Certain areas in the country are still unexplored, which need intensive exploratory survey.

- Conservation of FGR across the forest, including production forests and agro-forests, have not been developed for most of the species nor applied in India

- Research is needed to identify the best combination of approaches (*in situ*, *ex situ* and *circa situ*) for species that are important for livelihoods and subsistence in areas of high diversity and/or high poverty.

3.15. Ex-situ conservation

Ex-situ conservation is the conservation of components of biological diversity outside their natural habitats (CBD, 1992). Plant species and varieties can be preserved under artificial conditions away from the places where they naturally grow. Ex-situ plant collections have a number of uses for conservation and development, including for the re-vitalization of plant populations and associated economies and cultures (Hamilton and Hamilton, 2006). The following are the techniques generally employed for ex-situ conservation (Table 13 & 14).

The storage of seeds in seed banks has advantages for preserving species, but can only be used for species with seeds capable of remaining viable after long term storage (known as 'Orthodox' seeds). The typical techniques used for seed storage is to lower the moisture content of seeds to 2-6 percent or less and reduce temperature to around 0^0 C or lower.

Technique	Definition
Seed storage	The collection of seed sample at one location and their transfer to a gene bank for storage. The samples are usually dried to a suitable low moisture content and then kept at sub-zero temperatures.
Field gene bank	The collecting of seed or living material from one location and its transfer and planting at a second site. Large number of accessions of a few species are usually conserved.
Botanic garden/ arboretum	The collecting of seed or living material from one location and its transfer and maintenance at second location as living plant collections of species in an arboretum. Small numbers of accessions of a large number of species are usually conserved.
In-vitro storage	The collection and maintenance of ex-plants (tissue samples) in a sterile, pathogen free environment).
DNA/ Pollen storage	The collecting of DNA or pollen and storage in appropriate, usually refrigerated condition.

Table 13. Techniques of *ex-situ* conservation (Hawkes *et al.* 2000)

Technique	Definition
Seed storage	The collection of seed sample at one location and their transfer to a gene bank for storage. The samples are usually dried to a suitable low moisture content and then kept at sub-zero temperatures.
Field gene bank	The collecting of seed or living material from one location and its transfer and planting at a second site. Large number of accessions of a few species are usually conserved.
Botanic garden/ arboretum	The collecting of seed or living material from one location and its transfer and maintenance at second location as living plant collections of species in an arboretum. Small numbers of accessions of a large number of species are usually conserved.
In-vitro storage	The collection and maintenance of ex-plants (tissue samples) in a sterile, pathogen free environment).
DNA/ Pollen storage	The collecting of DNA or pollen and storage in appropriate, usually refrigerated condition.

Table 14. Techniques of *ex-situ* conservation (Hawkes *et al.* 2000)

4. Identification of research needs gaps and constraints

4.1. Identification of Research and Development Gaps

i.   **Identification of research needs:**

a.   Biodiversity documentation

b.   Global Net-Working in identification and monitoring forest biodiversity

c.   Strategies and Actions for Wild biodiversity

d.   Involvement of stakeholders in biodiversity conservation through participatory forest management and yield sustainable benefits to people

e.   Critical trends such as degradation/ fragmentation of habitat, extinction of species and destruction of unique habitats and its need for monitoring

f.   A possible REDD-mechanism (financial incentives for reducing emission from deforestation and forest degradation) under the post-2012 framework of the Kyoto Protocol should consider effects on local communities and poor people, and survive to ensure a fair sharing of benefits.

g.   Extent of invasion of Weeds and Invasive species in different Forest Types of India

h.   Identification of indicator species for understanding ecosystem health so as to evaluate the efficiency of management interventions

i.   Socio-economic variables and their interaction with biological component of various ecosystems with special attention to ethnology, tribal livelihood and dependence on forests

ii.   **Gaps and Constraints:** There are numerous gaps and constraints, which hamper in undertaking research activities. These gaps and constraints are there in the forms of lack of knowledge, lack of technologies in developing countries, unavailability of finance, non-availability of relevant data, data non-accessibility, data organization constraints, lack of well framed and effective policy work etc. The details of these gaps and constraints are summarized below

a.   *Knowledge Gaps:* Knowledge and information gaps on biodiversity-related issues and solutions at local levels. There is international and regional cooperation in areas of knowledge gap.

b.   *Financial Gaps :* There is an urgent need for action and investment planning for the research activities of forestry sector. It is important to make simple and easy approachable rules and regulation for funding the project work related to biodiversity conservation and ecological securities

c.   *Research Gaps:* There is a considerable gap in our knowledge of the natural resources of India. There are three broad areas, related to knowledge and data gaps that need to be addressed:

- First, there is much to learn about the potential magnitude and rate of extinction of plant biodiversity at the regional and local levels, and subsequent impacts on the full range of biodiversity endpoints and ecosystems.

- Second, there is no consolidated handbook of proven biodiversity conservation techniques, covering all the eco-regions of India.

- Third, detailed analysis need to be developed for each of the priority climate change threats to biodiversity and other resources.

- A further strategic approach is needed for detailed research on different ecosystem services and functions to estimate the potential impacts of climate change

d. *Technological and Capacity Building Gaps*: Today, science and technology is growing very fast. But it is not hard to see that the forestry sector of India has not been able to tap into the advances in technology to the optimal and uniform level. On the one hand, we have high and technical devices and on the other forestry sector lacks well equipped technical devices needed to research activity. Perhaps, one of the reasons why the status of forestry statistics has become a cause of concern because there is a vast gap in the current technology applications and their adaptation to the day-today working of forest research.

The local capacity to collect data at regional level is weak. Before implementing technological advances in statistical data reporting work, it is necessary to build adequate capacity for collection of data from primary sources. The primary data collector should be well versed not only in the terminology of the database, but also with the importance of such a database to ensure sincerity in the work. The capacity-building programmes should have a sustainable structure aiming at timely upgrading in tandem with the technology. Use of local and wide area networks is essential to ensure on-time data availability.

e. *Capacity strengthening*: The capacity to identify, collect and share data, use information and methods and build knowledge relevant for biodiversity conservation and ecological security is critical because of rapidly changing climatic, environmental and socio-economic conditions. Extension services and mechanisms have been weakened greatly over the last two decades. Extension will need to be strengthened substantially in order to address biodiversity conservation for providing an efficient interface between policy-makers and the forest community.

5. Road map for future research

The following research areas are suggested for conservation and sustainable utilization of forest biodiversity:-

- Species recovery research

- Establishing of new permanent preservation plots of representative forest types

- Locating of old permanent preservation plots, data collection & maintenance

- Establishing of Gene Pool Conservation Areas (GPCAs) outside protected areas, where there is concentration of endemic and RET species.

- Studies on plant-animal interactions/ associations

- Diversity of soil microflora and fauna

- Establishing of Gene Sanctuaries of endemic & RET species

- Preparation of Biodiversity Registers in a collaborative mode with the State Biodiversity Boards.

- Biodiversity assessment and updating for Forest Working Plans.

- Monitoring biodiversity of Sacred Groves and provide inputs for conservation.

- Studies on regeneration status of important primary and secondary timber species.

- Studies on reproductive biology and breeding systems of important tree species.

- Awareness creation on biodiversity, benefit sharing and other legal issues.

- Phenological studies on important tree species vis-à-vis climate change.

- Studies on intra-specific variations in important timber species.

- Monitoring the structural composition of different forest types vis-à-vis climatic factors.

- Studies on species yielding natural dyes, wild fruits, tubers and fodder.

- Identification, documentation and domestication of wild plant species of aesthetic value for Urban Forestry.

- Studies on usage of plants and animals in lesser known ethnic communities and validation of information.

- Studies on sustainability of over-exploited bioresources among forest dwelling communities.

- Eco-restoration of degraded forest areas.

- Studies on factors contributing towards forest degradation.

- Impact of natural calamities on forests and developing technologies for re-vegetating the affected areas.

- Studies on plant successions.

- Environment Impact Assessment Studies.

- Studies on reclamation of lands subjected to pollution and mining.

- Microfloral and faunal dynamics of forest litter vis-à-vis climate change

- Eco-restoration of degraded riparian and swamp ecosystems.

- Studies on aquatic biodiversity in hill streams and rivers.

- Studies on the impact of invasive weeds in forest ecosystem.

Abbreviations

FSI	---	Forest Survey of India
SG	---	Sacred Groves
CBD	---	Convention on Biological Diversity
SCBD	---	Secretariat of the Convention on Biological Diversity
DANIDA	---	Danish International Development Agency
MoEF	---	Ministry of Environment and Forest
IUCN	---	International Union for Conservation of Nature
TEEB	---	The Economics of Ecosystems and Biodiversity
IPCC	---	The Intergovernmental Panel on Climate Change
EIA	---	Environmental Impact Assessment
SEZ	---	Special Economic Zone
UNCED	---	The United Nations Conference on Environment and Development
PA's	---	Protected Ares (for Wildlife)
UT	---	Union Territory
SFD	---	State Forest Department
KFD	---	Kyasanur forest disease
KFRI	---	Kerala Forest Research Institute
REDD	---	Reducing emission of carbon from deforestation and forest degradation
FGR	---	Forest Genetic Resources

Author details

Vivek Khandekar* and Anita Srivastava*

*Address all correspondence to: vykhandekar@hotmail.com

Biodiversity Conservation, Indian Council of Forestry Research and Education, Ministry of Environment and Forests, Government of India, India

References

[1] Bahuguna V K. 2000. Forests in the Economy of the Rural Poor: An Estimation of the Dependency Level. *Ambio* 29(3):126-129

[2] Bharath Kumar L B, Patil B L, Basavaraja H, Mundinamani S M, Mahajanashetty S B, and Megeri S N. 2011. Participation Behaviour of Indigenous People in Non-Timber Forest Products Extraction in Western Ghats Forests. *Karnataka Journal of Agricultural Science.* 24(2): 170–172

[3] BSI, 2012. Botanical Survey of India. http://bsi.gov.in/ floristics.shtm (Accessed on 11.01.2012).

[4] Deb, P. & Sundriyal, R.C.2007. Tree species gap phase performance in the buffer zone area of Namdapha National Park, Eastern Himalaya, India. Tropical Ecology. 48(2): 209-225,

[5] Deb, S., Barbhuiya, A.R. Arunachalam, A. and Arunachalam, K. 2008. Ecological analysis of traditional agroforest and tropical forest in the foothills of Indian eastern Himalaya: vegetation, soil and microbial biomass. Tropical Ecology. 49(1): 73-78,

[6] Deshmush, S.V. 1991. Mangroves of India: Status report. In: *Proceedings of the project formulation workshop for establishing a global network of mangrove genetic resource centres for adaptation to sea level rise* (eds. Sanjay V. Deshmukh and Rajeswari Mahalingam), Proceedings No.2, CRSARD, Chennai, pp. 15-25.

[7] Emanuel, C. J. S. K., Kapoor, M.L. and Sharma, V.K. 1990. Achievements in tree improvement and their significance in Gene Conservation. *Jour. Econ. Bot. Phytochem.* 1:48-54.

[8] FSI, 2011, India State of Forest Report, Dehradun.

[9] Gupta, R., 1998, Life support species for medicinal use in India, In *Life Support Plant Species* (NBPGR, eds), New Delhi.

[10] Hawkes, J.G., Maxted, N. and Ford-Lloyd, B. 2000. *The ex situ conservation of plant genetic resources.* Kluwer Academic Publishers, USA.

[11] IPCC. 2007: *Climate Change 2007: Impacts, Adaptation and Vulnerability. Contribution of Working Group II to the Fourth Assessment Report of the Intergovernmental Panel on Climate Change,* M.L. Parry, O.F. Canziani, J.P. Palutikof, P.J. van der Linden and C.E. Hanson, Eds., Cambridge University Press, Cambridge, UK, 976 pp.

[12] Khan, T.I. and Frost, S. (2001). Floral biodiversity: a question of survival in the Indian Thar Desert. The Environmentalist, 21, 231–236.

[13] Khan, T.I., Dular, A.K. and Solomon, D.M., 2003, Biodiversity conservation in the Thar desert; with emphasis on endemic and medicinal plants. The Environmentalist, 23, 137–144, 2003

[14] Khoshoo, T. N. 1986. Environmental priorities in India and sustainable development (Presidential Address). Indian Science Congress Association, New Delhi, 224p.

[15] Khullar Pankaj, 1992. Conservation of biodiversity in natural forests through preservation plots- a historical perspective. *Indian For.* 118 (5): 327-337.

[16] Kunhikannan, C. and Gurudev Singh, B. 2005. *Strategy for Conservation of Sacred Groves*, Institute of Forest Genetics and Tree Breeding, Coimbatore.

[17] Mahapatra K and S. Kant. 2005. Tropical Deforestation: A Multinomial Logistic Model and Some Country Specific Policy Prescriptions. *Forest Policy and Economics* 7: 1-24

[18] Mc Neely, J. A. 1984. Conserving biological diversity – A decision maker's guide. *IUCN Bulletin* 20 (4-6): 6-7

[19] Ministry of Environment and Forests. 2006. *Report of the National Forest Commission.* New Delhi: Ministry of Environment and Forests, Government of India. 421 pp.

[20] Ministry of Environment and Forests. 2009, Asia-Pacific Forestry Sector Outlook Study II: India Country Report. Working Paper No. APFSOS II/WP/2009/06. Bangkok: FAO pp 78

[21] Mishra, B.P., Tripathi, O.P. and Laloo, R.C. 2005. Community characteristics of a climax subtropical humid forest of Meghalaya and population structure of ten important tree species. *Tropical Ecology.* 46(2): 241–251

[22] MoEF 2012. Protected Area Network in India. http://moef.nic.in/downloads/public-information/protected-area-network. (Accessed on 7.1.2012).

[23] MoEF, 2009a. India's Fourth National Report to the Convention on Biological Diversity, Ministry of Environment and Forests, Govt. of India, New Delhi.

[24] MoEF, 2009b. State of Environment Report, India, Ministry of Environment and Forests, Govt. of India, New Delhi, 179 p.

[25] Nageswara Rao, M. Ganeshaiha, K.N. and Uma Shaankar, R. 2001. Mapping genetic diversity of sandal (*Santalum album*) in south India: Lesson for *in-situ* conservation of sandal genetic resources. In: *Forest Genetic Resources: Status and Conservation Strategies* (eds. Uma Shaankar, R., Ganeshaiha, K.N. and Bawa, K. S.). Oxford & IBH Publishing Co. Pvt. Ltd, New Delhi.

[26] Nayar, M. P. and Sastry, A. R. K. 1990. *Red Data Book of Indian Plants.* Volume III. Botanical Survey of India. 271 p.

[27] Nayar, M. P. and Sastry, A. R. K. 1987. *Red Data Book of Indian Plants.* Volume I. Botanical Survey of India. 377 p.

[28] Nayar, M. P. and Sastry, A. R. K. 1988. *Red Data Book of Indian Plants.* Volume II. Botanical Survey of India. 268 p.

[29] NBA, 2004. The Biological Diversity Act, 2002 and Biological Diversity Rules, 2004. National Biodiversity Authority, Chennai, 57 p.

[30] NBAP, 2008. National Biodiversity Action Plan. Ministry of Environment and Forests, Govt. of India, New Delhi, 66p.

[31] Padmini, S., Nageswara Rao, M., Ganeshaiah, K.N. and Uma Shaanker, R. 2001. Genetic diversity of *Phyllanthus emblica* in tropical forests of South India: Impact of anthropogenic pressures. *Journal of Tropical Forest Science* 13(2): 297-310.

[32] Pandey, R.P. and Shetty, B.V., 1985, Rare and threatened plants of Rajasthan. *Proc. Nat. Symp. Evaluate Environ.* Zoology Deptt. J.N. Vyas University, Jodhpur. *GEOBIOS* 238-241.

[33] Pandey, R.P., Shetty, B.V. and Malhotra, S.K., 1983, A preliminary census of rare and threatened plants of Rajasthan. In *An Assessment of Threatened Plants of India* (Eds. S.K. Jain and R.R. Rao), pp. 52-62. BSI, Howrah.

[34] Parkinson, C. E.1923. A Forest Flora of Andaman Islands. Government Central Press, DehraDun.

[35] Quezel, O., 1965, Contribution a l'etude d'endemism chez les Phanerogames Sahariens. *C. R. Somm seanec. Soc. Biogeo,.* 359, 89–103.

[36] Ramakrishna, Raghunathan, C. and Sivaperuman, C. 2010. *Recent Trends in Biodiversity of Andaman and Nicobar Islands.* Zoological Survey of India, Kolkata, 542 p.

[37] Ravikanth, G., Ganeshaiha, K.N. and Uma Shaankar, R. 2001. Mapping genetic diversity of rattans in Central Western Ghats: Identification of hot-spots of variability for in-situ conservation. In: *Forest Genetic Resources: Status and Conservation Strategies* (eds. Uma Shaankar, R., Ganeshaiha, K.N. and Bawa, K. S.) Oxford & IBH Publishing Co. Pvt. Ltd, New Delhi.

[38] Ravikumar, K. 2010. Medicinal Plants Diversity in India and Conservation of Endangered and Threatened Medicinal Plants – Indian Perspective (http://www.apforgen.org/ FGR 20 Coimbatore /08072010/1.IFGTB-CBE-080710.pdf)

[39] Rodgers, W.A. 1991. Forest preservation plots in India-I – Management status and value. *Indian For.* 117 (6): 425-433.

[40] Sadashivappa P, Suryaprakash S, Vijaya Krishna V. 2006. Participation Behavior of Indigenous People in Non-Timber Forest Products Extraction and Marketing in the Dry Deciduous Forests of South India. Conference on International Agricultural Research for Development, Tropen tag University of Bonn, October 11–13

[41] Saha D and Sundriyal R C. 2012. Utilization of Non-Timber Forest Products in Humid Tropics: Implications for Management and Livelihood. *Forest Policy and Economics 14: 28–40*

[42] Saha, Amita and Guru B. 2003. Poverty in Remote Rural Areas in India: A Review of Evidence and Issues, GIDR Working Paper No 139, Ahmedabad: Gujarat Institute of Development Research. 69 pp

[43] Samati, H. and Gogoi, R., 2007, Sacred groves in Meghalaya. *Current Science.* 93(10): 1338-1339

[44] Sambandan, K. and Dhatchanamoorthy, N. 2012. Studies on the phytodiversity of a Sacred Grove and its traditional uses in Karaikal, District, U. T. Puducherry. *Journal of Phytology* 4(2): 16-21.

[45] SCBD (Secretariat of the Convention on Biological Diversity), 2010, Forest biodiversity-Earth's living treasure, Montreal.

[46] Sharma, S., 1983, A Census of rare and endemic flora of S. E. Rajasthan. *An Assessment of Threatened Plants of India* (Eds. S.K. Jain and A.R.K. Shastry), pp. 63-70.

[47] Singh, V.,1985, Threatened taxa and scope for conservation in Rajasthan. *J. Econ. Tax. Bot.* 7: 573-577.

[48] Subramanian K.N. and Sasidharan, K.R. 1992. Conservation of biological diversity-Forest trees. In: *Sustainable Management of Natural Ressources* (eds. T.N. Khoshoo and Manju Sharma). Malhotra Publishing House, New Delhi, pp. 217-242.

[49] Subramanian, K.N. and Sasidharan, K. R. 1996. The panorama of Indian Forests: A reservoir of plant and animal health. *Diversity* 12 (3): 26-29.

[50] Tikader, A., Anandarao, A. and Thangavelu, K.2001. Geographical distribution of Indian mulberry species. *Indian Jour. Pl. Gen. Res.* 15 (3):262-266.

[51] UNEP (United Nationa Environment Programme), 1992, Convention on Biological Diversity, (NA92-7807), New York, UNEP.

[52] Vasudeva, R., Raghu, H. B., Suraj, P.G., Uma Shaanker, R. and Ganeshaiah, K.N. 2002. Recovery of a critically endangered fresh-water swamp tree species of the Western Ghats. In: Proceedings Lake 2002. 9-13 December 2002. Bangalore

[53] Warrier K. C. S., Kunhikannan, C. and Sasidharan, K. R. 2008, Status and floristic diversity of Sacred Groves - The only remnants of natural forests in Alappuzha District, Kerala. Project Completion Report submitted to ICFRE, Dehradun.

[54] Wilson, E O, 1988, The current state of biological diversity, In Biodiversity edited by E O Wilson and F M peter, Washington D. C., National Academy Press.

Chapter 11

The Ecology and Species Richness of the Different Plant Communities Within Selected Wetlands on the Maputaland Coastal Plain, South Africa

M.L. Pretorius, L.R. Brown, G.J. Bredenkamp and
C.W. Van Huyssteen

Additional information is available at the end of the chapter

http://dx.doi.org/10.5772/58219

1. Introduction

The Maputaland Coastal Plain (MCP) is located on the eastern seaboard of southern Africa in KwaZulu-Natal. This area is renowned for its distinct geological history, rich biodiversity, diverse ecosystems, and internationally recognized wetlands (Figure 1). The KwaZulu-Natal Province has the second highest wetland surface area in South Africa [1], and the MCP itself contains a very rich collection of surface water bodies. This includes rivers, floodplains, estuaries, swamps, pans, and coastal lakes [2]. Land use on the MCP is mainly dominated by protected areas, agricultural practices and rural areas. There are currently few urbanized areas. Despite this few wetlands are still intact. Although wetlands play an important role for especially the local inhabitants on the MCP, its value is still underestimated, and little has been done for the promotion of conservation and sustainable utilization of these sensitive ecosystems.

Even though the vegetation of the MCP is remarkably diverse, few vegetation studies have been done on wetlands in the area. The major vegetation types of the MCP have been broadly described by Moll [3,4], and Morgenthal [5]. Tinley [6, 7, 8] conducted vegetation surveys along the coast, while Lubbe [9] conducted a detailed vegetation study of the coastal strip. Many detailed local vegetation studies have been conducted in the protected areas on the MCP but very little in the unprotected areas of the MCP. None of the studies mentioned above provide detailed descriptions of the wetland vegetation and their species richness. The only vegetation study focusing exclusively on wetlands is on the Mfabeni mire in the iSimangaliso Wetland Park [10].

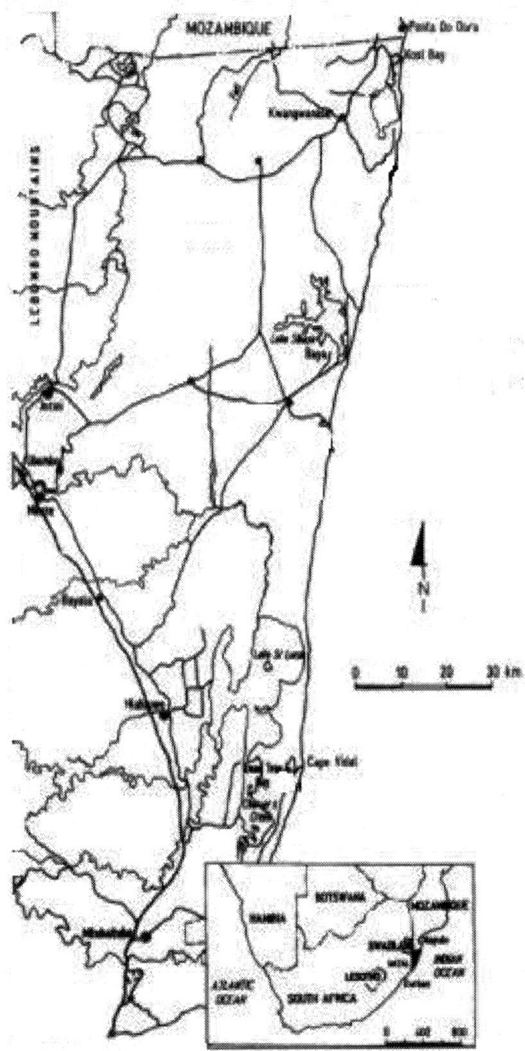

Figure 1. Locality of Maputaland within the South African context

Furthermore, the delineation of wetlands on the aeolian derived sandy soils associated with the MCP is regarded as problematic when using the soil form and-wetness indicators described by the Department of Water Affairs and Forestry (DWAF 2005) [11]. However, the distinct changes in plant species composition along the wetness gradient of a wetland provide an indication of wetland zoning [12, 11] and therefore guides the delineation procedure.

This study aims to elucidate the relationship between vegetation communities, plant species richness, and their environmental setting within the various wetland types on the MCP, in order to contribute to the understanding of wetland zones.

2. Study area

The MCP is demarcated by the Mozambican border in the north; the town Mtunzini in the south; the Indian Ocean in the east; the Lebombo Mountains in the north-west; and the N2 on the south-west. This study focuses on the northern parts of the MCP only (Figure 1).

The MCP has a subtropical climate with very hot summers and mild winters. The area receives 60% of its rainfall during summer and 40% during winter, with a mean annual precipitation of 963 mm [13]. Rainfall decreases sharply from east to west, with an approximate mean of 1200 mm at the coast, and 800 mm – 1000 mm at the crest of the Lebombo Mountains [13].

Aeolian distributed sands from the Tertiary and Quaternary period dominate most parts of the MCP. These sands are relatively infertile and of low-productivity [14]. The study area is characterized by undulating dune topography located up to roughly 70 m above sea-level [15]. In the east the Plain is separated from the Indian Ocean by an uninterrupted barrier dune system [14]. A long, relatively flat coastal plain stretches between the Lebombo Mountain Range and the coastal barrier dunes. Dune cordons occurring sporadically all over the MCP are interspersed with various wetland types such as floodplains, lakes, fens, swamp forests and pans [16]. Groundwater is the principal source of water for most of the lakes and wetlands in Maputaland [17], and moves rapidly through the system due to high permeability, high rainfall, and low water gradients. Two primary porosity aquifers are present on the MCP-a shallow, unconfined aquifer and a deeper, confined aquifer [18]. The shallow, unconfined aquifer is driven by rainfall which infiltrates and percolates through the sandy soil until it reaches the impermeable Kosi-Bay Formation, where after the water then moves laterally to exit the aquifer in the form of a surface water source.

In terms of biodiversity the MCP fall within the Maputaland Centre of Endemism Centre. This is one of Africa's most important biodiversity and endemism hotspots, and is located at the southern end of the African tropic where many plant and animal species reach the limit of their range. An assortment of diverse ecosystems and many broad ecological zones such as thicket, grassland, bushveld, forest, sand forest and swamp forest occur here [19].

Most of the wetlands occurring outside conservation areas are degraded. Local inhabitants of the area utilise the wetland areas extensively for subsistence agriculture due to the infertile nature of the sandy soil. A recent threat to the health of wetlands is the informal plantations that have sprung up all over the MCP during the past 20 years. These *Eucalyptus* plantations have a marked effect on the water table and the subsequent dynamics of the wetlands systems in the area. The MCP is rich in peatlands and contains about 60% of the estimated peat resources of South Africa [20]. This region contains the largest and highest density of peatlands of all the Peat Eco-Regions. It is estimated that 60 – 80% of these peatlands are currently being utilised by the local community for subsistence agriculture and other uses [21].

Five wetland types were identified and investigated in this study (Figure 2):

- Interdune-depression (IDD) System –

  - Scattered depression type wetlands between vegetated coastal dunes,

  - Linked with the regional water table,

  - Peaty soil in the pristine wetlands [33],

  - Intense local utilisation of the fertile peaty soils for subsistence.

- Muzi North Swamp System (MS) –

  - A linear valley-bottom system,

  - Linked with the regional water table,

  - The permanently wet areas of the system are peaty,

  - Clay lenses occur at 300 – 500 mm depth on the banks of the system,

- Perched Pan (PP) and Depression (DP) Systems –

  - A series of scattered seasonal pans occurring parallel to the MS system,

  - Inside the Tembe Elephant Park the pans occur as open areas surrounded by closed woodland (PP System),

  - Outside the Park the pans are open and degraded (DP System),

  - High clay content in the soil results in a perched water table for several months per year [31],

  - The pans are clay-rich, calcareous duplex soils.

- Upland Wetland (PL) System –

  - Located on the upland flat area between the Tembe Elephant Park and Manguzi,

  - Slightly undulating Lala Palm veld with interspersed spaces of open, moist grassland,

  - Depressions occur in large patches in the Palm Veld.

  - These wetlands are seasonal and water table fluctuation plays a prominent rolep [17].

3. Methods

Wetland areas occurring between the Tembe Elephant Park (TEP) and Kosi-Bay were identified using Google Earth and 1:3000 Orthophotos, and verified with a field visit to the area. The wetlands were selected based on accessibility, safety, land owner consent, data availability, and land use.

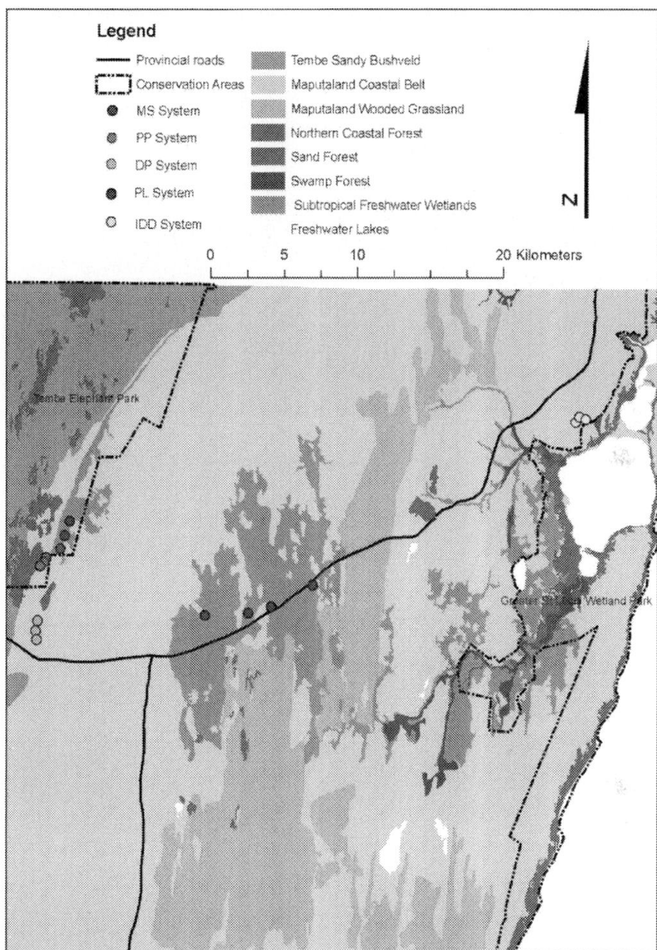

Figure 2. Wetland systems occurring in the northern parts of the Maputaland Coastal Plain.

Between three to five wetlands in each system were selected to be surveyed. These wetlands were first stratified into their various vegetation zones. Between three and five vegetation zones were identified in each wetland. For the purpose of the data collection in the field the different zones sampled were based on vegetation communities observed and not on hydrological regime. Therefore these zones were not termed 'permanent', 'seasonal', 'temporary', or 'terrestrial', but rather as Zone 1, Zone 2, Zone 3, etc (Figures 3 and 4). However after the data analysis the different zones were grouped into the different wetness zones as listed above and discussed accordingly under the discussion section.

Figure 3. The Interdune-depression (IDD) System with an example of the zone delineation.

Figure 4. The Upland Wetland (PL) System, with an example of zone delineation.

Vegetation surveys were conducted in March 2010 following the Zurich-Montpellier (Braun-Blanquet) School of total floristic compositions approach [22]. A total of 72 sample plots (2 m x 2 m) were placed within the different zones in a stratified random manner. Plant species were identified in the field, while the unknown plant species were collected and sent to the South African National Biodiversity Institute for identification.

The vegetation relevés were captured into TURBOVEG for Windows 1.97 [23] and exported to JUICE 6.5 [24]. A modified TWINSPAN was performed in JUICE using the Whittaker's beta-diversity, with the following pseudospecies cut-levels: 0, 1, 5, 25, 50, and 75. The Fisher Exact Fidelity Test at P<0.001 were used. The final classification was manually refined according to the Braun-Blanquet procedure [25]. No re-arrangements of clusters or relevés were done, but only species groups were manually re-arranged.

Six different ordination methods were applied to the plant community data in PCOrd [26]-the Bray-Curtis ordination, Canonical Correspondence Analysis, Weighted Averaging, Reciprocal Averaging, Detrended Correspondence Analysis (DCA), and Nonmetric Multidimensional Scaling (NMS). The DCA and NMS analyses gave the best results. The DCA ordination results are presented in this study, as it emphasized the variation and combination of the plant communities better than the NMS results. Various environmental factors thought to influence the distribution of the vegetation communities were superimposed on the ordination results. The overlay of the floristic communities identified by [27], the five wetland systems, and substrate type are included in the final results.

The Chi-Square Test [28] was performed on the data to determine whether significant differences exist between the species richness of the different plant communities.

4. Results

The modified TWINSPAN analysis [29] resulted in the identification of 11 plant communities that can be grouped into eight major communities and six sub communities. The results of the DCA ordination for all the plant communities are contained in Figure 5. From the DCA ordination axes 1 and 2 were selected as it was the most interpretable ordination. An Eigenvalue of 0.933 and 0.828 were obtained for Axis 1 and Axis 2 respectively.

The clay communities (communities 1–3) are positioned distinctly to the right of the ordination diagram. The communities which are located on predominantly sandy substrates (e.g. Community 4) are found on the extreme opposite end from the clay communities. The close proximity between Community 2 and sub community 6.2 is because both originate from the PP system. Sub communities 7.1, 7.2, and 7.3, all from the MS system, are affiliated with each other despite hydrological differences between the different zones. Sub community 5.1 and most of Community 8 originate from the PL System, explaining this association. The significant distance between sub community 5.1 (PL System) and 5.2 (IDD System) is as a result of the fact that they occur in different systems, despite similar environmental settings. Sub community 7.3 has a wide distribution, as some of its dominant species occur in other communities as well. Of these the graminoids *Stenotaphrum secundatum* and *Cynodon dactylon* are known to be variable in their habitat preference, and are not limited to a certain environment.

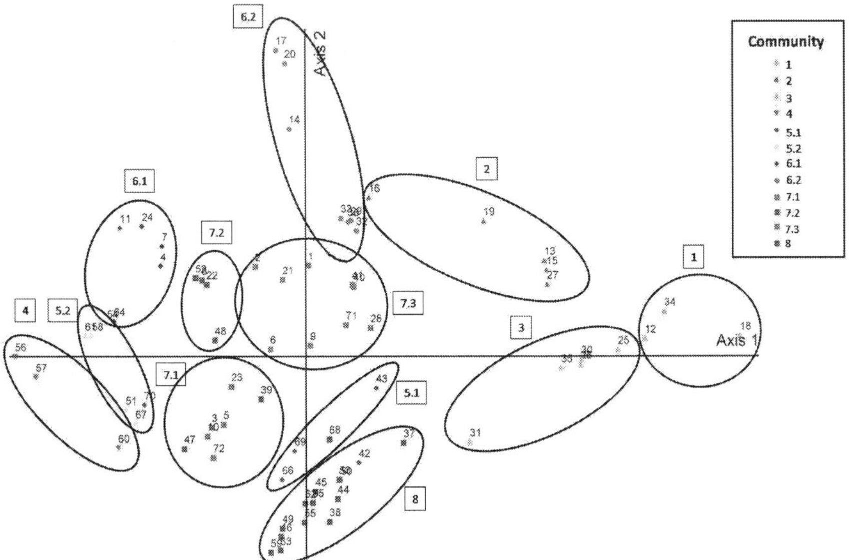

Figure 5. DCA ordination indicating clustering of communities.

4.1. Substrate overlay

Substrate has a strong influence on the spatial occurrence of vegetation communities and plant species. The clay communities (Communities 1 – 3 and 6.2) cluster strongly together with the communities occurring on duplex soil (sub communities 6.1 and some of 7.1) to a lesser degree so (Figure 6). A part of sub community 7.1 is found near the sand and high organic clusters because they are not only characterized by duplex soils, but also by higher organic matter content. The relationship between the plant species assemblages occurring on sandy and the high organic substrates are interesting, as the "High Organic" and the "Sand" communities form two overlapping clusters with a wide distribution. The "High Organic" cluster originates from Community 7 and is regarded as the "Organic MS System". This cluster is seasonally to permanently flooded, but because it originates from the MS System (which is characterized by clay lenses on the banks of the wetland) it occurs close to the cluster with the duplex substrates. Towards the bottom of Axis 2 the "Sand & High Organic" cluster contains Community 8 (seasonally and permanently flooded) (Figure 4) which originates from the sandy PL and IDD Systems.

4.2. Wetland system overlay

It was hypothesized that due to the divergent characteristics and environmental settings, the five wetland systems would contain plant communities entirely unique to each system. However, the dominant division was mainly between the two clay systems (PP and DP) and

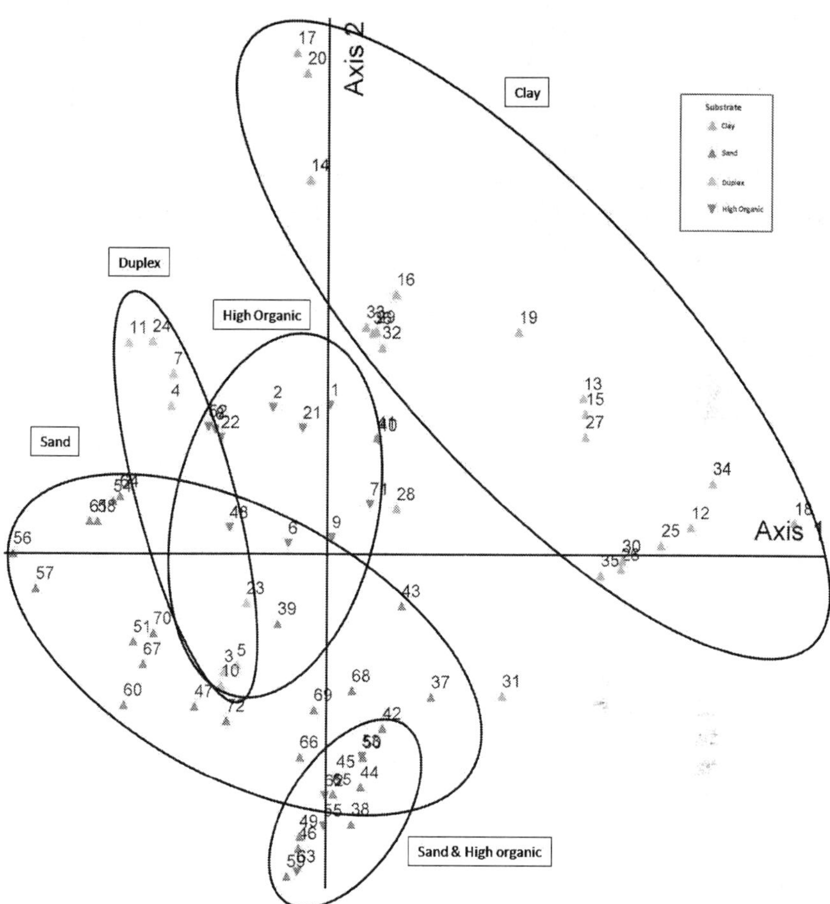

Figure 6. DCA ordination with substrate overlay.

the rest of the systems (MS, IDD & PL) (Figure 7), which obscured all other environmental distinctions between plant communities. Other differences in terms of vegetation composition could be ascribed to system characteristics such as substrate, geology, and hydrological regime. There exists no differentiation between the PL and IDD system, which is unexpected as these two systems are so distinct from each other.

In order to elucidate the relationship between the MS, IDD & PL systems, the clay PP and DP Communities (Communities 1-3 and 6.2) were eliminated, and the data analysed again. The communities on high organic substrates with a seasonally to permanently wet hydrological

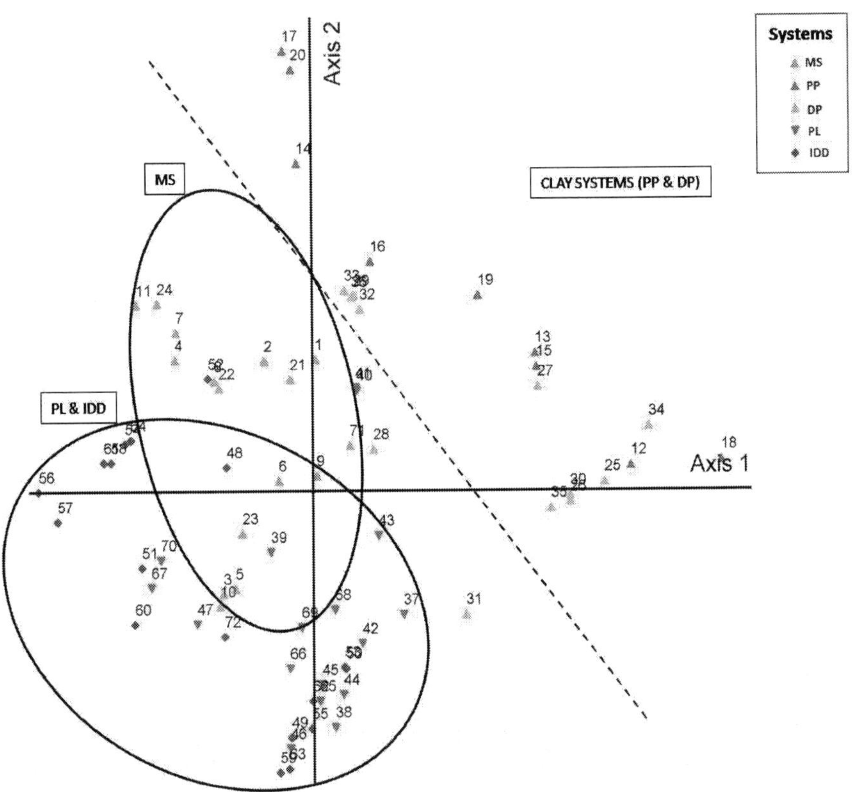

Figure 7. DCA ordination with Systems overlay.

regime (Communities 7 and 8) cluster to the left, while the communities on seasonal duplex soils and more terrestrial sandy soils (Communities 4 – 6.1) cluster to the right (Figure 8). In addition to the influence of the type of substrate on plant assemblages, there is therefore also a strong dry to wet influence.

4.3. Species richness

The species richness and average species per 4m^2 of each community is indicated in Table 1. The average species richness per 4m^2 is clearly lower in Communities 1 – 4 (the seasonal zone of the clay wetlands) than in the rest of the communities. There was a significant association between the number of plant species and plant communities present at X2(7)=382.35, p<0.0001. Sub communities 5.2 and 6.1 have exceptionally high species richness. The only environmental characteristic that these two sub communities have in common is that both communities are

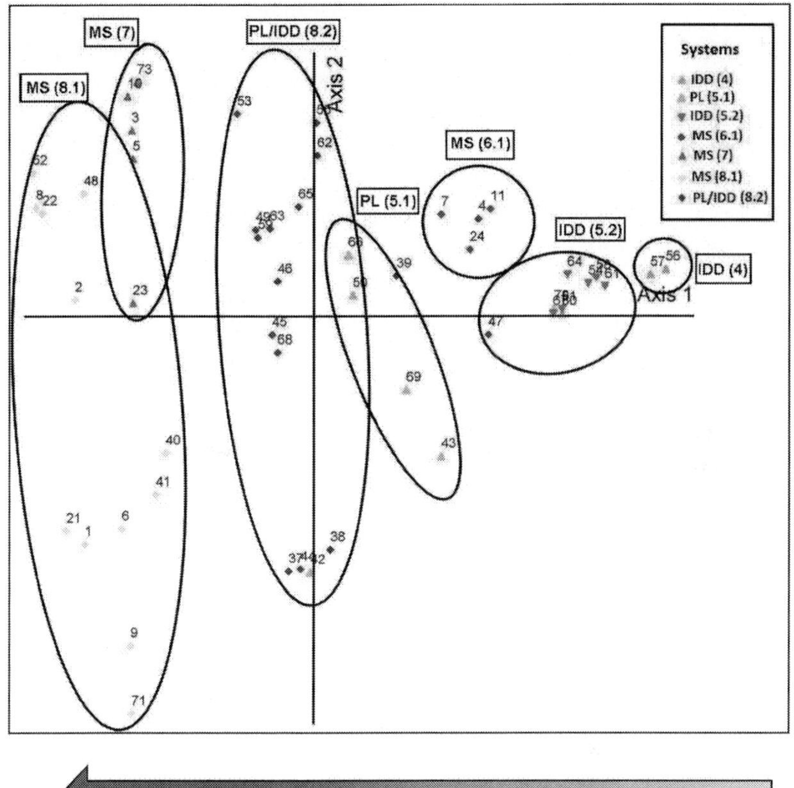

Figure 8. DCA ordination with community overlay of only the MS, PL, and IDD Systems which indicates a gradient from terrestrial to seasonally and permanently wet communities.

located well outside the wetland in the terrestrial zone. Based on the standardized residuals, species in Communities 1 to 4 are underrepresented (less than -1.96) and those in communities 5 to 8 are over represented (greater than 1.96).

The average species richness for the various systems is contained in Table 2. As in Table 1 it is clear that the clay PP and DP communities are more species poor that the rest of the wetland systems. The highly organic IDD system is significantly more species rich. Although the MS system also has high organic substrate, it contains many clay-related plant assemblages due to the clay lenses on the edges of the peat substrate.

The plant assemblages give a good indication of the wetness levels of the various zones. Additionally the peatlands on the MCP are permanently wet, and all connected to the groundwater table [17]. This was used to assign the various plant communities to the three wetness zones indicated in Table 3. Usually species richness decrease with increasing wetness levels as few plant species are adapted to waterlogged soil [12]. However, in this study species richness in the permanently wet (therefore high organic and peat substrate) zones is actually much higher than that of seasonally wet zones.

Community	Species richness	Average species per 4m²	Lower	Higher
1	9	3	2	4
2	32	6.4	4	10
3	22	4.4	3	7
4	19	6.33	5	7
5.1	58	9.7	5	13
5.2	103	17.17	14	24
6.1	59	14.75	12	19
6.2	75	10.7	8	17
7.1	33	6.6	4	10
7.2	26	6.5	2	10
7.3	79	8.78	4	14
8.1	70	11.7	6	17
8.2	67	7.44	4	10

Table 1. Species richness of each community.

System	Community	Average species per 4 m²
PP & DP	1; 2; 3; 6.2	6.1
MS	6.1; 7.1; 7.2; 7.3	9.2
IDD	8.1; 4; 5.2	11.7
PL	5.1; 8.2	8.6

Table 2. The average species richness for the various systems.

Zone	Community	Average species per 4 m²
Permanently wet	8.1; 7.3; 7.2	8.993
Seasonally wet	8.2; 7.1; 4; 1; 2; 3	5.695
Terrestrial	6.2; 6.1; 5.1; 5.2	13.08

Table 3. The average species richness for the various wetness levels.

5. Discussion

5.1. Muzi Swamp (MS) system

Characteristic plant species of the Muzi Swamp System

Terrestrial zone:*Acacia nilotica* & *Hyperthelia dissoluta*

Seasonal zone:*Imperata cylindrica*

Permanently wet zone:*Cladium mariscus, Phragmites australis, Stenotaphrum secundatum, Cynodon dactylon, Dactyloctenium aegyptium*

Typical plant communities: 6.1, 7.1, 7.2, 7.3

There is a distinct division between the terrestrial zones (sub community 6.1) and the permanently and seasonally wet zones of the MS System (Community 7). The grass *Imperata cylindrica* invariably characterizes the seasonal zone (sub community 7.1); even though this zone is very closely associated with the permanently wet zones (sub community 7.2 and 7.3). This community is described in Matthews *et al.* [31]. It also correlates with the "proximal-seasonally inundated floodplain" in Patrick & Ellery [30], in that it is functionally connected to the channel by being exposed to seasonal flood events and sedimentation.

The MS system is characterized by both a peat substrate which has a relatively high species richness as well as clay lenses on its edges which, in this study, has shown to have a lower species richness.

5.2. Perched Pans (PP) and Depressions (DP) clay systems

Characteristic plant species of the PP and DP (clay) systems (Figure 9)

Terrestrial zone:*Acacia nilotica, Acacia karroo, Justicia flava, Panicum maximum*

Seasonal zone:*Cyperus fastigiatus* (PP System) & *Echinochloa colona* (DP System)

Permanently wet zone:*Lemna gibba*

Typical plant communities: 1, 2, 3, 6.2

Matthews [30] describe two communities which occur on clay pans in the TEP-a "grassland on clay between thicket and pan marsh edges" Community, which does not correlate with what was found in the PP System; and a "*Nymphaea nouchali* aquatic vegetation in marshes and pans" which do correlate with the inundated zones found in the PP and DP Systems. There is a strong division between zone 1 (Community 1) and zone 2 (Community 2) of the PP System; as well as zone 1 (Community 1) and the seasonal zone 2 (Community 3) of the DP System. Community 3 is composed of many species that are regarded obligate hydrophytes such as *Marsilea sp., Pistia stratiotes,* and *Nymphaea nouchali,* yet it is regarded a seasonal zone. This classification of this community is as a result of the prominence of *Echinochloa colona* which didn't occur in open water, but in the area which is still waterlogged and able to host hydrophytic species such as those named above. *Echinochloa colona* is indicative of overgrazing and trampling [32], and occur in wetlands due to human influences. The PP System is utilized by animals of the Tembe Elephant Park as a water hole. The trampling of the pans by large animals decreases the open water zone and compacts the seasonal zone, destructing the habitat of the hydrophytes that could have occurred there.

The terrestrial zones cluster together into one community (Community 6), despite the differences that divide the inundated zones of the DP and PP System. Community 6 is far removed from Communities 1-3 (the wet and seasonal zones of the PP and DP Systems). The association of sub community 6.1 (MS System) with 6.2 (PP System) is based on the similarity of the substrate – a sandy topsoil underlain by a horizon with a significant increase in clay (duplex soil).

Figure 9. The floristic differences between the two clay systems with a) PP System and the b) DP System.

The species richness in Community 1 (the inundated community of both the PP and DP systems) is much lower than in Community 2 and 3 (the seasonal zones) and sub community 6.2 (the terrestrial zones). These results highlight the difference between species richness in high organic, fertile substrate versus the poor species richness on clay substrates, and supports the general rule of decreasing species richness with increasing wetness. As with the other systems the species richness is higher in the terrestrial zones, although still somewhat less in these clay systems as opposed to the other systems' terrestrial zones.

5.3. Interdune-depression (IDD) system

Characteristic plant species of the IDD system

The terrestrial zones of the IDD System have exclusive species assemblages. The disturbed community (Community 4) and the permanent-and seasonal zones, share many species with the PL System.

Terrestrial zone: *Themeda triandra, Trachypogon spicatus*

Permanently wet & seasonal zone:*Cladium mariscus, Cyperus natalensis, Hemarthria altissima, Thelypteris interrupta*

Disturbed seasonal zone:*Scleria sobolifer, Xyris capensis*

Typical plant communities: 8.1, 4, 5.2

The species composition of Community 4 illustrates the over-exploitation of the IDD system for the fertile, high organic substrate in the waterlogged areas. The permanently wet and seasonal zones are drained by trenches and drains through the wetland. The drainage lines stay visible, coining the term "fossil gardens" [33] (Figure 9 b). Only the seasonal zones from the disturbed wetlands form part of the disturbed Community 4. The permanently wet zone 1 doesn't floristically show disturbance as much as the seasonal zone, and seems to be buffered to some extent. A disturbance regime therefore has a much larger effect on the seasonal zones where the hydrological regime is variable.

The IDD and the PL Systems are generally grouped very closely together. The wet zones of both these systems occur close together in the ordination space, despite the IDD System having a permanently wet peat substrate and the PL System being periodically flooded open sandy plains. The only explanation that can be put forward for the similarity in species is that the peat might be shallow enough that the plants in the IDD System is rooted in the underlying sand substrate beneath the peat, and not necessarily in the peat itself. There is also a high similarity between the terrestrial zones of the PL (sub community 5.1) and IDD (sub community 5.2) systems, probably due to the sandy substrate of these terrestrial zones.

In the ordination results, however, these sub communities are far removed from each other. The zone differentiation in the IDD System is somewhat different to that of the other systems, as there is only a permanently wet zone and a terrestrial zone. Zones 1 and 2 cluster together in the highly organic, waterlogged community, and Zones 3 and 4 clusters together in the sandy grassland terrestrial community. This is because the transition between the permanently wet zone and the terrestrial zones is so sharp due to the steep slope of the depression, that the "seasonally wet zone" is just a small area at the slope foot. This zone is still high in organic carbon content, and therefore shares many species with the permanently wet zone.

The IDD system has the highest species richness of all the systems. This can be attributed to the high organic substrate that dominates this system. It is also devoid of clay, which seems to support a lower species diversity.

5.4. Upland Wetland (PL) System

Characteristic plant species of the PL system

Terrestrial zone:*Cyperus natalensis, Bulbostylis contexta*

"Wet" and seasonal zone:*Cyperus natalensis, Centella asiatica, Hemarthria altissima, Eragrosits heteromera*

Typical plant communities: 5.1, 8.2

Moist grasslands feature strongly in all vegetation studies done on the MCP, and are termed "hygrophilous grasslands". Various studies [9, 34, 35, 36] detail 'high water-table grassland' communities termed 'hygrophilous grasslands', which corresponds loosely to both The PL and IDD systems. All the above studies noted dominant occurrence of *Ischaemum fasciculatum*, which was not found abundantly in the PL System. The water table of the PL System (> 3 m deep) is in most areas not as high as the hygrophilous grassland communities described by the above authors. The absence of *I. fasciculatum* from the PL System might thus be a result of the variable hydrological regime. This argument is supported by Matthews [33] who states that *I. fasciculatum* is a species which reflects periods of inundation.

There is therefore a large overlap between the terrestrial zones and the wetter zones of the PL and IDD Systems (e.g. *Sorghastrum stipoides*, which occurs in high abundances and in Community 5 and sub community 8.2). The terrestrial zones of the PL System (sub community 5.1) are similar to the wet and seasonal zones of both the PL and IDD system (sub community 8.2), and not so much similar to the terrestrial zones of the IDD System (sub community 5.2). This is because the transition between the zones of the open PL System is much more gradual than that of the closed and sharply demarcated IDD System.

As a result of the hydrological regime and gradual zone transition, the zones of the PL System are difficult to delineate with certainty, and display a lot of species overlap. Still there is a strong division between the 'wet' zones and the terrestrial zones of the PL System. Zones 1 and 2 occur together as the 'Sandy Organic Grasslands' (sub community 8.2), and Zones 3 and 4 occur as the 'Terrestrial Sandy Grassland' (sub community 5.1). *Cyperus natalensis* occurred in most of the zones in the PL System, as well as in some IDD communities. *Centella asiatica* occurred abundantly in the wet zones of the PL and IDD Systems, but not at all in the terrestrial zones. These two species together seem to be indicative of some signs of 'wetland' conditions on sandy substrates (they were absent in the clay systems).

One of the biggest threats to seasonally wet, event-driven, rainwater-dependent, hygrophilous grasslands such as the PL System is a drop in the water table [9, 36]. This is mostly caused by afforestation, and can already be seen as the numerous informal plots of *Eucalyptus* trees (Figure 10 d). These hygrophilous grasslands are an essential and important part of the wetland catchment area of the Kosi Bay lake system and Lake Sibaya, and are also responsible for the recharge of the lower lying wetland areas such as the Muzi Swamp to the west and the numerous swamp forests occurring in the drainage lines to the east of the PL System [9, 17]. The drop of the water table over the past 20 years have had a significant effect on the PL System, and might explain the floristic and hydrological differences that exist between this system and the other described hygrophilous grasslands on the MCP.

Subsistence agriculture also poses a threat to the wetlands of the PL System (Figure 10 a). These gardens make use of the organic rich and moist soil in the wettest portions of the wetlands. No drainage lines are usually necessary, as the PL System is not permanently wet. Because it is mainly a sedge and grassland system, the vegetation removal to make space for crops is minimal. The effect of the gardens are thus less severe in this system, but a lot of the soil organic carbon still goes lost during the agriculture practices.

6. Conclusion

The results from this study indicate clear differences between the different wetland systems in terms of plant communities and species richness.

The clay systems (PP and DP) have three distinct zones:

- a wet zone (not permanently wet, but saturated for at least 6 months of the year);

- a seasonal zone; and

- a terrestrial zone.

The DP System has more vegetation zones than only three, but they cluster with the hydrological zones set out above.

The sandy and organic wetlands (including the duplex MS System) are characterized by more than three vegetation zones, which can be grouped into a permanently and seasonally wet, and a terrestrial hydrological zone:

- The permanently and seasonally wet zones were found to group together, with the terrestrial zones separately, due to the substrate type.

- The permanently and seasonally wet zones of the wetlands on the MCP are extremely high in organic carbon content, and thus have similar vegetation assemblages.

- The PL System also varies a bit, as there is a lot of overlap from zone 1 to zone 4.

Few of the communities, sub-communities and variants in this study are floristically associated with other vegetation communities described in the literature, probably due to the detailed scale of this wetland study. Although some vegetation studies have been conducted on the MCP, few have focused on wetlands specifically.

The statement by Matthews [31]: '...the important determinants of vegetation communities (are) the interconnected effects of water table (moisture), soil type and topography ....' is supported by this study. Although the specific type of wetland systems add to the various vegetation assemblages found, it does not account for all the differences encountered between vegetation communities. The main difference between vegetation compositions can be accounted for by the substrate type. In the ordination following the removal of the clay substrate type, the main division made was based on substrate (organic versus sand) and hydrological regime (a terrestrial group, and a combined seasonally and permanently wet group (Figure 8)). Although it is unclear at this stage which of these two factors is the main divisive factor, it is deemed unnecessary to investigate in detail as it is known that hydrological regime and organic content of soils are interlinked.

The specific type of system from which a relevé originates is the final classification factor. In certain instances the whole system is characterized by a specific substrate (such as the DP and PP systems), in which case it can be said that their vegetation types are limited to that specific system. The rest of the wetlands on the MCP occur on a predominantly sandy substrate, and species assemblages will therefore not be limited or exclusive to a specific wetland system (also the reason why the sandy IDD and PL systems are more associated with each other than with the somewhat duplex MS System). Vegetation composition of a specific wetland zone can therefore be influenced and driven on two levels:

1. by the substrate type and hydrological regime; and

2. by the wetland system it occurs in.

Plant species assemblages (communities) and species richness are therefore characteristic for the different wetland zones. However, zone delineation using vegetation composition varies between the different wetland systems in terms of amount and types of zones present, and should be evaluated according to the specific system in question. Not only can the different plant assemblages be used for the successful identification of the different zones within certain wetland types on the MCP, but all could be related to environmental conditions in the field.

Figure 10. Examples of wetland degradation on the MCP with a) cultivation in wetlands in the PL System; b) slash and burn and subsequent cultivation in the swamp forests; c) a destructed wetland now known as a 'fossil garden'; and d) one of numerous *Eucalyptus* plantations in a wetland on the MCP.

It is thought that the wetlands on the MCP are currently under stress as a result of drought and intensified forestation and agricultural practices on the MCP. These wetlands, especially the Upland Wetland (PL) System which act as a recharge area for the whole MCP [17], are extremely sensitive ecosystems. In the unprotected areas these wetlands are currently being exploited on a large scale for its goods and services (Figure 10). Human population increases are putting a demand on these resources which cannot be sustained. The Tonga community is dependent on the wetlands on the MCP. However, the current rate of uncontrolled utilization, with the added stress of the *Eucalyptus* plantations, could eventually cause these sensitive wetlands to become totally degraded with resultant loss of plant species and ecosystem functioning.

Author details

M.L. Pretorius[1], L.R. Brown[1*], G.J. Bredenkamp[1] and C.W. Van Huyssteen[2]

*Address all correspondence to: lrbrown@unisa.ac.za

1 Applied Behavioural Ecology and Ecosystem Research Unit, University of South Africa, Republic of South Africa

2 Department of Soil, Crop, and Climate Sciences, University of the Free State, Republic of South Africa

References

[1] South African National Biodiversity Institute, 2010, *Further Development of a Proposed National Wetland Classification System for South Africa*, The Freshwater Consulting Group (FCG), Pretoria.

[2] Grundling, A.T., 2009, 'Annexure A of Project Proposal', Agriculture Research Council – Institute for Soil, Climate and Water, Water Research Commission Project K5/1923.

[3] Moll, E.J., 1980, 'Terrestrial plant ecology', in M. N. Bruton & K. H. Cooper (eds.), *Studies on the Ecology of Maputaland*, Rhodes University Press and the Natal Branch of the Wildlife Society of Southern Africa, Rhodes University.

[4] Moll, E.J., 1978, 'The vegetation of Maputaland-a preliminary report of the plant communities and their present and future conservation status', *Trees in South Africa*, 29, 31-58.

[5] Morgenthal, T.L., Kellner, K., Van Rensburg, L., Newby, T.S. & Van der Merwe, J.P.A., 2006, *Vegetation and habitat types of the Umkhanyakude Node*, South African Journal of Botany, 72, 1 – 10.

[6] Tinley, K.L., 1985, 'Coastal dunes of South Africa. South African National Scientific Programmes Report', No. 109, CSIR, Pretoria.

[7] Tinley, K.L., 1976, 'The ecology of Tongaland' Wildlife Society of Southern Africa, Durban, pp. 1-40.

[8] Tinley, K.L., 1971, 'Determinants of coastal conservation: dynamics and diversity of the environment as exemplified by the Mozambique coast' Pretoria, SARCCUS, Proceedings SARCCUS symposium on Nature Conservation as a form of land use, pp125-153.

[9] Lubbe, R.A., 1997, 'Vegetation and flora of the Kosi Bay Coastal Forest Reserve in Maputaland, northern KwaZulu-Natal, South Africa', M.Sc., Dept of Botany, University of Pretoria, Pretoria.

[10] Venter, C.E., 2003, 'The vegetation ecology of Mfabeni peat swamp, St. Lucia, Kwazulu-Natal', M.Sc., University of Pretoria, Pretoria.

[11] Department of Water Affairs and Forestry, 2005, *A practical field procedure for identification and delineation of wetlands and riparian areas*, Pretoria, South Africa.

[12] Collins, N.B., 2005, *Wetlands: The basics and some more.* Free State Department of Tourism, Environmental and Economic Affairs.

[13] Kelbe, B. & Germishuyse, T., 2010, *Groundwater/surface water relationships with specific reference to Maputaland*, Water Research Commission Report No. 1168/1/10. Pretoria.

[14] Maud, R.R., 1980, 'The climate and geology of Maputaland', in M.N. Bruton & K.H. Cooper (eds.), *Studies on the Ecology of Maputaland*, pp. 1–7, Rhodes University and the Natal branch of the Wildlife Society of Southern Africa, Rhodes University.

[15] Mucina, L. & Rutherford, M.C., 2006, 'The vegetation of South Africa, Lesotho and Swaziland', Strelitzia 19, South Africa National Biodiversity Institute, Pretoria.

[16] Watkeys, M.K., Mason, T.R. & Goodman, P.S., 1993, '*The role of geology in the development of Maputaland, South Africa*' Journal of African Earth Sciences, 16, 205-221.

[17] Grundling, A.T. & Grundling, P., 2010, *Fourth Research Note: Groundwater Distribution on the Maputaland Coastal Aquifer (North-Eastern Kwazulu-Natal)*, Water Research Commission Project K5/1923.

[18] Colvin, C., 2007, Focus on CSIR research in water resource: Aquifer Dependent Ecosystems, 2007 Stockholm World Water Week, 13-17 August 2007.

[19] Van Wyk, A.E. & Smith, G.F., 2001, *Regions of floristic endemism in Southern Africa*, Umdaus Press, ISBN 1919766189, Pretoria.

[20] Grundling, P., Mazus, H. & Baartman, L., 1998, *Peat resources in northern KwaZulu-Natal wetlands: Maputaland*, Department of Environmental Affairs and Tourism Report no. A25/13/2/7, Pretoria.

[21] Grundling, P., 1996, 'Sustainable Utilisation of Peat in Maputaland, KwaZulu-Natal, South Africa', in G.W. Lottig (ed.), *Abstracts of the 10th International Peat Society Congress*, May 1996, Bremen.

[22] Westhoff V. & Van der Maarel E., 1978, 'The Braun-Blanquet approach' in R.H. Whittaker (ed.), *'The Braun-Blanquet approach'*, pp287-399, Dr W Junk Publishers, London.

[23] Hennekens, S.M., 1996, *TURBO(VEG): software package for input, processing and presentation of phytosociological data. User's guide*, IBN-DLO, Wageningen, and Lancaster University, Lancaster.

[24] Tichy, L., 2002, 'JUICE, software for vegetation classification' *Journal of Vegetation Science*, 13, 451-453.

[25] Brown, L.R., Du Preez, P.J., Bezuidenhout, H., Bredenkamp, G.J., Mostert, T.H.C. & Collins, N.B., 2013, 'Guidelines for phytosociological classifications and descriptions of vegetation in southern Africa', Koedoe 55(1), Art. #1103, 10 pages. http://dx.doi.org/10.4102/ koedoe.v55i1.1103

[26] Mccune, B. & Mefford, M.J., 1999, 'PC-ORD. Multivariate Analysis of Ecological Data, Version 4.0' MjM Software Design, Gleneden Beach, Oregon.

[27] Pretorius, M.L., 2011, 'A vegetation classification and description of five wetland systems and their respective zones on the Maputaland Coastal Plain', M.Sc., Dept of Environmental Science, University of South Africa, Pretoria.

[28] Welman C, Kruger F & Mitchell B. 2007. *Research Methodology* (3rd edition). Oxford University Press Southern Africa, ISBN 978-0-19-578901-0, Cape Town.

[29] Roleček, J., Tichý, L., Zelený, D. & Chytrý, M., 2009, *'Modified TWINSPAN classification in which the hierarchy respects cluster heterogeneity'*, Journal of Vegetation Science 20, 596–602. http://dx.doi.org/10.1111/j.1654-1103.2009.01062.

[30] Patrick, M.J. & Ellery, W.N., 2006, *Plant community and landscape patterns of a floodplain wetland in Maputaland*, Northern KwaZulu-Natal, South Africa, African Journal of Ecology, Vol 45, pp. 175–183.

[31] Matthews, W.S., Van Wyk, A.E., Van Rooyen, N. & Botha, G.A., 2001, *Vegetation of the Tembe Elephant Park, Maputaland, South Africa*, South African Journal of Botany, Vol. 67, pp. 573–594.

[32] Van Oudtshoorn, F., 2002, *Guide to grasses of Southern Africa*, Briza publications, Pretoria.

[33] Grundling, P., 2002, 'The quaternary peat deposits of Maputaland, northern KwaZulu-Natal, South Africa: Categorization, Chronology, and Utilization', MSc thesis, Dept Geology, Rand Afrikaanse Universiteit, Johannesburg.

[34] Matthews, W.S., Van Wyk, A.E. & Van Rooyen., 1999, *Vegetation of the Sileza Nature Reserve and neighbouring areas, South Africa, and its importance in conserving the woody grasslands of the Maputaland Centre of Endemism*, Bothalia, 29(1), 151-167.

[35] Van Wyk, G.F., 1991a, *Classification of the grassland communities of the Nyalazi State Forest*, Project 91513 (1990/91), Department of Water Affairs and Forestry.

[36] Van Wyk, G.F., 1991b, *The grass/shrubland communities of Sodwana State Forests and the Mosi State Land Complex*, Futululu Forestry Research Station, Mtubatuba.

[37] Kotze, D., 2011, '*The vegetation is changing, and what does it mean? Trying to understand wetland vegetation change in southern Africa*', National Wetland Indaba 2011, Drakensberg, South Africa.

Chapter 12

Marine Ecosystem Diversity in the Arabian Gulf: Threats and Conservation

Humood A. Naser

Additional information is available at the end of the chapter

http://dx.doi.org/10.5772/57425

1. Introduction

The Arabian Gulf is a marginal and semi-enclosed sea situated in the subtropical region of the Middle East between latitudes 24° and 30° N and longitudes 48° and 57° E (Figure 1). The Arabian Gulf constitutes part of the Arabian Sea Ecoregion, and represents a realm of the tropical Indo-Pacific Ocean (Spalding et al., 2007). It is a shallow sedimentary basin with an average depth of 35 m and a total area of approximately 240,000 km² (Barth and Khan, 2008).

Due to the high-latitude geographical position, the relative shallowness and the high evapo-ration rates, the Arabian Gulf is characterized by extreme environmental conditions. Sea temperatures are markedly fluctuated between winter and summer seasons (15 - 36°C). Salinity can exceed 43 psu and may reach 70-80 psu in tidal pools and lagoons. Therefore, marine organisms in the Arabian Gulf are living close to the limits of their environmental tolerance (Price et al., 1993).

Despite these harsh environmental conditions, the Arabian Gulf supports a range of coastal and marine ecosystems such as mangrove swamps, seagrass beds, coral reefs, and mud and sand flats (Naser, 2011a). These ecosystems contribute to the maintenance of genetic and biological diversity in the marine environment and provide valuable ecological and economic functions as they form feeding and nursery grounds for a variety of commercially important marine organisms.

However, these ecosystems are under ever-increasing pressure from anthropogenic activities that are associated with the rapid economic, social and industrial developments in the Arabian Gulf countries. The Arabian Gulf is considered among the highest anthropogenically impacted regions in the world (Halpern et al., 2008). The coasts of the Arabian Gulf are witnessing rapid industrialization and urbanization that contribute to the degradation of naturally stressed marine ecosystems. Coastal and marine environments are affected by intensive dredging and

reclamation activities, and several sources of pollution, including industrial waste, brine waste waters, ports and refiners, oil spills, and domestic sewage (Sheppard et al., 2010). These threats warrant the designation of the Arabian Sea Ecoregion, including the Arabian Gulf as 'critically endangered' by the International Union for the Conservation of Nature (IUCN) and the World Wildlife Fund (WWF) (http://wwf.panda.org).

Due to its unique environmental setting, the Arabian Gulf is increasingly receiving international scientific interest to study the effects of environmental extremes on marine organisms, and to investigate the potential impacts of future climate change on the ecological integrity of marine ecosystems (Riegl and Purkis, 2012; Feary et al., 2013). This chapter identifies valued ecosystem components in the Arabian Gulf, characterizes natural and anthropogenic impacts on these ecosystems, and suggests measures for conservation that might contribute to the protection of the fragile marine ecosystems in the Arabian Gulf.

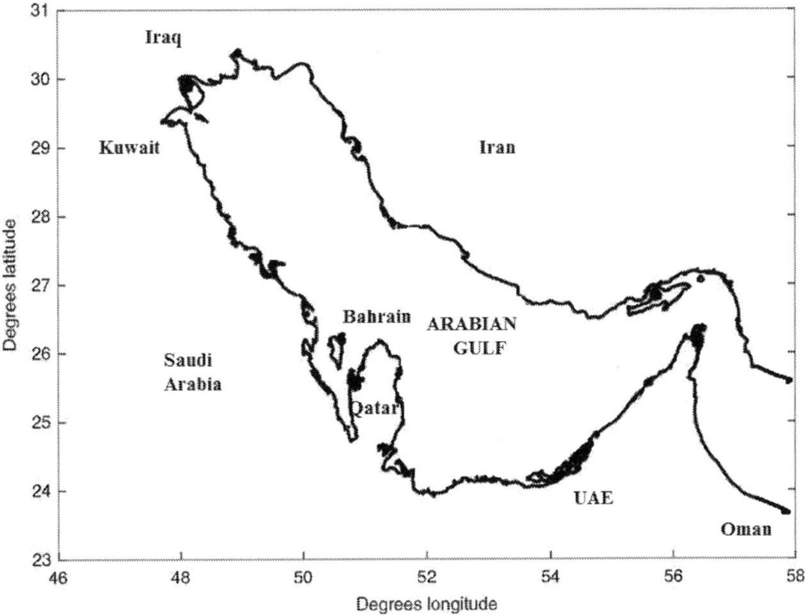

Figure 1. Map of the Arabian Gulf.

2. Valued ecosystem components

People of the Arabian Gulf are related economically, culturally and socially to the sea. Several ecosystems, including seagrass beds, coral reefs, mangroves, and mudflats contribute significantly to the productivity of marine resources in the Arabian Gulf (Khan et al., 2002). These ecosystems are considered Valued Ecosystem Components (VECs) because they provide important ecological goods and services (Treweek 1999). Most of these habitats are rich in varieties of fish, which are a major source of food for people in the Arabian Gulf. Ecosystem benefits in the Arabian Gulf are not limited to the direct consumptive value of seafood, but extend to other services ranging from primary production and nutrient cycling to erosion and sedimentation control.

2.1. Seagrass beds

Seagrass beds are highly productive ecosystems that provide important ecological functions and economic services (Sheppard et al., 1992). Ecologically, seagrass ecosystems provide food sources and feeding grounds for several threatened species in the Arabian Gulf such as turtles and dugongs (Abdulqader and Miller, 2012; Preen et al., 2012). They can also improve water quality by stabilizing loose sediment and by filtering some pollutants out of the water (Duffy, 2006). Economically, they serve as important nursery grounds for penaeid shrimps, pearl oysters and other organisms of importance to the Arabian Gulf's commercial fisheries (Erftemeijer and Shuail, 2012).

Three species of seagrass occur in the Arabian Gulf; namely, *Halodule uninervis*, *Halophila stipulacea* and *Halophila ovalis* (Phillips, 2003). These species are generally tolerant to salinity and temperature extremes of the Arabian Gulf. For instance, seagrasses are thriving in the extreme thermosaline conditions of Salwa Bay, south of Bahrain, where sea temperature exceeds 31 °C and salinity may reach 70 psu in summer.

Seagrass beds are distributed along most of the shores of the Arabian Gulf. According to Erftemeijer and Shuail (2012), around 7000 km² of seagrass habitat have been mapped in the Arabian Gulf to date. Extensive growth of seagrass is typically associated with sandy and muddy substrates in nearshores and shallow waters (less than 10 m). The largest areas of seagrass beds occur off the coasts of United Arab Emirates and between Bahrain and Qatar with estimated areas of 5500 and 1000 km², respectively (Erftemeijer and Shuail, 2012). These seagrass habitats are of critical importance as they support the largest population of dugongs known outside Australia (Preen, 2004).

2.2. Coral reefs

Coral reefs ecosystems provide a variety of ecological services such as renewable sources of seafood, maintenance of genetic, biological and habitat diversity, recreational values, and economic benefits such as utilizing reefs for creating land. Numerous nearshore reefs have been reclaimed along the coastline of the Arabian Gulf. Coral reefs are featured by both biological diversity and high levels of productivity. The high diversity of coral reefs provides a wide range of habitats for other reef species and fish. Coral reefs in the Arabian Gulf have traditionally been important habitats for fisheries.

Coral growth occurs in most of the Arabian Gulf with best development on offshore shoals. Additionally, fringing corals occur along the coastlines of United Arab Emirates, Qatar, Saudi Arabia and Bahrain (Riegl and Purkis, 2012). Extremes in temperature, salinity and other physical factors in the Arabian Gulf restrict the growth and development of corals to patchy forms (Sheppard et al., 2010). However, despite these harsh environmental conditions, corals in the Arabian Gulf exhibit remarkable resilience and vitality.

Recently, corals in the Arabian Gulf have been exposed to severe temperature anomalies at a recurrence faster than in any other coral regions in the world. Therefore, it is argued that corals in the Arabian Gulf already exist in a thermal environment that is equal to the 2099 projections of the Intergovernmental Panel on Climate Change (IPPC) in tropical oceans (Riegl and Purkis, 2012). This embarked regional and international interest in using the Arabian Gulf as a model ecosystem to understand the potential impact for future climate change (Burt, 2013). One aspect of that interest is the establishment the Mideast Coral Reef Society (MCRS) in 2013, with the aim of generating a deep understanding of these unique ecosystems and promoting their conservation and sustainable use.

Unfortunately, coral reefs in the Arabian Gulf have been severely affected by recent bleaching events as well as human impacts such as sediment runoff from dredging and reclamation activities and pollution from different land-based sources. Large-scale decline in coral reef has been observed. It is estimated that almost 70% of original reef cover in the Arabian Gulf may be considered lost and a further 27% threatened or at critical stages of degradation (Wilkinson, 2008).

2.3. Mangrove swamps

Mangrove habitats are ecologically important coastal ecosystems that provide food, shelter and nursery areas for a variety of terrestrial and marine fauna. Mangrove habitats of the Arabian Gulf support a variety of important species of fish, shrimps, turtles, and birds, and significantly contribute to the coastal productivity (Al-Maslamani et al., 2013). The Arabian Gulf coastlines are dominated by only one species of mangroves, *Avicennia marina*. The processes of osmoregulation and salt secretion allow *Avicennia marina* to cope with hypersalinity conditions in the Arabian Gulf (Dodd et al., 1999). Naser and Hoad (2011) investigated salinity tolerance and salt secretion in Bahrain and found that successful establishment of mangroves depends critically on the tolerance of these plants to salinity at the early stages of development.

Mangroves are largely distributed along the southern shores of the Arabian Gulf. Dense growth of mangroves is particularly confined to low-energy and sheltered coastal areas along the coastlines of United Arab Emirates, Saudi Arabia and Qatar. While mangroves are relatively widespread throughout the Arabian Gulf countries, there are variations in their distributions from one country to another. For instance, the total extent of mangroves in United Arab Emirates is estimated to be 38 km^2 (Dodd et al., 1999) compared with only 0.31 km^2 in Bahrain (Abido et al., 2011).

According to the IUCN Red List of threatened species, mangroves in the Arabian Gulf are classified as 'Least Concern'. The Red List assessment concluded that although there are overall range declines in many areas, they are not enough to reach any of the threatened category thresholds. However, it can be argued that this classification might not be applicable to certain countries in the Arabian Gulf, particularly Bahrain. It is recognized that the IUCN Red List categories and criteria provide objective framework for the classification of studied species. However, these categories and criteria are primarily designed for global taxon assessments (IUCN, 2012).

Bahrain is a good example to demonstrate the effect of rough resolution assessment (i.e. globally or regionally rather than locally). Based on the Red List, mangroves in Bahrain are classified as 'Least Concern'. However, mangrove stands in Bahrain are severely subjected to human impacts that might affect the existence of this important ecosystem. The marine area of Tubli Bay, which hosts the last remaining mangroves in Bahrain, has been reduced from to 25 to 12 km^2 in 2008 due to intensive reclamation activities. These activities significantly destroyed mangrove stands and reduced their spatial distribution in Bahrain. Due to the severe reduction in mangroves distribution accompanied by continuous anthropogenic threats, mangroves in Bahrain should arguably be classified within the threatened categories (i.e. Critically Endangered, Endangered or Vulnerable). Consequently, urgent conservation measures, including rehabilitation and restoration should be carried out to sustain the remaining of mangrove ecosystem in Bahrain.

2.4. Mudflats

Due to the sedimentary nature of the Arabian Gulf, sand and mud substrata are the most widespread habitats. They extend from intertidal salt marshes to the maximum depth and account for more than 97% of the bottom substrate of the Arabian Gulf (Al-Ghadban, 2002). Tidal mudflats are generally restricted to low energy environment associated with low water movement mainly along the coastlines of Kuwait, Saudi Arabia and United Arab Emirates. These habitats are favorable areas for the colonization by mangroves, algal and cyanobacterial mats, which play important roles in primary productions and food chains. Subtidal and tidal muddy habitats are extremely rich in macrobenthic assemblages, which form the largest and most diverse marine ecosystem in the Arabian Gulf.

Generally, biodiversity and distribution of macrobenthos in the Arabian Gulf are primarily governed by sediment type, temperature, salinity, primary productivity, depth and physical disturbance (Coles and MacCain, 1990). Macrobenthic assemblages through their high secondary productivity contribute significantly to the overall fisheries and marine productivity. Additionally, mudflat habitats provide feeding and roosting grounds for important shorebird populations. Some of these important bird areas are declared as wetlands of international importance (Ramsar Convention of Wetlands). Tubli Bay in Bahrain is an example of a Ramsar site that supports large numbers of wintering and migratory shorebirds (Al-Sayed et al., 2008).

3. Environmental stressors

The Arabian Gulf is facing multiple natural and anthropogenic environmental stressors. The unique physical and chemical settings represented by extremes in temperature and salinity accompanied by anthropogenic impacts may pose threats to marine species diversity and ecosystems integrity. The naturally stressed marine ecosystems are subjected directly or indirectly to human actions ranging from habitat destruction by coastal reclamation to pollution from a variety of land-based activities.

Environmental impacts on marine ecosystems could be generally attributed to natural or anthropogenic stressors. However, distinguishing between natural and anthropogenic stressors might be difficult due to the complexity of ecosystems responses to the variety of disturbances. For instance, anthropogenic impacts on ecosystems may not be detected until they are interacted with natural environmental stressors. Additionally, some environmental changes in ecosystems that appear to be natural may have been influenced by anthropogenic actions.

3.1. Natural stressors

Natural stressors in marine environment have a large number of forms and sources. Environmental extremes represent stressors that interfere with normal functioning of marine ecosystems (Breitburg and Riedel, 2005). The arid physical setting of the Arabian Gulf represented by extreme levels of salinity and temperature has pronounced effects on physiological aspects of marine organisms as well as their diversity, abundance and spatial distribution.

Physiological stresses are reflected in the dwarfism phenomena in fauna and flora inhabiting areas with high levels of salinity. For example, Price (2002) attributed the occurrence of dwarfism in echinoderms to the high salinity waters of the Arabian Gulf. Similarly, Naser and Hoad (2011) investigated morphological characteristics of mangroves in two distinctive coastal areas in Bahrain. The first area of mangroves receives input of low-salinity water from nearby farms, and underground seepage with salinity ranging between 5 psu in winter and 29 psu in summer. Salinity of the coastal water in the second area of mangroves can exceed 44 psu. This study reported differences in tree heights of mangroves ranging from 1.5 to 2.5 m in the coastal area compared with 4 to 5 m in the first area that receives low-salinity water.

Generally, extreme environmental conditions in the Arabian Gulf are reflected in reduced levels of species richness (Price, 2002). However, it is recognized that the Arabian Gulf has distinctive marine assemblages as well as habitats (Sheppard et al., 1992). Therefore, it can be argued that while species richness is relatively low, change in species composition along a spatial gradient is high (Price, 2002).

Stressors of biological sources such as invasive species and algal blooms could play important roles in ecosystems degradation in the Arabian Gulf. With more than 25 000 oil tankers navigating through the Strait of Hormuz each year (Literathy et al., 2002), the introduction of aquatic invasive species via ballast water is considered one of the major threats facing the marine environment in the Arabian Gulf. Hamza (2006) reported several exotic phytoplankton and zooplankton species in water samples collected from ballast water tanks of a gas tanker stopped along the United Arab Emirates coastal area. Some of these exotic species, particularly dinoflagellate organisms, are linked to the red tide and fish kill that frequently reported in recent years in Kuwait, Oman, Saudi Arabia and United Arab Emirates (Hamza and Munawar, 2009).

Extensive blooms (i.e. red tides) have been causing severe ecological and economical impacts in the Arabian Gulf. For instance, the massive blooms affected the Arabian Gulf from August 2008 to May 2009 caused widespread fish kills, damaged coral reefs, restricted fishing activities, impacted tourism industry, and interrupted desalination operations. The 2008-2009 harmful algal blooms were associated with the dinoflagellate species *Cochlodinium*, which was the first time to be observed in the Arabian Gulf waters (Richlen et al., 2010).

Effects of climate change and global warming on natural ecosystems and human well-being are major global concerns (Sheppard, 2006). Although climate change is directly or indirectly attributed to anthropogenic sources, it generally results from large-scale interactions for several variables over a very long time. There are three main features of the global climatic changes; namely, extreme sea-surface temperatures, marine acidification, and sea-level rise that may pose potential risks to marine ecosystems in the Arabian Gulf.

Although ecosystems in the Arabian Gulf are adapted to extreme environmental conditions, anomalous sea-surface temperatures due to climatic changes may result in severe impacts on the integrity of theses vulnerable ecosystems. The massive bleaching and subsequent mortality of corals in the Arabian Gulf occurred in 1996 and 1998 with maximum sea-surface temperatures reaching 37.3 °C and 38.0 °C, respectively (Sheppard and Loughland, 2002; Burt et al., 2011). Although most of the Arabian Gulf countries were affected by these beaching events, Bahrain was the worst affected with an estimated overall loss of 97% of live corals. Recovery of Bahrain reefs was limited in the subsequent years due to continuing coastal developments that are associated with intensive dredging and reclamation (Burt et al., 2013). Additionally, warmer waters can also lead to oxygen depletion and suffocation of marine organisms. Higher temperatures where attributed to the massive fish mortalities along the coasts of Qatar (Al-Ansi et al., 2002).

The harmful effects of increasing atmospheric levels of carbon dioxide (CO_2) and other greenhouse gases are reflected on the environment and human health. The Arabian Gulf is a major sink for atmospheric CO_2, which may lead to acidification of the marine environment. Measurements of pH concentration in surface waters of the Arabian Gulf over a four year period (2007-2010) indicated that waters are becoming increasingly acidic with time (Uddin et al., 2012). Increasing acidity in the marine environment is critical for several organisms, including corals, molluscs and calcareous phytoplankton.

Sea-level rise is another effect of global climatic change that poses threats on coastlines of the Arabian Gulf. Bahrain as a group of low land islands is particularly threatened by any sea-level rise due to global climatic change. Al-Jeneid et al. (2008) have predicted that 77 km^2 of Bahrain's land could be inundated with seawater in the case of a rise of the sea level by 0.5 m. These areas are harboring sensitive ecosystems such as mangroves, and intertidal mudflats.

3.2. Anthropogenic impacts

3.2.1. Reclamation and dredging

Coastal and marine environments in the Arabian Gulf are the prime target for most of the major housing, recreational, and economic developments (Naser et al., 2008). Coastal developments along the Arabian Gulf have accelerated at an unprecedented rate in the past decade to accommodate large-scale projects, including artificial islands, waterfront cities, ports and marinas (Khan, 2007).

Habitat destruction due to intensive reclamation and dredging activities is the prime threat for biodiversity loss and ecosystem degradation in the Arabian Gulf. It is estimated that more than 40% of the coasts of the Arabian Gulf have been developed (Hamza and Munawar, 2009). Examples of large-scale coastal developments in the Arabian Gulf include 'Palm Islands' and 'The World' in Dubai, United Arab Emirates, 'The Pearl' in Qatar, 'Al Khaleej' and 'Half Moon Bay' in Saudi Arabia 'Pearl City' in Kuwait, and 'Durrat Al Bahrain', 'Amwaj', and 'Dyar Al Muharraq' in Bahrain.

It is likely that reclamation will accelerate in the coming decades in order to secure land for large-scale projects as populations in the Arabian Gulf countries continue to grow. This is reflected in the long-term land use strategies and master plans in the Arabian Gulf. For instance, Bahrain National Land Use Strategy 2030 recognizes reclamation as the major option for securing the future needs for land, indicating that coastal environment will continue to be the major focus for developmental projects in the coming future (Naser, 2011b).

Given its limited land area (762 km^2), Bahrain has markedly been affected by coastal developments. Presently, reclamation activities in Bahrain resulted in the addition of around 95 km^2 representing an increase of 12% of the total land area (Naser, 2011b). Additionally, more than 80 % of Bahrain's coastline has extensively been modified due reclamation activities (Fuller, 2005).

Typically, reclamation in the Arabian Gulf is conducted by extracting sand and mud from designated marine borrow areas then dumping them into the coastal and shallow subtidal areas (Figure 2). Alternatively, reclamation could be carried out by infilling the coastline by rocks and sands extracted from quarries (Figure 3).

Dredging and reclamation processes are associated with short and long term biological, physical and chemical impacts. These activities involve the direct removal of macrobenthos and permanent modification of the physical environment. Deposition of dredged material during reclamation process may result in physically smothering the coastal and subtidal habitats and deoxygenating the underlining sediments (Allan et al., 2008). Reclaimed lands could also interfere with water circulation and subsequently alter the salinity (Al-Jamali et al., 2005). These physical and chemical alternations may reduce biodiversity, richness, abundance and biomass of marine organisms (Tu Do et al., 2012).

Additionally, dredging activities may contribute directly or indirectly to the loss of seagrass beds in the Arabian Gulf due to direct physical removal and burial, and the increase in turbidity levels (Al-Wedaei et al, 2011). Dredging and reclamation activities have resulted in the loss of many prime mudflats that support shorebird populations (Figure 4), and the degradation of coral reefs due to sediment runoff and the increase levels of turbidity (Al-Sayed et al., 2008).

Figure 2. Sand and mud materials are pumped from a marine burrow area into the reclamation site. 'Dyar Al Muharraq' development in Bahrain (2013).

3.2.2. Industrial effluents

The Arabian Gulf countries have witnessed a rapid industrial growth, mainly in the sectors of oil refining and petrochemical industries. These major industries are discharging wastewater containing a variety of chemicals, including heavy metals, hydrocarbon compounds, and nutrients (Sale et al., 2010). Petroleum refinery wastewaters are composed of different chemicals, which include oil and greases, phenols, sulphides, ammonia, suspended solids, and heavy metals like chromium, iron, nickel, copper, molybdenum, selenium, vanadium and zinc (Wake, 2005). Coastal and marine environments receiving intensive industrial effluents along the coastline of the Arabian Gulf are recognized as hotspots for high concentrations of hydrocarbons (De Mora et al., 2004; 2010) and heavy metals (Naser, 2012a; 2013a).

Naser (2013b) investigated the effects of industrial wastewater discharges that characterized by high inputs of heavy metals and hydrocarbons on crustacean assemblages along the eastern coastline of Bahrain. This coastline is heavily occupied by industrial facilities including the oil refinery, aluminum smelters and desalination plants. This study indicated that the analyzed heavy metals exhibited higher levels of concentrations in sediments influenced by industrial discharges. The study also argued that the synergistic effects of industrial effluents that contain

Figure 3. Rocks and sands extracted from nearby quarries are used to reclaim a coastal area along the eastern coastline of Bahrain (2012).

Figure 4. A mudflat along the northern coastline of Bahrain that supports wader birds is proposed to be reclaimed, which my result in a loss of important feeding grounds for bird populations (2012).

high levels pollutants, brine discharges, and sedimentation due to intensive dredging and reclamation activities were reflected on the reduced levels of crustacean diversity and abundance in the sampling stations.

The flushing time of seawater in the Arabian Gulf is ranging between 3 and 5 years. Therefore, pollutants, including heavy metals and hydrocarbons are likely to reside in the Arabian Gulf for a considerable time. Continuous inputs of industrial effluents from different anthropogenic sources in the Arabian Gulf could be critical for both marine ecosystems and humans who rely on marine resources for food, recreation and industry.

3.2.3. Desalination effluents

The Arabian Gulf countries are witnessing rapid industrial development and population growth, which increase the need for fresh water (Smith et al., 2007). Due to the low precipitation and high aridity in the Arabian Gulf countries, most of the fresh water needs are being obtained from seawater through the various processes of desalination, including multi-stage flash (MSF), and seawater/brackish reverse osmosis (RO) (Hashim and Hajjaj, 2005). It is estimated that the amount of desalinated water in the Arabian Gulf countries accounts for more than 60% of the world's total production (Lattemann and Hopner, 2008).

Large quantities of reject water from desalination plants are being discharged on a daily basis to coastal and subtidal areas in the Arabian Gulf. Therefore, Hypersaline water discharges from desalination plants are increasingly becoming a serious threat to marine ecosystems in the Arabian Gulf (Areiqat and Mohamed, 2005).

Coastal and marine environments receiving these discharges are typically subject to chemical and physical alterations. Desalination effluents are commonly associated with harmful chemical components, including heavy metals, antiscaling, antifouling, antifoaming, and anticorrosion additive substances (Lattemann and Hopner, 2008). Additionally, discharges from desalination processes may alter physically and chemically the characteristics of receiving seawater, including water temperature and salinity. These alterations in seawater quality, temperature, dissolved oxygen and salt concentration may severely affect several marine organisms and assemblages.

Naser (2013c) investigated the effects of tow major desalination plants that use MSF and RO technologies on macrobenthic assemblages. The study found reduced levels of biodiversity and abundance in areas adjacent to the outlet of MSF reflecting severe impacts on macrobenthic assemblages caused by brine effluents that are associated with high temperatures, salinities, and a range of chemical and heavy metal pollutants (Figure 5).

The demand for desalinated water in the Arabian Gulf will increase in the coming future (Dawoud and Al-Mulla, 2012). This may result in cumulative impacts from the brine discharges leading to substantial fluctuations in salinity levels. It is forecasted that brine discharge will increase the salinity of the Arabian Gulf (Smith et al., 2007). Bashititalshaaer et al. (2011) predicted that brine discharge will increase the salinity of the Arabian Gulf by 2.24 g l^{-1} in 2050. This marked increase in seawater salinity could arguably be critical to the naturally stressed marine ecosystems in the Arabian Gulf.

Figure 5. Marine organisms at the proximities of desalination plants are influenced by chemical, physical and thermal pollution. Reduced levels of diversity and abundance of macrobenthos were recorded adjacent the outlet of this MSF desalination plant along the eastern coastline of Bahrain (2012).

3.2.4. Sewage discharges

Sewage effluents are considered one of the most common anthropogenic disturbances of marine ecosystems in the Arabian Gulf. Despite high standards of sewage treatment (i.e. secondary or tertiary) (Sheppard et al. 2010), large quantities of domestic effluents are discharged to coastal and marine environments in the Arabian Gulf. These effluents are characterized by high-suspended solids and high loads of nutrients such as ammonia, nitrates and phosphates (Naser, 2011a). Sewage effluents are generally accompanied by biological and chemical pollutants, including pathogen microorganisms and heavy metals (Shatti and Abdullah, 1999). Bioaccumulation and biomagnification of pathogenic organisms and chemical contaminants due to sewage discharges affect the quality of human food and subsequently pose threat to human health.

Shallow subtidal areas and semi-enclosed embayments are the receiving environments for most of the sewage discharges in the Arabian Gulf, which can cause localized eutrophication, nutrient enrichment and oxygen depletion. Kuwait Bay in Kuwait and Tubli Bay in Bahrain have witnessed several eutrophication conditions and fish mortality phenomena due to excessive sewage discharges (Al-Ansi et al., 2002; Glibert et al., 2002). Naser (2013b) studied crustacean assemblages influenced by sewage discharges from a major treatment station in Bahrain. The study reported a reduction in biodiversity, richness and evenness of crustaceans reflecting severe habitat degradation in the nearby marine environment. Additionally, influenced areas were characterized by marked increase in organic enrichment, mainly ammonia and phosphate.

3.2.5. Oil pollution

The Arabian Gulf is considered the largest reserve of oil in the world (Literathy et al., 2002). Consequently, coastal and marine environments in the Arabian Gulf are under permanent threat from oil related pollution. Oil exploration, production, and transport have been major contributors to pollution in the Arabian Gulf. Sources of oil spills in the Arabian Gulf include offshore oil wells, underwater pipelines, oil tanker incidents, oil terminals, loading and handling operations, weathered oil and tar balls, illegal dumping of ballast water, and military activities (Sale et al., 2010).

The Arabian Gulf has been a scene for major oil spill incidents in the world. Bahrain experienced one of the earliest major oil spills in the Arabian Gulf in 1980. A large oil slick (20,000 barrels) invaded the north and west coasts of Bahrain causing severe ecological and economical damages (Brown and James, 1985). This major incident was a precursor for establishing the first governmental authority concerned with the protection of environment in Bahrain in 1980; namely, the Environmental Protection Committee (EPC). The most notorious oil spill reported in the Arabian Gulf occurred during the 1991 Gulf War. An estimated 10.8 million barrels of oil were spilled in the Arabian Gulf waters (Massoud et al., 1998).

Oil pollution adversely affects marine ecosystems by reducing photosynthetic rates in phytoplankton and marine algae, accumulating toxic chemicals in several benthic organisms, and contaminating human food chains with carcinogenic substances. Seabirds and intertidal waders are predominantly vulnerable to oil pollution. For instance, several seabirds suffered severe mortality (22–50%) during the 1991 oil spill in the Arabian Gulf (Evans et al., 1993). Environmental consequences of long-term chronic oil pollution include degradation of sensitive ecosystems such as seagrass bed, coral reefs and mangroves, which may subsequently lead to decline in fish stocks and other renewable marine resources.

4. Conservation and management

Conservation biology is an integrated, multidisciplinary scientific field that has developed in response to the challenge of preserving species and ecosystems. Valued ecosystem components in the Arabian Gulf are facing several challenges due to habitat destruction, fragmentation, degradation and pollution. These are reflected in the decline in regional coral reefs due to natural and anthropogenic stressors, the loss of prime mudflats and mangroves swamps and seagrass beds due to intensive dredging and reclamation activities and anthropogenic effluents. Therefore, effective conservation and management of marine ecosystems in the Arabian Gulf is becoming an urgent need in order to protect and sustain these vulnerable ecosystems. Additionally, effectively managed ecosystems provide a range of essential environmental services that contribute to economic, social and cultural aspirations in the

Arabian Gulf (Al-Cibahy et al., 2012). This section therefore suggests conservation approaches and management strategies that might contribute to the protection of the fragile marine ecosystems in the Arabian Gulf, including marine protected areas, Environmental Impact Assessment (EIA), environmental regulations, ecological restoration, and environmental monitoring and scientific research.

4.1. Marine protected areas

A marine protected area (MPA) is defined by the International Union for Conservation of Nature (IUCN) as "any area of intertidal or subtidal terrain, together with its overlying water and associated flora, fauna, historical and cultural features, which has been reserved by law or other effective means to protect part or all of the enclosed environment" (Dudley, 2008).

Marine protected areas (MPAs) are globally recognized as the most important tool for in situ conservation (Chape et al., 2005). MPAs contribute significantly to both preservation and conservation of genetic characteristics, species, habitats and cultural diversity in coastal and marine environments. They can help in preventing or reducing the ongoing declines in marine biodiversity, habitats and fisheries productivity. MPAs can also improve ecosystem functions and services through maintaining ecological structure and processes that support economic and social uses of marine resources (Agardy, 1994). Additionally, MPAs can contribute towards climate change adaptation by protecting ecosystem resilience and protecting essential ecosystem services (McLeod et al., 2009).

Various relevant international conventions including Convention on Biological Diversity, Convention on Wetlands of International Importance (Ramsar Convention), and World Heritage Convention serve to advance the number and coverage of MPAs worldwide (Green et al., 2011). Similarly, regional conventions may promote the conservation benefits of marine protected areas in the Arabian Gulf. For instance, the Convention on the Conservation of Wildlife and their Natural Habitats in the Gulf Cooperation Council Countries (Bahrain, Kuwait, Oman, Qatar, Saudi Arabia, and United Arab Emirates) provides the basis for integrating protected areas into national and regional environmental strategies and polices (GCC, 2010). This convention aims to effectively conserve ecosystems and wildlife habitats. It is also concerned with the protection of threatened species on a regional levels, especially when the distribution of these species exceed the international borders of two or more neighboring countries or when these species migrate across the boundaries of the member states.

The coastal and marine areas extended from Gulf of Bahrain to the United Arab Emirates have been identified as a potential transboundary marine protected area (Knight et al., 2011). These areas are shared by four countries (Saudi Arabia, Bahrain, Qatar, and United Arab Emirates) and characterized by high levels of species and habitat diversity.

Higher priority of conservation is typically given for areas that characterized by distinctiveness, endangerment, and utility features (Primack, 2010). These characteristics are reflected on the proposed transboundary marine protected area. This area supports distinctive species, population and habitats, including vulnerable mega-fauna such as dugongs, turtles and dolphins, and ecological complex of seagrass beds, coral reefs and fisheries. However, it is susceptible to anthropogenic threats due to coastal developments and pollution from a variety of land-based sources. Further, this area provides economic, touristic, cultural and educational benefits to the local people in the Arabian Gulf.

The archipelago of Hawar Islands, which is located in the Gulf of Bahrain, is characterized by varied coastal habitats, including muddy, sandy, and rocky shores as well as saline wetlands known locally as 'sabkha'. These islands are surrounded by shallow waters, which promote the growth of extensive seagrass beds and algal mats. These habitats support large populations of dugongs, green turtles and dolphins (Preen, 2004). Additionally, Hawar Islands provide undisturbed habitats for a variety of avian fauna. These islands host the largest breeding colonies of the endemic Socotra Cormorant (*Phalacrocorax nigrogularis*) in the world, with a winter population of 200,000 individuals (King, 1999). Due to the remarkable diversity in habitats and their associated fauna and flora, Hawar Islands were declared nationally as a wildlife sanctuary in 1996, and internationally as a Ramser site (convention on wetlands of international importance) in 1997.

Designation and implementation of MPAs are arguably critical for the protection of naturally stressed coastal and marine ecosystems in the Arabian Gulf. Toward this, about 38 officially designated MPAs covering around 18,180 km² have been established in the Arabian Gulf (Van Lavieren et al., 2011). However, number and coverage of MPAs may not provide an indication of the effectiveness of these MPAs in achieving their conservation goals (Chape et al., 2005).

Van Lavieren and Klaus, (2013) evaluated the management effectiveness of MPAs in the Arabian Gulf and revealed variable levels of performance. Several weaknesses in the MPAs in the Arabian Gulf were identified, notably the limitation in regulation enforcement, the lack of management plans, and the weak communication with local stakeholders, traditional communities, and local marine resource users (Van Lavieren and Klaus, 2013).

Local communities are recognized as the key focus for the success of conservation initiatives (Kideghesho et al., 2007). Public understanding, support and participation are important for the success of marine protected areas as a conservation management tool (Jameson et al., 2002). This could be promoted through reviving the concept of 'Hima' in the Arabian Gulf. Hima is considered a community based environmental resource management system that could help in building understanding and acceptance of protected areas and promoting the need to conserve and use marine resources wisely (Knight et al., 2011; Van Lavieren et al., 2011).

4.2. Environmental Impact Assessment

Environmental Impact Assessment (EIA) is a systematic process of identifying, predicting, evaluating and mitigating the environmental consequences of a proposed project on the biological and physical environments. EIA aims at integrating environmental considerations in the decision-making system, minimizing or avoiding adverse impacts, protecting natural systems and their ecological processes, and implementing principles of sustainable development (Glasson et al., 2005).

EIA is considered a standard tool for decision-making in most countries throughout the world. It ensures that authorities are provided with necessary knowledge relating to any likely significant effects of a proposed project on the environment prior to the decision-making process. The integration of environmental considerations may result in a rational and structured decision-making process that maintains a balance of interest between the development action and the environment (Glasson et al., 2005). EIA minimizes or avoids the adverse effects of a proposed development on the environment by addressing effective designs, alternatives, mitigations, cumulative impacts, and monitoring (Cooper and Sheate, 2002).

Since the early stages of incorporating EIA in The National Environmental Policy Act in 1969, in the USA, considerations to protect ecosystems and biodiversity of natural resources and habitats have been an integrated part of the EIA process (Gontier et al., 2006).

EIA is one of the frequently used approaches in coastal planning and management (Kay and Alder, 2005). It is considered as an effective tool to minimize anthropogenic impacts and to induce the implementation of protection measures of coastal environment. The importance of EIA in protecting biodiversity and promoting the sustainable use of coastal and marine resources is represented in its fundamental role as a process for predicting the environmental effects of projects or programmes in coastal and marine areas. Therefore, EIA can be used to ensure that necessary measures needed to protect biodiversity and its sustainable use are addressed in the process of development planning (Khera & Kumar, 2010).

Additionally, EIA involves in facilitating consultation between various stakeholders as well as the public, considering alternatives for projects and locations, ensuring early identification of potential effects on coastal and marine environments, and implementing mitigation and compensation measures (Badr et al, 2004). Consequently, an effective EIA can contribute to the protection of biodiversity and to the sustainable use of coastal and marine environments in the Arabian Gulf.

Recognizing the role of EIA in protecting environment from degradation and pollution associated with rapid economic developments, Arabian Gulf countries have adopted EIA in their environmental policies (El-Fadl and El-Fadel, 2004). Coastal development projects, including reclamation and dredging activities, are required to be subjected to EIA in the Arabian Gulf. However, effectiveness of EIA in coastal and marine environments is constrained by many factors that are also common in many other regions in the world. These include lack of adequate legal and regulatory frameworks, limited public participation, inadequate guidelines on procedural EIA, and lack of provisions related to cumulative impacts and strategic environmental assessment (Van Lavieren et al., 2011; Naser, 2012b).

Multiple anthropogenic stressors can lead to cumulative impacts on marine ecosystems (Crain et al., 2008). Coastlines of the Arabian Gulf are witnessing a rapid increase in the number and scale of coastal developments. The negative effects of these several separate developments may synergistically combine, additively or multiplicatively, to destroy biodiversity and marine ecosystems in the Arabian Gulf.

Planning of dredging and reclamation activities is typically carried out in the Arabian Gulf at a project-by-project basis, without assessing environmental impacts strategically. This approach may ignore the cumulative impacts of coastal reclamation on the valued ecosystem components in the Arabian Gulf. For instance, several reviewed EIA reports indicated that the allocated sites of their projects were already impacted or degraded due to surrounding existing and ongoing projects (Naser, 2012b). Therefore, maintaining sustainable use of coastal and marine natural resources in the Arabian Gulf requires measures to holistically address the interactions among the several reclamation and dredging activities and their additive and cumulative impacts on valued ecosystem components.

Strategic Environmental Assessment (SEA), a tool to integrate environmental considerations into decision-making, may contribute toward achieving environmentally sound and sustainable development. SEA is defined as the process of evaluating the environmental consequences of proposed policies, plans and programmes, and addressing them into higher-level decision-making systems (Lamorgese and Geneletti, 2013). SEA has emerged as an important element in environmental decision-making process in developed countries, including Europe and North America. However, SEA is still relatively new or need to be introduced in the Arabian Gulf countries (Rachid and El-Fadel, 2013).

The need for a more strategic approach to environmental assessment in the Arabian Gulf can be illustrated by reference to coastal developments that associated with dredging and reclamation. The project-level EIA has been criticized for failing to ensure adequate considerations for potentially severe indirect, cumulative and synergistic environmental impacts on coastal and marine ecosystems (Naser et al., 2008).

Several reclamation and dredging activities are increasingly taking place within a relatively small geographical range on coastlines of the Arabian Gulf countries, which could have several cumulative consequences on the coastal and marine environments. SEA has the potential to promote sustainable development in coastal and marine environments through identifying cumulative impacts of exiting or planned projects, investigating feasible alternatives to coastal developments, and implementing effectively mitigation and compensation measures (Duisk et al., 2006).

SEA also has the advantages of integrating the coastal concerns into planning policies, facilitating consultation between various organizations as well as the public. Additionally, SEA can identify social, economic, and environmental issues associated with coastal development in the Arabian Gulf, and subsequently assist in the implementation of an important principle of sustainability (Barker, 2006).

Nonetheless, similar to many countries in the world (Liou and Yu, 2004), there are difficulties and challenges associated with the implementation of SEA in the Arabian Gulf. These include introducing and enforcing SEA law provisions, producing SEA related guidelines, clarifying administrative and procedural responsibilities of concerned bodies in SEA, institutionalizing networks, and encouraging public participation.

4.3. National, regional and international environmental regulations

Environmental legislations related to pollution prevention and biodiversity protection in the Arabian Gulf are based on a range of national laws and regulations as well as regional and international agreements. Nationally, there are several framework laws with respect to protecting the wildlife and their environment and combating environmental pollution in each country. These general environmental laws facilitate the implementation of related regional and international regulations and agreements (Khan and Price, 2002).

Several laws and regulations dealing with protection of environment and biodiversity have been developed in the countries of the Arabian Gulf. These national instruments include laws with respect to environment, exploitation and protection of living marine resources, protection of wildlife and natural environment, environmental quality standards, environmental assessment, prevention of oil pollution, banning of catching endangered species, and establishment of marine protected areas. Although these national laws can, directly or indirectly, contribute to the protection of marine ecosystems in the Arabian Gulf, their effectiveness might be restricted by the lack of enforcement (Al-Awadhi, 2002).

Regionally, the Kuwait Regional Convention for Cooperation on the Protection of the Marine Environment from Pollution (Kuwait Convention), which was established in 1978, provides the basis for an integrated regional response to protecting biodiversity and combating pollution (Khan, 2008). The Regional Organization for the Protection of the Marine Environment (ROPME) was established under the Kuwait Convention to act as a focal point for regional cooperation (Khan and Price, 2002).

Currently, there are four Protocols under the Kuwait Convention; namely, protocol concerning regional cooperation in combating pollution by oil and other harmful substances in cases of emergency, protocol concerning marine pollution resulting from exploration and exploitation of continental shelf, protocol for the protection of the marine environment against pollution from land-based sources, and protocol on the control of marine transboundary movements and disposal of hazardous wastes. These protocols collectively address the pollution of marine environment and propose criteria for protection and management of ecosystems and marine resources (Khan and Price, 2002; Khan, 2008).

Internationally, the Convention on Biological Diversity (CBD) provides a legal, scientific and practical mechanism for biodiversity conservation. The CBD requires member states to develop national strategies and action plans for the conservation and sustainable use of biodiversity, integrate biodiversity into the relevant plans, programs and policies, identify activities likely to have significantly adverse impacts on biodiversity, develop a system of protected areas to conserve biodiversity, integrate consideration of conservation into national decision-making systems, and introduce environmental impact assessment to avoid or minimize adverse impacts of proposed projects on biodiversity.

Additionally, the Arabian Gulf countries have accepted or ratified several international agreements that can contribute to the protection of the coastal and marine environments. These include, among others, Convention on Wetlands of International Importance (Ramsar Convention), World Heritage Convention, United Nations Convention on the Law of the Sea, United Nations Framework Convention on Climate Change, International Maritime Organization (IMO) conventions, CITES (the Convention on International Trade in Endangered Species of Wild Fauna and Flora). These international agreements provide mechanisms for dealing with many aspects and concerns relating to the marine environment, and consequently contributing to the protection and conservation of marine ecosystems in the Arabian Gulf.

4.4. Ecological restoration

The Arabian Gulf is experiencing a substantial loss of productive habitat and ecosystem services (Price et al., 2012). Dredging and reclamation have transformed extensive coastal areas into artificial environments. Additionally, several types of pollution are contributing to ecosystem degradation. Ecological restoration is an approach that could help in minimizing or reversing the decline in ecosystem integrity in the Arabian Gulf.

Ecological restoration is described as an assisted recovery of degraded, damaged or destroyed ecosystems (Clewell and Aronson, 2006). Ecological restoration is increasingly playing an important role in conservation biology (Young, 2000). Ecological restoration provides the opportunity to conduct experiments to understand the structures and functions of ecosystems. Insight from such research can be invaluable for the conservation and the management of natural resources.

Although marine restoration lags behind terrestrial and freshwater counterparts (Elliott et al., 2007), restoration activities are increasingly conducted in coastal and marine environments worldwide. Similarly, several restoration projects have been conducted in the Arabian Gulf (Weishar, 2008). Planting projects to restore mangrove ecosystems have been conducted in most of the Arabian Gulf countries. The success of planting mangroves depends critically on the topographical and hydrological conditions of the selected site, including low energy shorelines with stable and non-eroding soil, gentle slop, sufficient depth, quantity and quality of water entering the site, and the requirement of low-salinity water (Field 1998).

Successful establishments and growth of mangrove plants have been reported in Qatar (Abdel-Razik, 1990; Al-Khayat and Jones, 1999) and Kuwait (Bhat et al., 2004; Bhat and Suleiman, 2005; Al-Nafisi et al., 2009; Almulla, 2013). However, only limited success of mangrove plantation has been reported in a sheltered bay in Bahrain (Al-Sayed et al., 2008). Therefore, there is an urgent need to investigate alternatives to restore the critically threatened mangrove ecosystems in Bahrain.

Planting mangroves in suitable intertidal areas in the Arabian Gulf is considered a sound approach for increasing coastal and marine productivity and supporting a wide range of biodiversity. For instance, coastal areas associated with planted and natural mangroves along the Qatari coastline support similar levels macrobenthos diversity (Al-Khayat and Jones, 1999). Likewise, Al-Nafisi et al. (2009) reported positive impacts for planted mangroves on coastal environment in Kuwait. Therefore, ecological and environmental benefits provided by planted mangroves may contribute positively to enhancing the overall productively of coastal and marine environments in the Arabian Gulf (Almulla et al., 2013).

Coral restoration could be carried out by creation of artificial reefs or translocation of healthy coral fragments to damaged reefs (Edwards and Gomez, 2007). Artificial reefs are frequently deployed in marine environment of the Arabian Gulf in an attempt to restore or enhance biodiversity and productivity of marine ecosystems. Generally, abundant and diverse communities of reef fish, coral and benthos organisms can develop on artificial structures (Feary et al., 2011). However, these communities may differ structurally and functionally from those in natural reefs (Burt et al., 2009).

Translocations of healthy corals have been conducted in some cases in the Arabian Gulf to avoid their distraction by large-scale marine projects. For instance, around 4,500 coral colonies from pipeline corridors which would have been affected by proposed expansion projects were relocated to another suitable location in Qatar (O'Donovan and McDonald, 2008). However, the success of such environmental initiatives in protecting affected corals remains to be investigated. Coral culture and transplantation within the Arabian Gulf is proposed as a feasible approach to maintain coral populations and preserve their adaptive capacities to future thermal stress events due to climate change (Coles and Riegl, 2013).

4.5. Integrated environmental monitoring and scientific research

Monitoring can be described as systemic observations and measurements of physical, chemical and biological variables to detect environmental changes over time (Lovett et al., 2007). Monitoring can provide decision makers with information on the state of biodiversity, and consequently, assist in identifying management goals and assessing priorities for conservation (Collen et al., 2013).

The key to protecting and managing biodiversity and marine resources is to characterize the structures of coastal and marine ecosystems (i.e. species and populations involved) and functions (i.e. flow of energy, growth and productivity). This could be achieved by adopting a holistic environmental monitoring approach that investigates, spatially and temporally, the physical, chemical and biological aspects of the valued ecosystem components in the Arabian Gulf (Naser, 2011a).

Several logistical and technical limitations may restrict the effectiveness of environmental monitoring in the Arabian Gulf. Van Lavieren and Klaus (2013) indicated that ecological monitoring and surveys in the Arabian Gulf are poorly designed and do not provide adequate information for decision-making systems.

Developing necessary plans and mechanisms for population and habitat conservation requires adequate knowledge and description of species. Therefore, there is a need to promote taxonomic research in the Arabian Gulf. Environmental impacts can be detected in a coarser level of taxonomic identification such as genus and family levels of biotic assemblages (Naser, 2010). However, effective conservation can only be achieved if the state of the environment is fully documented and understood, including species diversity. Therefore, it could be argued that while coarser taxonomic levels can be logistically useful in routine environmental monitoring, species-level is critically important to assess the biodiversity and to understand the structure and function of marine ecosystems in the Arabian Gulf.

Transboundary monitoring in the Arabian Gulf is needed to ensure that representatives of marine communities and habitats are included in the conservation measures. This could be addressed by increasing the cooperation between local and regional institutions and organizations concerned with ecological research and monitoring in the Arabian Gulf. The Regional Organization for the Protection of the Marine Environment (ROPME) may play an important role in strengthening the coordination of environmental monitoring and ecological surveys in the Arabian Gulf.

A key research need for marine conservation is to understand the individual and cumulative impacts of human disturbances on marine ecosystems. Therefore, monitoring should also be extended to processes and activities that are likely to have significant adverse impacts on the valued ecosystem components in the Arabian Gulf.

Feary et al. (2013) identified research topics that are considered to be the highest priority areas for future coral reef research in the Arabian Gulf, which could be extend to the other valued ecosystem components. These research areas include marine protected areas development, biological and ecological processes structuring marine ecosystems, climate change impacts on ecology and biology of ecosystems, effects of anthropogenic activities on marine ecosystems, connectivity of coral reef communities, disease biology, economic evaluation of ecosystems functions and services, monitoring and ecological surveys of species and communities, coral reef restoration and management, and mechanisms governing ecosystems' resistance and adaptation to environmental extremes. Strengthening cooperation between national, regional and international universities and scientific institutions in field of environment and conservation could facilitate the development and implementation of long-term research pogroms in the Arabian Gulf.

Building capacity toward scientific research in the field of environment and conservation biology is important in order to effectively conserve and mange marine ecosystems in the Arabian Gulf. Therefore, there is a need for significant improvement in the number and quality of programs related to marine sciences in the Arabian Gulf universities (Burt, 2013).

5. Conclusions

The Arabian Gulf is one of the world's most enclosed, small-scale marine environments. It is characterized by shallow depth and restricted water exchange with the wider Indian Ocean. The Arabian Gulf represents one of the harshest marine environments in the world due to marked fluctuations in seawater temperatures and high levels of salinities. These environmental extremes may interfere with normal functioning of marine ecosystems and affect physiological aspects of marine organisms and their diversity, abundance and spatial distribution.

The Arabian Gulf hosts some of the world's most critically endangered species such as dugongs, green and hawksbill turtles, and supports a variety of marine ecosystems, including seagrass bed, mangroves, coral reefs and mudflats that are uniquely adapted to environmental extremes. These ecosystems are under ever-increasing pressure from anthropogenic activities that are associated with the rapid economic, social and industrial developments in the Arabian Gulf countries.

Marine environment of the Arabian Gulf is severely impacted. The coasts of the Arabian Gulf are witnessing rapid industrialization and urbanization that contribute to degradation of naturally stressed marine ecosystems. Coastal development associated with dredging and reclamation is particularly damaging to coastal and marine ecosystems. This is combined with several anthropogenic factors, including industrial and domestic effluents, brine wastewater discharges and oil pollution.

Conserving species and communities and maintaining healthy ecosystems are important priorities in the marine environment of the Arabian Gulf. These could be achieved by adopting conservation approaches and management strategies that might contribute to the protection of the fragile marine ecosystems in the Arabian Gulf, including marine protected areas, environmental impact assessment, environmental regulations, ecological restoration, and environmental monitoring and scientific research.

Designation and implementation of marine protected areas are arguably critical for the protection of coastal and marine ecosystems in the Arabian Gulf. Although several marine protected areas have been established, lack of comprehensive management plans may hinder their effectiveness.

Environmental impact assessment can play an important role in the protection of biodiversity and in the sustainable use of coastal and marine environments in the Arabian Gulf. However, its effectiveness is constrained by the lack of adequate legal and regulatory frameworks, limited public participation, inadequate guidelines on procedural EIA, and lack of provisions related to cumulative impacts. Therefore, there is a need for a more strategic approach to environmental assessment that identifies environmental consequences of proposed policies, plans and programmes, and integrates environmental considerations into higher-level decision-making systems in the Arabian Gulf.

The Arabian Gulf countries have extensive national regional and international environmental legislations in place. Strengthening the implantation and the enforcement of the current regulations and agreements can substantially contribute to the protection of marine environment in the Arabian Gulf.

Ecological restoration principles could be adopted to minimize or reverse the decline in ecosystem integrity in the Arabian Gulf. Several restoration projects have been conducted in the coastal and marine environments of the Arabian Gulf. Planted mangroves provide several ecological and environmental benefits that may contribute to the productivity of coastal and marine habitats. However, the true impact of some restoration projects such as coral restoration remains to be investigated.

A holistic environmental monitoring and scientific research in the fields of marine sciences and conservation biology are integral part of any effort to conserve and manage biodiversity and marine resources in the Arabian Gulf. Improvement in both number and quality of academic programs related to marine sciences in the Arabian Gulf universities can contribute to building the long-term research capacity in the region.

Acknowledgements

Support provided by the Department of Biology and the Library and Information Services at University of Bahrain is greatly appreciated.

Author details

Humood A. Naser

Department of Biology, College of Science, University of Bahrain, Kingdom of Bahrain

References

[1] Abdulqader, E., Miller, J. (2012). Marine turtle mortalities in Bahrain territorial waters. Chelonian Conservation and Biology, 11:133-138

[2] Abido, M., Abahussain, A., Abdel Munsif, H. (2011). Status and composition of mangrove plant community in Tubli Bay of Bahrain during the years 2005 and 2010. Arabian Gulf Journal of Scientific Research, 29:100-111.

[3] Abdel-Razik M (1990). Towards a prospective transplantation of the local mangrove Avicennia marina (Frossk.) Vierh. growing in coast of Qatar. Qatar University Science Bulletin 10: 199-211.

[4] Agardy, T. (1994). Advances in marine conservation: The role of marine protected areas. Trends in Ecology and Evolution, 9: 267-270.

[5] Al-Ansi, M., Abdel-Moati, M., Al-Ansari (2002). Causes of fish mortality along the Qatari waters (Arabian Gulf). International Journal of Environmental Studies, 59: 59-71.

[6] Al-Awadhi, B. (2002). Strengthening environmental law in state of Kuwait. Global judges symposium on sustainable development and the role of law. Johannesburg, South Africa, 18-20 August 2002.

[7] Al-Cibahy, A., Al-Khalifa, K., Boer, B., Samimi-Namin, K. (2012). Conservation of marine ecosystems with a special view to coral reefs in the Gulf. In: B. Riegal, S. Purkis (eds.) Coral reefs of the Gulf: Adaptation to climatic extremes. Springer, Dordrecht.

[8] Al-Ghadban, A. (2002). Geological oceanography of the Arabian Gulf. In: The Gulf ecosystem Health and Sustainability, N. Khan, M. Munawar, A. Price (Eds.), pp. 23-39, Backhuys Publishers, Leiden.

[9] Al-Jamali, F., Bishop, J., Osment, J., Jones, D., LeVay, L. (2005). A review of the impacts of aquaculture and artificial waterways upon coastal ecosystems in the Gulf (Arabian/ Persian) including a case study demonstrating how future management may resolve these impacts. Aquatic Ecosystem Health and Management, 8: 81–94.

[10] Al-Jeneid, S. Bahnassy, M., Nasr, S., El Raey, M. (2008). Vulnerability assessment and adaptation to the impacts of sea level rise on the Kingdom of Bahrain. Mitigation and Adaptation Strategies for Global Change, 13: 87-104.

[11] Al-Khayat, J. Jones, D. (1999). A comparison of macrofauna of natural and replanted mangroves in Qatar. Estuarine, Coastal and Shelf science, 49: 55-63.

[12] Allan, S. Ramirez, C. & Vasquez, J. (2008). Effects of dredging on subtidal macrobenthic community structure in Mejillones Bay, Chile. International journal of Environment and Health, 2: 64-81.

[13] Al-Maslamani, I., Walton, M., Kennedy, H., Al-Mohannadi, M., Le Vay, L. (2013). Are mangroves in arid environments isolated systems? Life-history evidence of dietary contribution from inwelling in a mangrove-resident shrimp species. Estuarine, Coastal and Shelf Science, 124: 56-63.

[14] Almulla, L. (2013). Soil site suitability evaluation for mangrove plantation in Kuwait. World Applied Sciences Journal, 22: 1644-1651.

[15] Almulla, L., Bhat, N., Thomas, B., Rajesh, L., Ali, S., George, P. (2013). Assessment of existing mangrove plantation along Kuwait coastline. Biodiversity Journal, 4: 111-116.

[16] Al-Nafisi, R., Al-Ghadban, A., Gharib, I., Bhat, N. (2009). Positive impacts of mangrove plantations on Kuwait's coastal environment. European Journal of Scientific Research, 26: 510-521.

[17] Al-Sayed, H. Naser, H. & Al-Wedaei, K. (2008). Observations on macrobenthic invertebrates and wader bird assemblages in a tropical marine mudflat in Bahrain. Aquatic Ecosystem Health & Management, 11: 450-456.

[18] Al-Wedaei, K., Naser, H., Al-Sayed, H., Khamis, A. (2011). Assemblages of macrofaun associated with two seagrass beds in Kingdom of Bahrain: Implications for conservation. Journal of the Association of Arab Universities for Basic and Applied Sciences, 10: 1-7.

[19] Areiqat, A. Mohamed, K. (2005). Optimization of the negative impact of power and desalination plants on the ecosystem, Desalination, 185: 95-103.

[20] Badr, E., Cashmore. M., Cobb, D. (2004). The consideration of impacts upon the aquatic environment in environmental impact statements in England and Wales. Journal of Environmental Assessment Policy and Management, 6: 19-49.

[21] Barker, A. (2006). Strategic environmental assessment (SEA) as a tool for integration within coastal planning. Journal of Coastal Research, 22: 946-950.

[22] Barth, H., Khan, N. (2008). Biogeophysical setting of the Gulf. In: A. Abuzinada, H. Barth, F. Krupp, B. Boer, T. Abdessalam. Protecting the Gulf's marine ecosystems from pollution. Brikhauser Verlag/Switzerland.

[23] Bashitialshaaer, R., Persson K., Aljaradin M. (2011). Estimated future salinity in the Arabian Gulf, the Mediterranean Sea and the Red Sea consequences of brine discharge from desalination. International Journal of Academic Research, 3: 133-140.

[24] Bhat, N., Shahid S., Suleiman. M. (2004). Mangrove (Avicennia marina) establishment and growth under the arid climate of Kuwait. Arid Land Research and Management, 18: 127-139.

[25] Bhat, N., Suleiman, M. (2005). Classification of soils supporting mangrove plantations in Kuwait. Archives of Agronomy and Soil Science, 50: 535 - 551.

[26] Breitburg, D., Riedel, G. (2005). Multiple stresses in marine ecosystems. In: Marine conservation biology. E. Norse, L. Crowder (eds). Island Press, London, pp.167-182.

[27] Brown, D. James, J. (1985). Major oil spill response coordination in the combat of spills in Bahrain waters. Journal of Petroleum Technology, 37: 131-136.

[28] Burt, J., Bartholomew, A., Usseglio, P., Bauman, A., Sale, P. (2009). Are artificial reefs surrogates of natural habitats for coral fish in Dubai, United Arab Emirates? Coral Reefs, 28: 663-675.

[29] Burt, J. (2013). The growth of coral reef science in the Gulf: A historical perspectives. Marine Pollution Bulletin, 72: 289-301.

[30] Burt, J., Al-Khalifa, K., Khalaf, E., Alsuwaikh, B., Abdulwahab, A. (2013). The continuing decline of coral reefs in Bahrain. Marine Pollution Bulletin, 72: 357-363.

[31] Burt, J., Al-Harthi, S., Al-Cibahy, A. (2011). Long-term impacts of coral bleaching events on the world's warment reefs. Maine Environmental Research, 72: 225-229.

[32] Chape, S., Harrison, J., Spalding, M., Lysenko, I. (2005). Measuring the extent and effectiveness of protected areas as an indicator for meeting global biodiversity targets. Philosophical Transactions of the Royal Society B: Biological Sciences, 360: 443-455.

[33] Clewell, A., Aronson, J. (2006). Motivations for the restoration of ecosystems. Conservation Biology, 20: 420-428.

[34] Coles, S. & McCain, J. (1990). Environmental factors affecting benthic communities of the western Arabian Gulf. Marine Environmental Research, 29: 289-315.

[35] Coles, S., Riegl, B. (2013). Thermal tolerances of reed corals in the Gulf: A review for the potential for increasing coral survival and adaptation to climate change through assisted translocation. Marine Pollution Bulletin, 72: 323-332.

[36] Collen, B., Pettorelli, N., Baillie, J., Durant S. (2013). Biodiversity monitoring and conservation: bridging the gap between global commitment and local action. John Wiley & Sons, London.

[37] Cooper, L., Sheate, W. (2002). Cumulative effect assessment: a review of UK environmental impact statements. Environmental Impact Assessment Review, 22: 415-439.

[38] Crain, C., Kroeker, K., Halpern, B. (2008). Interactive and cumulative effects of multiple human stressors in marine systems. Ecology Letters, 11: 1304-1315.

[39] Dawoud, M., Al-Mulla, M. (2012). Environmental impacts of seawater desalination: Arabian Gulf case study. International Journal of Environment and Sustainability, 1: 22-37.

[40] De Mora, S. Fowler, S. Wyse, E. & Azemard, S. (2004). Distribution of heavy metals in marine bivalves, fish and coastal sediments in the Gulf and Gulf of Oman. Marine Pollution Bulletin, 49: 410-424.

[41] De Mora, S. Tolosa, I. Fowler, S. Villeneuve, J. Cassi, R. & Cattini, C. (2010). Distribution of petroleum hydrocarbons and organochlorinated contaminants in marine biota and coastal sediments from the ROPME Sea Area during 2005. Marine Pollution Bulletin, 60: 2323-2349.

[42] Dodd, R., Blasco, F., Rafii, Z., Torquebiau, E. (1999). Mangroves in the United Arab Emirates: ecotypic diversity in cuticular waxes at the bioclimatic extreme. Aquatic Botany, 63: 291-304.

[43] Dudley, N. (2008). Guidelines for applying protected area management categories. The International Union for Conservation of Nature (IUCN), Gland.

[44] Duffy, J. (2006). Biodiversity and the functioning of seagrass ecosystems. Marine Ecology Progress Series, 311: 233-250.

[45] Duisk, J., Fischer, T., Sadler, B., Steiner, A., Bonvoision, N. (2006). Benefits of a strategic environmental assessment. London: REC.

[46] Edwards, A., Gomez, E. (2007). Reef Restoration Concepts and Guidelines: making sensible management choices in the face of uncertainty. Coral Reef Targeted Research & Capacity Building for Management Programme: St Lucia, Australia. 38 pp.

[47] El-Fadl, K., El-Fadel, M., 2004. Comparative assessment of EIA systems in MENA countries: challenges and prospects. Environmental Impact Assessment Review, 24: 553-593.

[48] Elliott. M., Burdon, D., Hemingway, K., Apitz, S. (2007). Estuarine, costal and marine ecosystem restoration: confusing management science- A revision of concepts. Estuarine, Costal and Shelf Science, 74: 349-366.

[49] Erftemeijer, P., Shuail, D. (2012). Seagrass habitats in the Arabian Gulf: distribution, tolerance thresholds and threats. Aquatic Ecosystem Health and Management. 15(S1): 73-83.

[50] Evans, M., Symens, P., Pilcher, C. (1993). Sort-term damage to coastal bird populations in Saudi Arabia and Kuwait following the 1991 Gulf War marine pollution. Marine Pollution Bulletin, 27: 157-161.

[51] Feary, D., Burt, J., Bartholomew, A. (2011). Artificial marine habitats in the Arabian Gulf: Review of current use, benefits and management implications. Ocean & Coastal Management, 54: 742-749.

[52] Feary, D., Burt, J., Bauman, A., Al Hazeem, S., Abdel-Moati, M., Al-Khalifa, K., Anderson, D., Amos, C., Baker, A., Bartholomew, A., Bento, R., Cavalcante, G., Chen, C., Coles, S., Dab, K., Fowler, A., George, D., Grandcourt, E., Hill, R., John, D., Jones, D., Keshavmurthy, S., Mahmoud, H., Tapeh, M., Mostafaci, P., Naser, H., Pichon, M., Purkis, S., Riegal B., Samimi-Namin, K., Sheppard, C., Samiei, J., Voolstra, C., Wiedenmann, J. (2013). Critical research needs for identifying future changes in Gulf coral reef ecosystems. Marine Pollution Bulletin. 72: 406-416.

[53] Field, C. (1998) Rehabilitation of mangrove ecosystem: an overview. Marine Pollution Bulletin, 37: 383-392.

[54] Fuller, S. (2005). Towards a Bahrain national report on the Convention on Biological Diversity. Report prepared for the General Directorate of Environment and Wildlife Protection, Kingdom of Bahrain and Unite Nations Development Program country office, Manama, Bahrain.

[55] GCC, Gulf Cooperation Countries, (2010). Guideline for the convention on the conservation of wildlife and their natural habitats in the Gulf Cooperation Council countries. Secretariat General, Riyadh.

[56] Glasson, J., Therivel R., Chadwick, A. (2005). Introduction to Environmental Impact Assessment. Spon Press, London.

[57] Glibert, P., Landsberg, J., Evans, J., Al-Sarawi, M., Faraj, M., Al-Jaeallah, M., Haywood, A., Ibrahem, S., Klesius, P., Powell, C., Shoemaker, C. (2002). A fish kill of massive proportion in Kuwait Bay, Arabian Gulf 2001: the roles of bacterial disease, harmful algae, and eutrophication. Harmful Algae, 1: 215-231.

[58] Gontier, M., Balfors, B. and Mortberg, U. (2006). Biodiversity in environmental assessment-current practice and tools for prediction. Environmental Impact Assessment Review, 26: 268-286.

[59] Green, S., White, A., Christie, P., Kilarski, S., Blesilda, A., Meneses, T., Samonte-Tan, G., Karrer, L., Fox, H., Campbell, S., Claussen, J. (2011). Emerging marine protected area networks in the Coral Triangle: Lessons and way forward. Conservation and Society, 9: 173-188.

[60] Halpern, B. Walbridge, S. Selkoe, K. Kappel, C. Micheli, F. D'Agrosa, C. Bruno, J. Casey, K. Ebert, C. Fox, H. Fujita, R. Heinemann, D. Lenihan, H. Madin, E. Perry, M. Selig, E. Spalding, M. Steneck, R. & Watson, R. (2008). A global map of human impact on marine ecosystems. Science, 319: 948-952.

[61] Hashim, A., Hajjaj, M. (2005). Impact of desalination plants fluid effluents on integrity of seawater, with the Arabian Gulf in perspective, Desalination, 182 373-393.

[62] Hamza, W. (2006). Observations on transported exotic plankton species to UAE coastal waters by gas tankers ballast water. In: A. Tubielewicz (Ed.), Living marine

resources and coastal habitats. EuroCoast-Littoral, pp. 47–53. Gdansk University of Technology, Poland.

[63] Hamza. W., Munawar, M. (2009). Protecting and managing the Arabian Gulf: Past, present and future. Aquatic Ecosystem Health & Management, 12: 429-439.

[64] Jameson, S., Tupperb, M., Ridley, J. (2002). The three screen doors: can marine protected areas be effective? Marine Pollution Bulletin, 44: 117-1183.

[65] IUCN (2012). IUCN Red List Categories and Criteria: Version 3.1. Second edition. Gland, Switzerland and Cambridge, UK: IUCN. 32 pp.

[66] Kay, R., Alder, J. (2005) Coastal planning and management. Taylor & Francis, London.

[67] Khan, N. (2007). Multiple stressors and ecosystem-based management in the Gulf. Aquatic Ecosystem Health & Management, 10: 259-267.

[68] Khan, N. (2008). Integrated management of pollution stress in the Gulf. In: A. Abuzinada, H-J. Barth, F. Krupp, B. Boer, T. Al-Abdessalaam (Eds.), Protecting the Gulf's Marine Ecosystems from Pollution, pp. 57–92. Birkhauser, Basel.

[69] Khan, N. Price, A. (2002). Legal and institutional frameworks In: N. Khan, M. Munawar, A. Price (Eds.), The Gulf Ecosystem Health and Sustainability, pp. 399–424. Backhuys, Leiden.

[70] Khan, N., Munawar, M., Price, A. (Eds.) (2002). The Gulf ecosystem: Health and sustainability. Backhuys, Leiden.

[71] Khera, N., Kumar, A. (2010). Inclusion of biodiversity in environmental impact assessment (EIA): a case study of selected EIA reports in India. Impact Assessment and Project Appraisal, 28: 189-200.

[72] Kideghesho, J., Roskaft, E., Kaltenborn, B. (2007). Factors influencing conservation attitudes of local people in Western Serengeti, Tanzania. Biodiversity Conservation, 16: 2213-2230.

[73] King, H. (1999). The breeding birds of Hawar. Ministry of Housing, Bahrain.

[74] Knight, A., Seddon, P., Al-Midfa, A. (2011). Transboundary conservation initiatives and opportunities in the Arabian Peninsula. Zoology in the Middle East. 54 (sup3): 183-195.

[75] Lamorgese, L., Geneletti,D. (2013). Sustainability principles in strategic environmental assessment: a framework for analysis and examples from Italian urban planning. Environmental Impact Assessment Review, 42: 116-126.

[76] Lattemann, S., Hoepner, T.,(2008). Environmental impact and impact assessment of seawater desalination. Desalination, 220: 1-15.

[77] Literathy, P. Khan, N. & Linden, O. (2002). Oil and petroleum industry. In: The Gulf ecosystem: Health and Sustainability, N. Khan, M. Munawar & A. Price (Eds.), pp. 127-156, Backhuys Publishers, Leiden.

[78] Liou, M., Yu, Y. (2004). Development and implementation of strategic environmental assessment in Taiwan. Environmental Impact Assessment Review, 24: 337-350.

[79] Lovett, G., Burns, D., Driscoll, C., Jenkins, J., Mitchell, M., Rustad, L., Shanley, J., Likens, G., Haeuber, R. (2007). Who needs environmental monitoring? Frontiers in Ecology and the Environment, 5: 253-260.

[80] Massoud, M., Al-Abdali, F., Al-Ghadban, A. (1998). The status of oil pollution in the Arabian Gulf by the end of 1993. Environment International, 24: 11-22.

[81] McLeod, E., Salm, R., Green, A., Almany, J. (2009). Designing marine protected area networks to address the impacts of climate change. Frontiers in Ecology and the Environment, 7: 362–370.

[82] Naser, H. (2010). Testing taxonomic resolution levels for detecting environmental impacts using macrobenthic assemblages in tropical waters. Environmental Monitoring and Assessment, 170: 435-444.

[83] Naser, H. (2011a). Human impacts on marine biodiversity: macrobenthos in Bahrain, Arabian Gulf. In: The importance of biological interactions in the study of Biodiversity, J. Lopez-Pujol (ed.), INTECH Publishing. pp. 109-126.

[84] Naser, H. (2011b). Effects of reclamation on macrobenthic assemblages in the coastline of the Arabian Gulf: A microcosm experimental approach. Marine Pollution Bulletin, 62: 520-524.

[85] Naser, H. (2012a). Metal concentrations in marine sediments influenced by anthropogenic activities in Bahrain, Arabian Gulf. In: Metal contaminations: sources, detection and environmental impacts, Shao Hong-Bo (Editor), NOVA Science Publishers, Inc. New York, pp. 157-175.

[86] Naser, H. (2012b). Evaluation of the environmental impact assessment system in Bahrain. Journal of Environmental Protection, 3: 233-239.

[87] Naser, H. (2013a). Assessment and management of heavy metal pollution in the marine environment of the Arabian Gulf: A review. Marine Pollution Bulletin, 72:6-13.

[88] Naser, H. (2013b). Soft-bottom crustacean assemblages influenced by anthropogenic activities in Bahrain, Arabian Gulf. In: G. Sisto (ed.) Crustaceans: Structure, Ecology and Life Cycle. NOVA Science Publishers, Inc. New York. pp 95-112.

[89] Naser, H. (2013c). Effects of multi-stage and reverse osmosis desalinations on benthic assemblages in Bahrain, Arabian Gulf. Journal of Environmental Protection, 4: 180-187.

[90] Naser, H. Bythell, J. & Thomason, J. (2008). Ecological assessment: an initial evaluation of the ecological input in environmental impact assessment reports in Bahrain. Impact Assessment and Project Appraisal, 26: 201-208.

[91] Naser, H., Hoad, G. (2011). An investigation of salinity tolerance and salt secretion in protected mangroves, Bahrain. Gulf II: an international conference. The state of the Gulf ecosystem: Functioning and services. Kuwait City, Kuwait. 7-9 February 2011.

[92] O'Donovan, D., McDonald, I. (2008). Coral reef initiatives in the Middle East. Wildlife Middle East, 3 (2): 8

[93] Phillips, R. (2003). The seagrasses of the Arabian Gulf and Arabian Region. In: World Atlas of seagrasses, E. Green & F. Short. (Eds.), pp. 74-81, UNEP-WCMC.

[94] Preen, A. (2004). Distribution, abundance and conservation status of dugongs and dolphins in the southern and western Arabian Gulf. Biological Conservation, 118: 205-218.

[95] Preen, A., Das, H., Al-Rumaidh, M., Hodgson, A. (2012). Dugongs in Arabia; In: E. Himes, J. Reynolds III, L. Aragones, A. Mignucci-Giannoni, M. Marmontel (Eds.) Sirenian conservation: Issues and strategies in developing countries. University Press of Florida, Gainesville.

[96] Price, A. (2002). Simultaneous hotspots and coldspots of marine biodiversity and implications for global conservation. Marine Ecology Progress Series, 241: 23- 27.

[97] Price, A., Donlan, M., Sheppard, C., Munawar, M. (2012). Environmental rejuvenation of the Gulf by compensation and restoration. Aquatic Ecosystem Health & Management, 15 (S1): 7-13.

[98] Price, A. Sheppard C. & Roberts, C. (1993). The Gulf: Its biological setting. Marine Pollution Bulletin, 27: 9-15.

[99] Primack, R. (2010). Essential of conservation biology. Sinauer Associates. New York.

[100] Rachid, G., El Fadel, M. (2013). Comparative SWOT analysis of strategic environmental assessment systems in the Middle East and North Africa region. Journal of Environmental Management, 125: 85-93.

[101] Richlen, M., Morton, S., Jamali, E., Rajan, A., Anderson, D. (2010). The catastrophic 2008-2009 red tide in the Arabian Gulf region, with observations on the identification and phylogeny of the fish-killing dinoflagellate Cochlodinium polykrikoides. Harmful Algae, 9: 163–172.

[102] Riegl, B, Purkis, S. (2012). Coral reefs of the Gulf: adaptation to climatic extremes. Springer, Dordrecht Heidelberg.

[103] Sale, P., Feary, D., Burt, J., Bauman, A., Cavalcante, G., Drouillard, K., Kjerfve, B., Marquis, E., Trick, C., Usseglio, P., Van Lavieren, H. (2010). The growing need for

sustainable ecological management of marine communities of the Persian Gulf. Ambio, 40: 4-17.

[104] Shatti, J., Abdullah, T. (1999). Marine pollution due to wastewater discharge in Kuwait. Water Science and Technology, 40: 33-39.

[105] Sheppard, C., Price, A., Roberts, C. (1992). Marine ecology of the Arabian region: patterns and processes in extreme tropical environments. Academic Press, London.

[106] Sheppard, C. (2006). Longer-term impacts of climate change on coral reefs. In: coral reef conservation, I. Cote, J. Reynolds. Cambridge University Press, Cambridge, pp: 264-290.

[107] Sheppard, C., Loughland, R. (2002). Coral mortality and recovery in response to increasing temperature in the southern Arabian Gulf. Aquatic Ecosystem Health & Management, 5: 395-402.

[108] Sheppard, C. Al-Husiani, M. Al-Jamali, F. Al-Yamani, F. Baldwin, R. Bishop, J. Benzoni, F. Dutrieux, E. Dulvy, N. Durvasula, S. Jones, D. Loughland, R. Medio, D. Nithyanandan, M. Pilling, G. Polikarpov, I. Price, A. Purkis, S. Riegl, B. Saburova, M. Namin, K. Taylor, O. Wilson, S. & Zainal, K. (2010). The Gulf: A young sea in decline. Marine Pollution Bulletin, 60: 3-38.

[109] Smith, R. Purnama, A., Al-Barwani, H. (2007). Sensitivity of hypersaline Arabian Gulf to seawater desalination plants. Applied Mathematical Modelling, 31: 2347-2354.

[110] Spalding, M., Fox, H., Allen, G., Davidson, N., Ferdana, Z., Finlayson, M., Halpern, B., Jorge, M., Lombana, A., Lourie, S., Martin, K., McManus, E., Molnar, J., Recchia, C., Robertson, J. (2007). Marine ecoregions of the world: A bioregionalizaton of coastal and shelf areas. BioScience, 57: 573-583.

[111] Treweek, J. (1999). Ecological Impact Assessment. Blackwell Science Ltd, Oxford.

[112] Tu Do, V., Montaudouim, X., Blanchet, H., Lavesque, N. (2012). Seagrass burial by dredged sediment: benthic community alteration, secondary production loss, biotic index reduction and recovery possibility. Marine Pollution Bulletin, 64: 3340-2350.

[113] Uddin, S., Gevao, B., Al-Ghadban, A., Nithyanandan, Al-Shamroukh, D. (2012). Acidification in the Arabian Gulf – Insights from pH and temperature measurements. Journal of Environmental Monitoring. 14: 1479-1482.

[114] Van Lavieren, H., Burt, J., Feary, D., Cavalcante, G., Marquis, E., Benedetti, L., Trick, C., Kjerfve, B., Sale, P. (2011). Managing the growing impacts of development on fragile coastal and marine ecosystems: lessons from the Gulf. A policy report, UNU-INWEH, Hamilton, ON, Canada.

[115] Van Lavieren, H., Klaus, R. (2013). An effective regional marine protected area network for ROPME Sea Area: Unrealistic vision or realistic possibility? Marine Pollution Bulletin, 72: 389-405.

[116] Wake, H. (2005). Oil refineries: a review of their ecological impacts on the aquatic environment. Estuarine, Coastal and Shelf Science, 62: 131-140.

[117] Weishar, L., Watt, I., Jones, D., Aubrey, D. (2008). Evaluation of arid salt marsh restoration techniques. In: A. Abuzinada, H. Barth, F. Krupp, B. Boer, T. Abdessalam. Protecting the Gulf's marine ecosystems from pollution. Brikhauser Verlag/Switzerland.

[118] Wilkinson, C. (2008). Status of Coral Reefs of the World. Australian Institute of Marine Science, Townsville.

[119] Young, T. (2000). Restoration ecology and conservation biology. Biological Conservation, 92: 73-83.

Chapter 13

Threats of Mining and Urbanisation on a Vulnerable Ecosystem in the Free State, South Africa

L.R. Brown and P.J. Du Preez

Additional information is available at the end of the chapter

http://dx.doi.org/10.5772/57589

1. Introduction

The Free State province is located in the central part of South Africa (Figure 1) and is approximately 1 300 m above sea level. The northern boundary is formed by the Vaal River with the Orange River forming the southern border. The Province covers an area of approximately 130 000 square kilometers comprising 10.6% of the total area of South Africa and has a population of almost 3 million people (southafrica.info 2012). Bloemfontein is the capitol with almost 370 000 residents and is located in the southern part of the Province.

Mining and agriculture are the major contributors to the province's economy. Various, coal, diamonds and bentonite mining activities occur throughout the province while approximately 120 000 square kilometres of land is used by the agricultural section for crop production and grazing purposes (southafrica.info 2012). These activities as well as the continued increase in human population numbers with resultant development of new infrastructure places stress on the natural environment.

A country's ability to conserve and sustainably manage its natural vegetation and water resources is reflected by its industrial potential and the standard of living of its people. Any injudicious utilisation of these natural resources will disturb the balance between the different components of the ecosystem and can have disastrous results for both humans and animals (Aucamp & Danckwerts 1989).

The environment consists of complex ecosystems within which a balance exists. Any disturbance in an ecosystem will affect the interactions between different species and therefore the natural resources available to different organisms. Vegetation is the most physical representation of the environment (Kent & Coker 1992; Kent 2012). Any changes environment whether

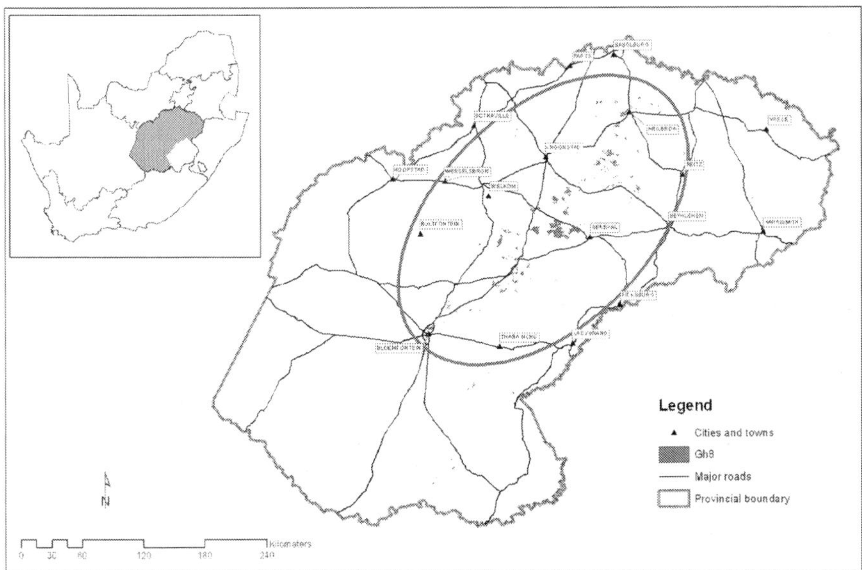

Figure 1. Locality of the Free State Province and the distribution of the Bloemfontein Karroid shrubland vegetation type (Gh8) in the Province.

it is as a result of pollution, development, droughts etc. is first seen in the vegetation and its species diversity and composition.

The term biodiversity refers to the diversity and number of plant and animal species on earth. Biodiversity conservation not only refers to the protection of all species, but also habitats, ecosystems and biomes (Brower *et al.* 1990; Van As *et al.* 2012). The diversity of species within an ecosystem is partly a reflection of the diversity of the physical environment. The more diverse the environment the higher the species richness is expected to be due to different microhabitats available for different plant and animal species (Van As *et al.* 2012). A diverse ecosystem contains a variety of genetic material that will ensure long-term stability and survival and also are less likely to be invaded by alien or pioneer species. Biodiversity conservation is very important for the survival of humans on earth. Each and every species on earth is important and crucial in an ecosystem. The loss of one species could lead to the loss of various others that in turn will have a chain reaction of events that could cause the destruction of one or many ecosystems.

The impact of humans on the environment is widespread and a cause for concern (Botha 2003). As the human population increased over time, people started to exert a bigger influence on nature (Grime 1997). The demand for land for housing, agriculture, mining and industries are increasing and so is demand for more food and water. The depletion of our natural resources to sustain our life styles causes large scale destruction of the environment. According to Van As *et al.* (2012) and Keddy (2007) the average ecological footprint of humans has reached a critical level and humans are destroying Earth at an unprecedented rate. Humans need to be aware of their actions and the effects it is having on our ecosystems and ultimately survival as a species.

The vegetation of the Free State Province of South Africa falls within the Grassland, Savanna and Nama-karoo biomes with grasslands forming the largest component. The Province has 34 different vegetation types (Mucina & Rutherford 2006). As previously mentioned large areas of these vegetation units are threatened and degraded due to various human actions (e.g. mining, development, agriculture). One of these vegetation units, the Bloemfontein Karroid Shrubland vegetation type (Gh8) (Mucina & Rutherford 2006) occurs as small islands scattered throughout the Province comprising a total area of 473.09 km^2 ha (0.004% of the Province). This very small vegetation unit's existence is under threat from mining, road construction and residential developments. If not properly protected and managed these areas and its unique plant species associations will be permanently destroyed.

The purpose of this chapter is to provide a broad description of the plant species associations and the species diversity of the Bloemfontein Karroid Shrubland vegetation type (Gh8) (Mucina & Rutherford 2006) and to provide guidelines to conserve this sensitive ecosystem. In this chapter we follow a broad plant phytosociological and floristic approach to describe the unique plant species and assemblages within this vegetation type.

2. Study area and methods

The Bloemfontein Karroid shrubland vegetation type (Gh8) (Mucina & Rutherford 2006) occurs as an archipelago of isolated patches on shallow dolerite outcrops within the Highveld grassland region of the Free State Province of South Africa (Figure 1). The vegetation is characterised by small-leaved dwarf karroid and succulent shrubs underlain by dolerite sheets of igneous origin (Figure 2). The soil is very shallow and gravelly with exposed rock outcrops prominent. In-between the rock crevices slightly deeper and less gravelly soil occur. A large proportion of the soil present on the rock sheets and those formed from the weathering of the rocks is washed into the adjacent lower-lying areas and depressions (Dingaan & Du Preez 2002).

The province is located within the summer rainfall area of South Africa and experiences warm to hot summers and cool to cold winters. Maximum temperatures are experienced in December and January (30.2°C) while June and July are the coldest months when the average daily temperature could drop to -1.6°C (Dingaan & Du Preez 2002). The eastern areas are prone to snowfalls especially on the higher-lying mountains while the western areas are more arid. The province receives approximately 580 mm of rain per annum with the highest rainfall between November and February.

In order to obtain a representative sample of the Bloemfontein Karroid Shrubland, a total number of 68 relevés (16 m^2) were surveyed within randomly stratified units of this vegetation type in various parts of the province. The data obtained is representative of five different stands of this vegetation type stretching from Bloemfontein in the south-west to the Willem Pretorius Nature Reserve in the north-east. The plot data were grouped into the five groups namely the Bloemfontein stand, the Winburg stand, the Willem Pretorius Nature Reserve stand, the Skoongesig stand and the Kareefontein stand. Habitat as well as floristic data was captured using TURBOVEG (Hennekens 1996; Hennekens & Schaminee 2001) and exported to JUICE (Tichø 2002) from where a raw table (Table 1-Annexure 1) was created for basic floristic interpretation. No formal phytosociological classification was done since the purpose of this study was not to obtain a formal classification, but to compare the different groups in terms of species richness and diversity.

Many people regard species richness and diversity as similar to species diversity. Species richness however refers to the number of species within an area or community (Kent & Coker 1992; Magurran 1988; Magurran 2005; Spellerberg & Fedor 2003). For this study species richness was calculated by determining the total number of species in each stand surveyed.

Species diversity refers to the diversity that occurs within a plant community or area and incorporates both species richness and the evenness of species' abundances (McGinley 2013). Species diversity is one component of the concept of biodiversity and is influenced by the relative abundances of different plant species present within the community. Various indices exist that measure both evenness and species richness into a single measure of species diversity (Stirling & Wilsey 2001). For the purpose of this study the Simpsons Index (Simpson 1949) as well as the Shannon-Wiener Species Diversity Index (Smith & Wilson 1996) was used to determine species diversity for each stand of the Bloemfontein Karroid shrubland surveyed in this study as expressed in the following formulas:

Simpson Index:

$$D = \sum \left(\frac{ni\,[ni - 1]}{N\,[N - 1]} \right) \tag{1}$$

Shannon-Wiener Index:

$$H' = -\sum_{i=1}^{S} pi(\ln pi) \tag{2}$$

Species richness (S), Simpson index if diversity (-ln (D)) and the the Shannon-Wiener index of diversity (H') were calculated for each stand:

S=Richness (Number of species per community)

pi=is the proportion of individuals of a species (relative proportion)

D=Simpson's index of diversity. It represents the likelihood that two randomly chosen individuals will be different species.

A the Chi-Square Test (Welman *et al.* 2007) was performed on the data to determine whether significant associations exist between the different stands.

3. Results and discussion

3.1. Habitat and growth forms

In 1937, Potts and Tidmarch published an article recognizing a vegetation type near Bloemfontein which has "marked affinities with the Karoo". In 1991, Du Preez and Bredenkamp (1991) named this vegetation unit the *Oropetium capense* community on rock sheets and classified it as a separate vegetation class. Dingaan and du Preez (2002) surveyed this vegetation unit near Bloemfontein and identified three different plant communities, namely the *Eragrostis trichophora–Aristida congesta*, *Heliophylla carnosa–Senecio radicans* and the *Stomatium braunsii–Avonia ustulata* Communities. In Mucina and Rutherford (2006) this vegetation unit, although small in size, is recognized as a separate vegetation type and has been described as the Bloemfontein Karroid shrubland (Gh8). Due to the presence of the scattered dolerite sheets this vegetation type has an archipelago appearance that occurs mainly in the Dry Highveld grassland.

Dolerite is of igneous origin and forms extensive sheets which vary in thickness. During the cooling process various horizontal cracks develop (Duncan and Marsh 2006; Holmes 2012). These cracks create areas where water infiltrates the rock. This allows chemical weathering to take place which in turn allows more water infiltration. Eventually the cracks develop into crevices into which soil and organic matter accumulates. Areas with deeper soil (50mm – 250mm) accommodate deep rooted species such as shrubs, perennial grasses and geophytes. Depressions occur on the exposed dolerite sheets where soil accumulates, These areas, although very shallow (10mm – 50mm), house a few species especially succulents and annuals. The two main environmental factors that differentiate the different plant communities on these dolerite outcrops are soil depth and soil moisture availability.

These dolerite sheets create an unusually arid habitat in a relatively high rainfall area due to the high loss of potentially available water. The loss of rainwater is caused by the poor water retention abilities of the coarse textured soil, poor infiltration, high evaporation tempos and high runoff. This unique habitat creates physiological drought conditions (Snyman 1984). The presence of these archipelagos of dolerite sheets with their shallow soils in a "sea" of deep soil and grass covered plains create a mosaic of scattered and isolated patches of arid habitats (Figure 2) which are relatively hostile environments for typical grassland species.

The physiological drought environment that is being created by the dolerite sheets and the shallow soil, is unsuitable most of the Grassland biome species but for a number of Nama-karoo biome species, which can tolerate the high temperatures and arid conditions present on these dolerite sheets, it creates a suitable habitat. This habitat can therefore be regarded as unique and in a certain sence can be regarded as an azonal vegetation type, Due to the lack of water, it is deviating strongly from the typical surrounding zonal vegetation (Mucina & Rutherford 2006)

The percentages of the number of plant species present per growth form recorded on these rocky sheets are indicated on Figure 3.

Figure 2. Typical appearance of the Bloemfontein Karroid shrubland vegetation type (Gh8). Succulents are limited to shallow crevices in the dolerite sheets while the grasses and low shrubs occur on slightly deeper soil.

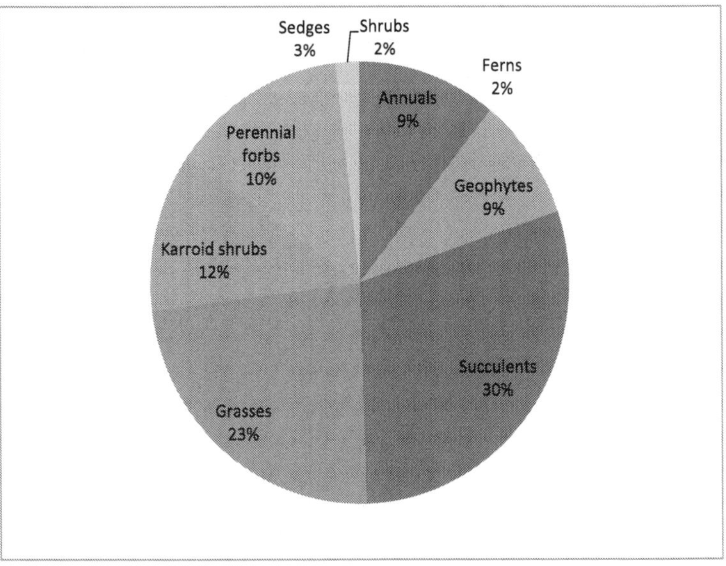

Figure 3. The percentages of the number of plant species per growth form present on the dolerite sheets.

It is interesting to note that number of succulents (30%) is the highest out of a total of 111 plant species noted on these dolerite sheets. This differs from the surrounding grasslands where less than 10% of the species are succulents. The second most important growth form is grasses (23%) followed by Karroid shrubs (12%). Perennial forbs make up 10% of the total species list while annuals and geophytes represent almost the same percentage (9%). Shrubs (3%), sedges and ferns (2%) struggle to survive on these arid habitats and are not well represented.

A number of species are endemic to this arid habitat. They are the succulents *Anacampserus filamentosa, A. telephiastrum, Avonia ustulata, Crassula tetragona, Euphorbia catervifolia, Hereroa species, Othonna protecta, Rabiea species, Ruschia unidens, Stomatium braunsii,* the *geophytes Brachystelma dimorphum subsp. gratum, Strumaria tenella subsp. orientalis* and the drought tollerant sedges *Cyperus bellus* and *Mariscus indecorus* (Annexure 1). It is only Brachystelma dimorphum subsp. gratum which is currently listed as a Red data species. According to POSA (2009) its status is rare.

3.2. Species richness and diversity

The Bloemfontein stand of the Bloemfontein Karroid Shrubland vegetation type (Gh8) has the highest species richness (81) with the Willem Pretorius Nature Reserve stand the second most namely 68 different species. That is followed by the Skoongesig stand with the Winburg and Kareefontein stands the lowest (Figure 4).

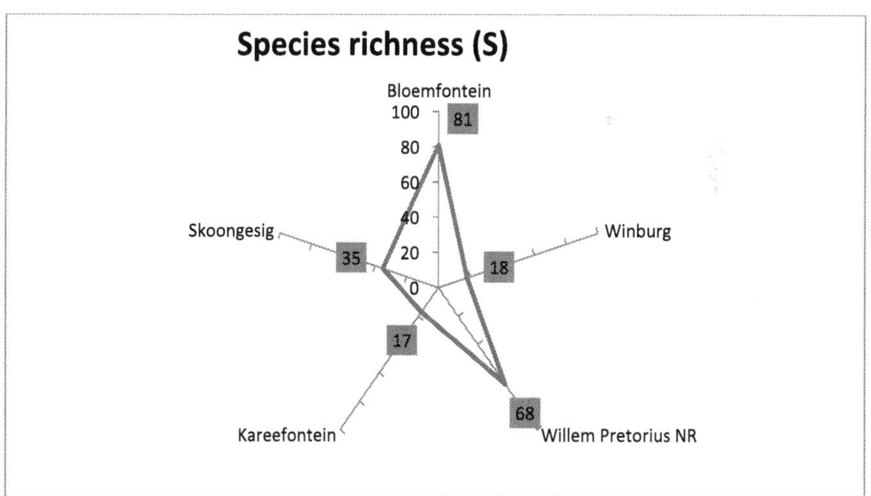

Figure 4. Species richness of the different stands of the Bloemfontein Karroid Shrubland (Gh8).

According to the Shannon-Wiener Index (Smith & Wilson 1996) values the Willem Pretorius Nature Reserve population of the Bloemfontein Karroid Shrubland vegetation type (Gh8) has the highest diversity followed by the Bloemfontein population (Figure 5). They are significantly different from the other three populations with the Kareefontein population the third most diverse. These results are confirmed in the Simpsons Index (Figure 6).

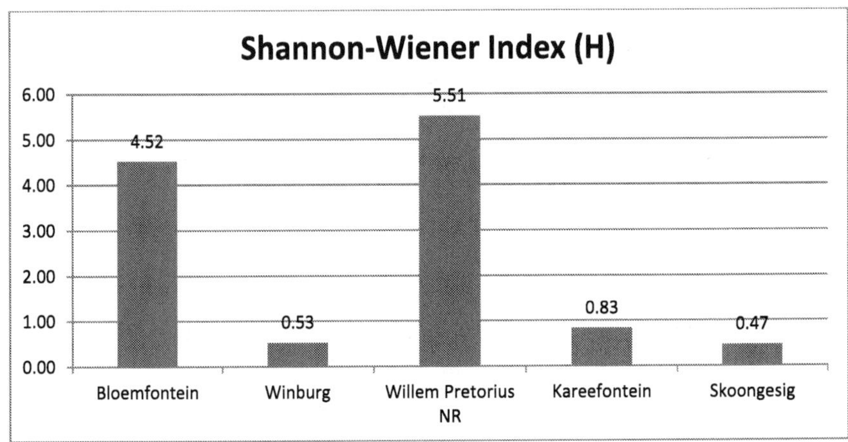

Figure 5. Shannon-Wiener index values for the five stands of the Bloemfontein Karroid Shrubland vegetation.

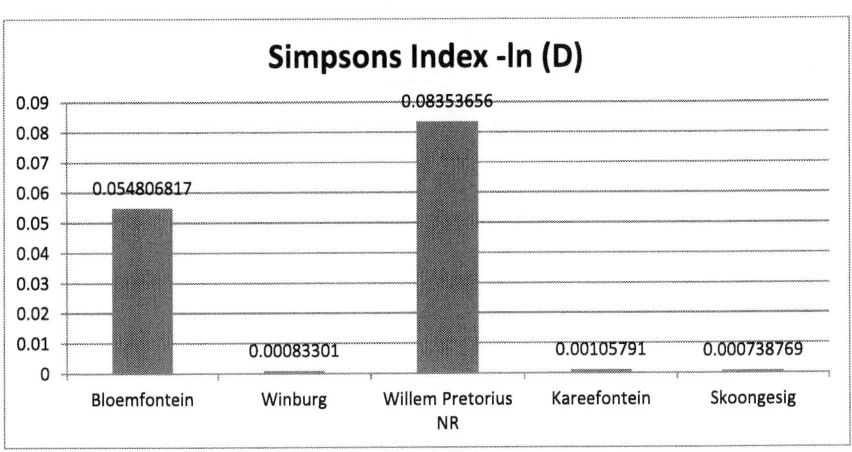

Figure 6. Simpson index values for the five stands of the Bloemfontein Karroid Shrubland vegetation.

There was a significant association between Shannon-Wiener Diversity Index values and areas surveyed X2(4)=10.06, p=0.039. Based on the standardized residuals, Willem Pretorius Nature Reserve is over represented (+2.04) and the main contributor to the association.

The species diversity of a plant community refers to the different plant species and the relative abundance of each species. It is widely believed that the more diverse a community the more complex it is and the higher its production will be (Brower *et al.* 1990). Some ecologists also believe that the more diverse a community, the more stable it would be, but this assumption would not be true in all cases (Brower *et al.* 1990) for example a degraded community could have a high diversity of pioneer species, but would not be regarded as a mature or stable ecosystem. Thus the total species composition and ecological status of the species should be evaluated when interpreting species diversity indices.

In the case of the Bloemfontein karroid shrubland (Gh8) the species comprises mostly of climax and secondary succesional species indicating the vegetation to be in a mature and stable condition.

The Willem Pretorius Nature Reserve and Bloemfontein stands have the highest species richness and are also the two most diverse stands of this vegetation type (Figures 4 & 5). Not only are these two stands the largest in area size compared to the other three, but the Willem Pretorius stand occurs within the Willem Pretorius Nature Reserve which is a protected area that fall within the jurisdiction of the Free State Department of Economic, Tourism and Environmental Affairs (DETEA). Although the Bloemfontein stand has a higher species richness (81 species - figure 4) it is not as diverse as the Willem Pretorius Nature Reserve stand with 68 different species (Figures 5 & 6). The higher number of species of the Bloemfontein stand is most probably the result of the different habitats surrounding the area which have resulted in the presence of single individuals of different species. Both stands do however have a high biodiversity and species richness compared to the other stands.

The Skoongesig and Winburg stands have the lowest diversity values (Figures 4 & 5). The Skoongesig stand however, has 35 different species compared to the 18 of the Winburg stand and the 17 of the Kareefontein stand. Both the Skoongesig and the Winburg stands are located in areas that are subjected to degradation. The Winburg stand is surrounded by local communities that utilise the whole area for grazing by domestic stock, while the N1 highway also passes through this vegetation type. The Skoongesig stand is also a small area that is surrounded by formal agricultural land with deep soil that has mostly been ploughed and the rest are grazed. Both these stands are small in size and are affected by surrounding human associated impacts that has led to low diversity and isolation of these stands.

In contrast the Kareefontein stand is very small with a low species richness (Figure 4) but a somewhat higher species diversity compared to the Winburg and Skoongesig stands (Figures 5 & 6). This stand is located within a private nature reserve with various game species utilising the vegetation. Whereas private game reserves are focused on tourism and hunting to ensure it to be economically viable compared to large provincial nature reserves, these reserves do many times have a higher number of animals stocked on their farms. Thus although protected it could be slightly more trampled than similar areas in larger nature reserves. Thus the higher diversity compared to the degraded stands is expected, but the lower species richness can be attributed to its small size in the private reserve as well as the effect of grazing by antelope.

The diversity and species richness of the three smaller stands (Kareefontein, Skoongesig and Winburg) were significantly lower than those of the larger stands (Bloemfontein and Willem Pretorius Nature Reserve) (Figures 4, 5 & 6). Factors that could contribute to the lower species richness as discussed above include small size and degradation of the habitat and surrounding areas. The larger areas although also surrounded in same places by various human related activities and degraded areas seem to have a more stable species composition and higher diversity. Thus these areas are better adapted to withstand and survive any threat to their ecosystem. The impacts of fragmentation and human related activities has influenced the smaller and more isolated areas of this vegegtation type.

3.3. Threats to the Bloemfontein Karroid Shrubland (Gh8) vegetation type

To survive, humans need continuous access to clean water, air, food and shelter (Van As *et al.* 2012). This can only be assured if the environment is utilised and managed in a sustainable way. If the environment is managed and utilised in an unsustainable way both our renewable an non-renewable resources would become depleted that could cause total degradation of our ecosystems. That in turn could lead to the mass extinction of all organisms on earth including humans (Van As *et al.* 2012). One of the results of habitat exploitation and degradation is the fragmentation of habitats. Franklin *et al.* (2002) maintain habitat fragmentation to be a primary concern in conservation biology. The disruption of large sections of an ecosystem into smaller intact units, usually as a result of human activity is also referred to as fragmentation (Franklin *et al.* 2002). Humans are responsible for large scale habitat fragmentation due to pollution, urban development, agriculture, the introduction of alien species, forest plantations and especially mining activities (MacDonald 1989; Hogan 2013). Although not true in all cases it is generally regarded that the larger an area the more diverse and sustainable it will be. The general view is that the ecological effects of habitat destruction and fragmentation are negative (Franklin *et al.* 2002). From the results of this study the larger areas are more diverse and species rich, however the smaller areas contain certain species not present within the larger areas (Table 1 – Annexure 1).

Due to the extensive dolerite layers associated with the Bloemfontein Karroid Shrubland they are frequently mined for road building material (Figure 7) while other areas are used for the development of houses. Most of these areas are left unrehabilitated causing further degradation of the ecosystem. As a result these abandoned areas are either developed or left to become species poor transformed areas.

Another threat to the existence of this unique vegetation type is severe grazing of the area (Figure 8). Due to the sparse vegetation cover and shallow soil layers in some areas on rock sheets, overgrazing by domestic and other animals results in a reduction of the vegetation cover and the trampling of the shallow soil. High rainfall events therefore leads to erosion washing all the soil to the adjacent vegetation communities on the lower-lying valley bottom areas. This in turn leads to even lower vegetation cover and recruitment of plant species that leads to the exposed rock areas becoming larger with more pioneer species present.

(a)

(b)

Figure 7. (a) Mining of the dolerite sheets of the Bloemfontein Karroid Shrubland vegetation type (Gh8) leads to total destruction and transformation of the area. (b) Mining of the dolerite hills of the Bloemfontein Karroid Shrubland vegetation type (Gh8).

The low vegetation structure and open rocky areas in-between the different plant species also leaves the impression that this vegetation type is generally degraded. As a result people often develop on these areas or use it for grazing without regard for the sensistivity ofthis ecosystem.

Figure 8. Degradation and reduction of vegetation cover of Bloemfontein Karroid Shrubland vegetation type (Gh8) as a result of overgrazing by domestic animals in the Winburg area..

4. Conclusion

Changes in ecosystems throughout the world are done to increase the flow of energy to one species only namely humans. Human populations continue to increase at the expense of other species (Keddy 2007). Not only has that resulted in large scale destruction of habitats, but also in immeasurable loss of species and ecosystem functions.

The results of this study provide more information on the distribution as well as species composition of the Bloemfontein Karroid Shrubland (Gh8) as described in Mucina & Rutherford (2006). A number of plant species unique to this vegetation type occurs while one red data plant was also found to be present. The effect of degradation and fragmentation on this vegetation type is clearly illustrated by the lower species diversity of the isolated and overgrazed patches. The important role that large nature reserves and the conservation of large sections of this vegetation type play in the conservation of the species diversity is also illustrated by the high species richness of the Bloemfontein and Willem Pretorius Nature Reserve stands. Although degraded and low in species diversity in some areas, these islands all are important and contribute to the larger community composition. Bond (1989) states that in some cases smaller islands of vegetation cover a wider area than a nature reserve. In such a case their combined species total could be greater than the smaller section conserved in a nature reserve. These smaller islands may also act as refugia for formerly widespread species even from the surrounding threatened ecosystems (Crawley 1997). The results from this study also indicate that the smaller stands contain plant species not present in the larger stands studied, thus their conservation and ecosystem value should not be underestimated.

It is important that these islands are conserved as natural communities to ensure continued existence of these unique species assemblages and related ecosystem processes. If uncontrolled development of these areas are allowed it will not only lead to local destruction of this sensitive ecosystem but also to further fragmentation that will lead to the total loss of this ecosystem and related plant species. Keddy (2007) states that the greater the loss of species and related ecosystem services, the more human survival is at risk. This chapter focused on the negative effect of humans on a unique vegetation type in the Free State Province of South Africa, however these negative effects is also applicable to other ecosystems in other parts of the world. It is important that nature conservation organisations consider all aspects related to an ecosystem (structure, species assemblages, ecosystem processes and functions, fragmentation, condition of the ecosystem etc.) before decisions are made on whether development can be allowed or not.

The increasing amount of environmental research provides a better understanding of human impact on ecosystems. Further studies are needed to fully understand the extent and distribution of this vegetation type. Every taxon has a specific geographical range of distribution (Van Wyk & Smith 2001). Endemism is a scale related concept and the term "endemic" refers to a taxon that is geographically limited in its distribution, while "near-endemic" refers to a taxon that is marginally present elsewhere (Van Wyk & Smith 2001). For the purpose of this study these terms are applied to the Bloemfontein Karroid Shrubland vegetation type (Gh8) (Mucina & Rutherford 2006). It is proposed that based on the data from this study, the Bloemfontein Karroid Shrubland vegetation type (Gh8) is regarded as an endemic vegetation unit within the Free State Province in need of a high conservation status. In view of the fragmented nature of this vegetation type and the threats such as urbanisation, mining and overgrazing this vegetation unit as a whole must be listed a threatened ecosystem and no development or mining activity may be allowed unless detailed vegetation and ecosystem functioning studies have been conducted. Without a strict policy to protect these fragments, it would be difficult to control the destruction and the eventual loss of an unique vegetation type.

Annexure 1

Total number of releves: 68																																																																					
Stands		Bloemfontein																																					Kareefontein					Winburg				Skoongesig					Willem Pretorius Nature Reserve															
Releve numbers		4	4	4	0	4	4	4	0	0	4	4	4	4	0	0	4	0	0	4	4	4	4	4	4	4	4	4	4	4	4	0	0	4	4	4	4	4	4	4	4	4	4	4	4	4	4	4	4	4	4	4	4	4	4	4	4	4	4	4	4	4	4	4	4	4	4	4	4
		7	7	7	1	7	2	2	0	0	5	5	5	5	0	0	2	0	1	2	2	2	2	2	2	2	2	2	5	5	2	0	0	2	2	2	2	2	2	2	2	9	9	9	3	3	3	3	2	2	2	2	2	2	2	2	6	6	6	6	6	6	6	6	6	6	6	6	6
		2	3	1	6	4	4	3	2	5	2	3	4	1	1	7	0	1	5	8	9	0	1	7	5	7	9	6	8	7	2	3	4	3	4	6	5	7	8	2	1	7	5	0	2	1	3	0	1	2	3	5	4	9	6	8	2	1	3	5	1	4	7	6	9	0	8	2	3
Species	**Growth form**																																																																				
Albuca setosa	Geophyte	+	r	+	.	r	r	r	r	+	+	+	+	r	r	r	+	r	.	.	r	r	+	.	+	.	+	.	.	.	r	+	.	r	+	r	+	.	.	.	.	.	.	.	.	.	.	.	.	.	.	.	.	.	.	.	+	1	1	1	.	+	2	.	.	.	.	1	1
Aristida congesta	Grass	.	.	r	.	.	.	.	r	.	.	.	.	.	.	.	.	r	.	.	.	.	.	.	.	.	.	.	.	.	.	r	.	.	+	.	.	.	.	.	.	+	.	+	+	+	+	1	.	.	.	+	.	.	.	.	.	.	.	.	.	.	.	.	.	.	.	.	.
Avonia ustulata	Succulent	.	r	+	.	+	r	.	+	+	r	r	r	.	r	r	+	r	.	.	r	.	r	.	.	+	+	.	.	.	.	.	.	.	.	r	.	.	+	.	.	.	.	.	r	r	r	r	+	+	+	.	+	+	+	.	.	.	.	.	.	.	.	.	.	.	.	.	.
Chasmatophyllum mustellinum	Succulent	r	r	.	.	r	.	.	.	.	.	.	.	.	.	.	.	r	.	.	.	.	.	.	.	.	.	.	.	.	.	.	.	.	.	+	.	.	.	.	.	.	+	+	.	r	.	.	.	.	.	+	r	.	.	.	.	.	.	.	.	.	.	.	.	.	.	.	.
Eragrostis nindensis	Grass	+	r	1	.	+	+	+	1	+	2	+	.	2	+	+	1	r	.	2	r	+	1	1	+	2	1	1	1	+	.	1	.	2	.	+	+	+	+	1	r	.	+	+	1	+	+	+	+	+	.	+	.	+	1	+	3	3	1	3	1	3	2	2	3	2	+	.	4
Jamesbrittenia pristicepala	Karroid shrub	.	.	.	.	.	.	.	.	.	r	.	.	r	.	+	.	r	.	.	.	.	.	.	.	.	.	.	.	.	.	.	.	.	.	.	+	.	.	.	.	.	.	.	.	.	.	.	.	.	.	.	.	.	.	.	.	.	.	.	.	.	.	.	.	.	.	.	.
Ledebouria luteola	Geophyte	.	.	.	.	.	+	r	r	+	r	.	r	.	r	.	.	r	.	.	r	.	.	r	.	.	.	+	r	+	.	.	.	.	.	.	+	.	+	.	.	.	.	+	r	r	r	.	.	.	.	.	.	.	.	.	.	.	.	.	.	.	.	.	.	.	.	.	.
Melinis repens	Grass	r	r	+	.	r	.	.	1	+	r	+	+	r	+	+	+	r	+	.	.	+	.	.	.	.	.	.	.	.	2	4	.	.	+	.	.	.	.	.	.	.	.	.	.	.	.	.	.	.	.	.	.	.	.	+	.	.	.	.	.	.	.	.	.	.	.	.	.
Oropetium capense	Grass	+	r	r	.	r	+	+	1	1	2	2	2	2	1	2	+	+	+	+	+	+	+	1	r	1	+	.	.	.	.	.	.	+	.	+	+	+	+	.	.	+	+	+	+	+	+	1	+	+	+	1	+	+	+	.	.	.	.	.	.	.	.	.	2	.	2	.	.
Stomatium braunsii	Succulent	.	.	.	+	+	.	.	1	+	+	2	1	1	1	1	1	+	+	+	1	1	+	+	.	+	1	.	.	.	.	.	.	.	1	+	+	1	+	+	r	.	.	.	.	.	.	.	.	.	.	.	.	.	.	.	.	.	.	.	.	.	.	.	.	.	.	.	.
Microchloa caffra	Grass	+	r	+	.	r	.	.	1	.	.	.	.	.	.	.	.	.	.	.	.	.	.	.	.	.	.	.	.	.	.	.	.	.	.	.	.	.	.	.	.	.	+	.	.	.	.	.	.	.	.	.	.	.	.	.	.	.	.	.	.	.	.	.	.	.	.	.	.
Ruschia spinosa	Succulent	.	.	.	.	.	+	+	r	.	.	.	.	.	.	.	.	.	.	1	1	+	.	1	+	.	+	.	.	.	.	+	+	.	.	+	.	+	+	.	.	.	.	.	.	.	.	.	.	.	.	.	.	.	.	.	3	3	2	3	2	3	3	1	1	3	1	.	2
Anacampseros filamentosa	Succulent	r	r	r	.	.	r	r	.	.	.	.	.	.	.	.	.	.	.	+	r	r	.	+	r	.	.	.	.	.	.	r	.	.	+	r	.	.	.	.	.	.	.	.	r	r	.	+	r	+	+	.	.	.	+	.	.	.	.	.	.	.	.	.	.	.	.	.	.
Aristida canescens	Grass	.	.	.	.	.	.	.	.	.	.	.	.	.	.	.	.	.	.	.	.	.	.	.	.	.	.	.	.	.	.	+	+	.	.	.	.	.	.	.	.	.	.	.	.	.	.	.	.	.	.	.	.	.	.	.	.	.	.	.	.	.	.	.	.	.	.	2	.
Bidens bipinnata	annual	.	.	.	.	.	.	.	.	.	.	.	.	.	.	.	.	.	.	.	.	.	.	.	.	.	.	.	.	.	.	+	r	.	.	.	.	.	.	.	.	.	.	.	.	.	.	.	.	.	.	.	.	.	.	.	.	.	.	.	.	.	.	.	.	.	.	.	.
Cheilanthes eckloniana	Fern	r	r	r	.	.	.	.	.	.	.	.	.	.	.	.	.	.	.	.	.	1	.	+	+	.	.	+	.	+	+	1	+	+	.	.	+	.	.	1	1	+	+	+	.	.	.	.	.	.	.	.	.	.	.	1	.	.	.	.	.	.	.	.	.	.	.	.	.
Crassula nudicaulis	Succulent	+	.	r	.	.	.	.	.	.	+	.	.	.	r	+	+	.	.	r	1	r	1	.	+	r	r	+	.	.	.	+	.	r	.	.	+	.	r	.	.	.	.	.	.	.	.	.	.	.	.	.	.	.	.	.	2	1	+	2	1	1	+	1	2	1	+	+	1
Digitaria eriantha	Grass	.	+	.	.	.	+	+	.	.	.	.	.	.	.	.	.	.	.	+	+	.	.	.	r	.	.	.	1	2	+	1	+	1	2	.	2	.	.	.	.	+	.	.	.	.	.	.	.	.	.	.	.	.	.	.	4	4	4	2	3	3	3	.	3	3	2	.	.
Euphorbia mauritanica	Succulent	.	.	.	.	.	.	.	.	.	r	.	.	.	.	.	+	.	.	.	1	.	+	.	.	.	.	+	.	.	1	+	2	.	+	.	.	.	.	.	.	r	.	.	.	.	.	.	.	.	.	.	.	.	.	.	2	2	4	4	3	3	3	.	4	.	3	2	1
Euryops subcarnosa	Karroid shrub	.	.	.	.	.	.	.	.	.	.	.	.	.	.	.	.	.	.	.	.	.	.	.	.	.	.	.	.	.	.	+	.	.	.	.	.	.	.	.	.	.	.	.	.	.	.	.	.	.	.	.	.	.	.	.	.	1	.	.	4	.	.	.	1	3	1	4	2
Monsonia angustofolia	annual	.	.	+	.	.	.	.	.	r	.	.	.	.	.	.	.	.	.	.	.	.	.	.	.	.	.	.	.	.	.	r	.	.	.	.	.	.	.	+	.	.	.	.	.	.	.	.	.	.	.	.	.	.	.	.	.	.	.	.	.	.	.	.	.	.	.	.	.
Nidorella resedifolia	annual	.	.	.	.	.	.	.	.	.	.	.	.	.	.	.	.	.	.	.	.	.	.	.	.	.	.	.	+	+	.	r	.	.	.	.	.	.	.	.	.	.	.	.	.	.	.	.	.	.	.	.	.	.	.	.	2	.	.	.	.	.	.	.	.	.	.	.	.
Schkuhria pinnata	annual	.	.	.	.	.	.	.	.	.	.	.	.	.	.	.	.	.	.	.	.	.	.	.	.	.	.	.	.	.	.	r	.	.	.	.	.	.	.	.	.	.	.	.	.	.	.	.	.	.	.	.	.	.	.	.	.	.	.	.	.	.	.	.	.	.	.	.	.
Strumaria tenella subsp. orientalis	Geophyte	.	.	.	.	.	+	+	.	1	.	.	.	.	.	.	+	.	.	+	r	.	+	+	+	1	+	.	.	.	.	r	+	+	+	+	.	+	+	.	.	.	.	.	.	.	.	.	.	.	.	1	+	+	.	.	1	.	2	2	.	1	1	1	.	.	.	.	.
Themeda triandra	Grass	.	.	.	.	.	+	+	.	.	.	.	.	.	.	.	.	.	.	.	.	.	.	.	.	.	.	1	4	4	+	1	+	.	+	.	1	.	.	.	.	+	.	.	.	.	.	.	.	.	.	.	.	.	.	.	3	.	2	2	3	.	3	.	2	2	3	3	4
Tragus koeleroides	Grass	.	.	.	.	.	r	+	.	.	.	r	.	.	.	.	.	.	.	+	.	.	.	.	r	.	+	.	r	+	.	+	.	+	.	.	.	.	+	.	+	+	+	r	r	.	r	+	+	+	.	.	.	.	+	.	.	.	3	2	3	3	2	3	3	2	2	3	4

Species	Growth form	Cover-abundance values (reading order)
Aristida diffusa	Grass	1 1 + 1 + + r + 2 + + + + + 1 + 1 + + 2 3 3 2
Bonatea speciosa	Geophyte	r r
Diospyros lyciodes	Shrub	r
Eriocephalus ericoides	Karroid shrub	+ + 1 + 1 1 1 1 1 1 + 2 2 2 3 2 1
Eustachys paspaloides	Grass	r + + + 2 3 3 2
Heliophia subcarnosa	annual	+ 3 3 3 1 3 3 1 1
Searsia ciliata	Shrub	1 2 3 3 2
Sarcostemma viminale	Succulent	+ + r + + + + + r + + + 2 4 4 2 3 3 3 2 2 1
Senecia radicans	Succulent	+ + 1 + r 1 + 1 + + + r r 4 5 4 5 4 3 3 3 3 3 2
Crassula tetragona	Succulent	+ r + + r + + + r r + r r r r + + + + 1 1 1 + 1 2 + 2 2 + 1
Ruschia unidens	Succulent	+ 1 2
Aloe grandidentata	Succulent	+ r + 2 1 1 2
Asparagus suaveolens	Karroid shrub	+
Cotyledon orbiculata	Succulent	+ r r r r r r r r + + + 1 1 +
Heteropogon contortus	Grass	+ + r 1 + 1 + + 1 + 4
Pellaea callomelanos	fern	+
Phyllanthus parvulus	perennial forb	+ + + + r r r r 1
Tephrosia capensis	perennial forb	r
Trachyandra saltii	Geophyte	r r r + + r + + + r + 1 1 r 1 + + 1 1 +
Anacamperos telephiastrum	Succulent	r + r r + r
Crassula lanceolata	Succulent	r + +
Crassula setosa	Succulent	r r r
Euryops empetrifolius	Karroid shrub	+ 1 1 2 3 2 4
Opuntia ficus-indica	Succulent	r r
Commelina africana	perennial forb	r r r r + r r r + 2 2 3 2 2 2 1 3 1
Mariscus indecorus	sedge	r r 1 1 + 1 2 2 3 3
Pollichia campestris	perennial forb	+ + r
Selago albida	Karroid shrub	r 3 2 +
Tripteris aghillana	perennial forb	r
Cyperus bellus	sedge	r 1 + + r + + +
Eragrostis trichophora	Grass	r + +
Eragrostis lehmanniana	Grass	1 2 1 + + 3 1 3 2 2 3 3 3 4
Bulbostylis burchellii	sedge	r + r
Euphorbia caterviflora	Succulent	+ + + 1 1 1 + + 2 1 1 1 1
Euphorbia clavaroides	Succulent	r r
Geigeria filifolia	perennial forb	r + + + r r + r 3 2 1

Species	Growth form	Cover-abundance values (left to right)
Hereroa species	Succulent	r . . + + + +
Cynodon hirsutus	Grass	1 1
Rabiea species	Succulent	+ +
Chamaecyce prostrata	perennial forb	+
Sutera caerulea	perennial forb	r + . r . + . + . + . 1 . 3 . 3
Crassula capitella	Succulent	+ . + + + . + . r
Eragrostis chloromelas	Grass	r . +
Nenax microphylla	Karroid shrub	+ . r . r r . r . +
Pterodiscus speciosus	Succulent	r
Othonna protecta	Succulent	+ r . r
Felicia filifolia	Karroid shrub	+ r 1 . +
Chascanum pinnatifidum	perennial forb	r
Conyza podocephala	perennial forb	+ . + + . +
Cymbopogon pospischillii	Grass	1 . 2
Trichodiadema barbatum	Succulent	1 . + . 3 . 3 . 1
Brachystelma dimorphum subsp. g	Geophyte	r . r . +
Stapelia grandiflora	Succulent	+ . 1
Orbeopsis lutea	Succulent	+ . r
Pachypodium succulentum	Succulent	+ +
Adromischus tryginus	Succulent	+
Crassula coralina	Succulent	+ + +
Bulbostylis humilis	sedge	+
Kalanchoe thyrsifolia	Succulent	+ r +
Oxalis corniculata	annual	r +
Pseudognaphalium oligandrum	annual	r
Sporobolus fimbriatus	Grass	+ . 2 . 1
Opuntia lindheimeri	Succulent	r
Raphionacme hirsutus	Geophyte	r
Duvalia corderoyii	Succulent	+
Eragrostis obtusa	Grass	1 2 . 2 2 3
Lessertia annularis	Karroid shrub	3 2 3 3 . 3 3 3 3 2 4
Oxalis depressa	annual	4 2 5 2 3 2 4 3 2 2 4 3 1
Scilla species	Geophyte	4 2 4 2 3 3 2 2 3 2 3 2 1
Senecio inaquidens	annual	4 3 . 3 3 2 2 . 1
Tragus berteronianus	Grass	3 2
Senecio burchellii	annual	2 . 2 . + 2 1

http://dx.doi.org/10.5772/57589

Selago densiflora	Karroid shrub	.	.	.	.	.	.	.	.	.	.	.	.	.	.	.	.	.	.	.	.	.	.	.	.	.	.	.	.	.	.	.	.	.	.	.	.	.	.	.	.	.	.	.	.	.	.	.	.	.	.	.	.	.	.	.	1	.	.	.	.	.	1	.	+	1	1	.	.
Anthospermum rigidum	Karroid shrub	.	.	.	.	.	.	.	.	.	.	.	.	.	.	.	.	.	.	.	.	.	.	.	.	.	.	.	.	.	.	.	.	.	.	.	.	.	.	.	.	.	.	.	.	.	.	.	.	.	.	.	.	.	.	.	.	.	3	1	.	.	.	.	.	.	.	.	3
Eriospermum species	Geophyte	.	.	.	.	.	.	.	.	.	.	.	.	.	.	.	.	.	.	.	.	.	.	.	.	.	.	.	.	.	.	.	.	.	.	.	.	.	.	.	.	.	.	.	.	.	.	.	.	.	.	.	.	.	.	.	.	.	3	.	.	.	.	.	.	.	.	.	.
Pelargonium minimum	Karroid shrub	.	.	.	.	.	.	.	.	.	.	.	.	.	.	.	.	.	.	.	.	.	.	.	.	.	.	.	.	.	.	.	.	.	.	.	.	.	.	.	.	.	.	.	.	.	.	.	.	.	.	.	.	.	.	.	.	.	.	.	.	.	.	+	.	r	+	.	.
Eragrostis curvula	Grass	.	.	.	.	.	.	.	.	.	.	.	.	.	.	.	.	.	.	.	.	.	.	.	.	.	.	.	.	.	.	.	.	.	.	.	.	.	.	.	.	.	.	.	.	.	.	.	.	.	.	.	.	.	.	.	.	.	.	.	3	.	.	.	.	3	.	3	3
Aristida junciformis	Grass	.	.	.	.	.	.	.	.	.	.	.	.	.	.	.	.	.	.	.	.	.	.	.	.	.	.	.	.	.	.	.	.	.	.	.	.	.	.	.	.	.	.	.	.	.	.	.	.	.	.	.	.	.	.	.	.	.	.	.	2	.	.	.	.	.	.	.	.
Ipomoea oenotheroides	geophyte	.	.	.	.	.	.	.	.	.	.	.	.	.	.	.	.	.	.	.	.	.	.	.	.	.	.	.	.	.	.	.	.	.	.	.	.	.	.	.	.	.	.	.	.	.	.	.	.	.	.	.	.	.	.	.	.	.	.	.	2	.	.	.	.	.	.	.	.
Chrysocoma ciliaris	Karroid shrub	.	.	.	.	.	.	.	.	.	.	.	.	.	.	.	.	.	.	.	.	.	.	.	.	.	.	.	.	.	.	.	.	.	.	.	.	.	.	.	.	.	.	.	.	.	.	.	.	.	.	.	.	.	.	.	.	.	.	.	.	.	.	.	.	.	.	1	.
Digitaria argyrograpta	Grass	.	.	.	.	.	.	.	.	.	.	.	.	.	.	.	.	.	.	.	.	.	.	.	.	.	.	.	.	.	.	.	.	.	.	.	.	.	.	.	.	.	.	.	.	.	.	.	.	.	.	.	.	.	.	.	.	.	.	.	.	.	.	.	.	.	.	2	.
Eragrostis superba	Grass	.	.	.	.	.	.	.	.	.	.	.	.	.	.	.	.	.	.	.	.	.	.	.	.	.	.	.	.	.	.	.	.	.	.	.	.	.	.	.	.	.	.	.	.	.	.	.	.	.	.	.	.	.	.	.	.	.	.	.	.	.	.	.	.	.	.	2	.
Indigofera alternans	perennial forb	.	.	.	.	.	.	.	.	.	.	.	.	.	.	.	.	.	.	.	.	.	.	.	.	.	.	.	.	.	.	.	.	.	.	.	.	.	.	.	.	.	.	.	.	.	.	.	.	.	.	.	.	.	.	.	.	.	.	.	.	.	.	.	.	.	.	2	.
Cymbopogon excavatus	Grass	.	.	.	.	.	.	.	.	.	.	.	.	.	.	.	.	.	.	.	.	.	.	.	.	.	.	.	.	.	.	.	.	.	.	.	.	.	.	.	.	.	.	.	.	.	.	.	.	.	.	.	.	.	.	.	.	.	.	.	.	.	.	.	.	.	.	.	1
Enneapogon scoparius	Grass	.	.	.	.	.	.	.	.	.	.	.	.	.	.	.	.	.	.	.	.	.	.	.	.	.	.	.	.	.	.	.	.	.	.	.	.	.	.	.	.	.	.	.	.	.	.	.	.	.	.	.	.	.	.	.	.	.	.	.	.	.	.	.	.	.	.	.	3

Table 1. Releves, species composition and growht form of the Bloemfontein Karroid Shrubland (Gh8) (Values included in the table are according to the Modified Braun-Blanquet cover abundance scale)

Author details

L.R. Brown[1] and P.J. Du Preez[2]

*Address all correspondence to: lrbrown@unisa.ac.za

1 Applied Behavioral Ecology and Ecosystem Research Unit, University of South Africa, Republic of South Africa

2 Department of Plant Sciences, University of the Free State, Republic of South Africa

References

[1] Aucamp, A. & Danckwerts, J.E. 1989. *Weiding: 'n Strategie vir die toekoms.* Agriforum '89. Departement van Landbou en Watervoorsiening, Pretoria.

[2] Bond, W.J. 1989. Describing and conserving biotic diversity. In B.J. Huntley (ed). *Biotic Diversity in southern Africa.* pp 2-18. Oxford University Press, ISBN 0195705491. Oxford.

[3] Botha, A.M. 2003. A plant ecological study of the Kareefontein Nature Reserve, Free State Province. MSc Dissertation, University of the Free State.

[4] Brower, J.E., ZAR, J.H. & Von Ende, C.N. 1990. *Field laboratory methods for general ecology.* Wm.C. Brown Publishers, ISBN 0679051455. Dubuque.

[5] Crawley, M.J. 1997. Biodiversity. In *Plant Ecology*, M.J. Crawley (ed). Blackwell Sciences, ISBN 0632036387. Oxford.

[6] Dingaan, M.N.V. & Du Preez, P.J. 2002. The phytosociology of the succulent dwarf shrub communities that occur in the "Valley of Seven Dams" area, Bloemfontein, South Africa. *Navorsinge van die Nasionale Museum Bloemfontein* 18: (3) 33-48.

[7] Du Preez, P.J. & Bredenkamp, G.J. 1991. .Vegetation classes in southern and eastern Orange Free State (South Africa) and the Highlands of Lesotho. *Navorsinge van die Nasionale Museum Bloemfontein.*7(10): 477-526.

[8] Duncan, A.R. & Marsh, J.S. 2006. The Karoo Igneous province In: Johnson, M.R., Anhaeusser, C.R. & Thomas, R.J. (eds). *The Geology of South Africa.* pp 1 – 691. Council for Geoscience. ISBN 1919908773. Pretoria.

[9] Franklin, A.B., Noon, B.R. & George, T.L. 2002. What is habitat fragmentation? *Studies in Avian Biology* (25):20-29.

[10] Grime, P.J. 1997. Climate change and vegetation. In: *Plant Ecology*, pp. 582-594, ed. M.J. Crawley. University Press, ISBN 0632036387. Cambridge.

[11] Hennekens, S.M. & Schaminée, J.H.J. 2001. TURBOVEG, a comprehensive database management system for vegetation data. *Journal of Vegetation Science* 12: 589-591 ISSN.1100-9233

[12] Hennekens, S.M. 1996. *TURBOVEG: A software package for input, processing, and presentation of phytosociological data.* User's guide, version 2001. University of Lancaster, IBN-DLO.

[13] Hogan, C. 2013. *Habitat fragmentation.* Retrieved from http://www.eoearth.org/view/article/153225

[14] Holmes, P. 2012. Lithological and structural controls on landforms. In Holmes, P. & Meadows, M. (eds) *Southern African Geomorphology - Recent trends and new directions.* Sun Press, ISBN 9781920382025. Bloemfontein.

[15] Keddy, P.A. 2007. *Plants and vegetation: Origins, processes, consequences.* Cambridge University Press, ISBN 139780521864800. Cambridge.

[16] Kent, M & Coker, P. 1992. *Vegetation description and analysis: A practical approach.* Belhaven Press, ISBN0849377560. London.

[17] Kent, M. 2012. *Vegetation description and data analysis: A practical approach.* Wiley-Blackwell, ISBN 9780471490937. Oxford.

[18] Macdonald, I.A.W. 1989. Man's role in changing the face of southern Africa. In Huntley B.J. (ed): *Biotic diversity in southern Africa,* pp 51-78. Oxford University Press, ISBN 0195705491. Oxford.

[19] Magurran, A.E. 1988. *Ecological Diversity and its Measurement.* Princeton University Press, ISBN: 0691084912. Princeton, NJ.

[20] Magurran, A.E. 2005. *Measuring biological diversity.* Blackwell Publishing, ISBN 0632056339. Malden, Australia.

[21] Mcginley, M. (2013). *Species diversity.* Retrieved from http://www.eoearth.org/view/article/156211

[22] Mucina, L. & Rutherford, M.C. (eds) 2006. Vegetation of South Africa, Lesotho and Swaziland. *Strelitzia* 19. South African National Biodiversity Institute (SANBI). ISBN 101919976213. Pretoria.

[23] Plants Of South Africa (POSA)(online checklist). 2009. South African National Biodiversity Institute (SANBI). Pretoria.

[24] Potts, G. & Tidmarch, C.E. 1937. An ecological study of a piece of Karroo-like vegetation near Bloemfontein. *Journal of South Afican Botany.* 2: 51 – 92.

[25] Simpson, E. H. 1949. Measurement of diversity. *Nature,* 163, 688.

[26] Smith, B. & Wilson, J.B. 1996. A Consumer's Guide to Evenness Indices. *Oikos,* 76: 70-82

[27] Snyman, H.A. 1984. Die invloed van verskillende suksessiestadia van natuurlike veld op effektiewe reënval. *Glen Agric* 13(2): 13-15.

[28] Spellerberg, I.F. & Feror, P.J. 2003. A tribute to Claude Shannon (1916 - 2001) and a plea for more rigorous use of species richness, species diversity and the 'Shannon-Wiener' Index. *Global Ecology &Biogeography*, 12: 177-179.

[29] Stirling, G. & Wilsey, B. 2001. Empirical Relationships between Species Richness, Evenness, and Proportional Diversity. *The American Naturalist*, 158: 286-299

[30] Tichⓞ, L., 2002. JUICE, software for vegetation classification. Journal of Vegetation Science 13, 451–453.

[31] Van As, J., Du Preez, J., Brown, L. & Smit, N. 2012. The story of life and the environment: An African perspective. Randomhouse Struik, ISBN 9781770075856. Cape Town.

[32] Van Wyk, A.E. & Smith, G.F. 2001. *Regions of floristic endemism in southern Africa*. Umdaus Press, ISBN 1919766189 Hatfield.

[33] Welman C, Kruger F & Mitchell B. 2007. *Research Methodology* (3rd edition). Oxford University Press Southern Africa, ISBN 9780195789010. Cape Town

Chapter 14

The Nile Fishes and Fisheries

Waleed Hamza

Additional information is available at the end of the chapter

http://dx.doi.org/10.5772/57381

1. Introduction

The Nile basin extends from 4° S to 31° N and includes ten African countries: Burundi, Egypt, Eritrea, Ethiopia, Kenya, Rwanda, Sudan, Tanzania, Uganda and Democratic Republic of Congo (Fig. 1). The source of the Nile water is one of the upper branches of the Kagera River in Rwanda. The Kagera follows the boundary of Rwanda northward, and continues along the border of Tanzania before draining into Lake Kyoga between Lake Victoria and Lake Albert, the Nile rushes along its course for 483 km within rocky valleys and over rapids and cataracts. The section between the two lakes is called Lake Kyoga and is part of the Victorian Nile, which name is used for the river section till its confluence with the Blue Nile in Sudan. The Nile discharges from the northern end of Lake Albert and flows through northern Uganda to the Sudan border where it becomes the Bahr El-Jebel. At its conjunction with Bahr Al-Ghazal, the river becomes the White Nile. At Khartoum, the White Nile is joined by the Blue Nile.

The 1529 km long Blue Nile has its source in Lake Tana in the Ethiopian Highlands. From Khartoum the Nile flows 322 km north to its junction with the Atbara River. Downstream from its confluence with the Atbara River, the Nile traverses the Nubian Desert, and is marked by two profound bends in its course, the first north of Khartoum and the second near Aswan in the Egyptian territory. The Nile flows towards the Mediterranean Sea through the Nile delta that splits into the two main Nile delta branches, the Rosetta and Damietta branches. The Nile River flows generally north to the Mediterranean Sea for a distance of 5,584 km. From its remotest head-stream, the Ruvyronza River in Burundi, the river is 6,671 km long, and it has a drainage area of more than 2,590,000 km² [1].

With a length of 6,695 km from source to Mediterranean outfall, the Nile River is the longest river in the world. The uniqueness of the Nile is not only related to morphometric features of its basin, but to other many facts. No other river traverses such a variety of landscapes, and such a spectrum of cultures and peoples as the Nile. No other river has historically had such

Figure 1. The Nile River trajectory from source to outfall

a profound material effect upon those who dwell along its banks, representing the difference between plenty and famine, between life and death, for multitudes since the earliest stages of human history [2]. The course of the Nile flows from highland regions with abundant moisture to lowland plains with semi-arid to arid conditions [1].

For Nile basin countries with no access to the coast for economic activities, the Nile River is their artery of life and economic sustenance. Its fertile sediments supply agriculture soil with necessary elements for plant growth, and the fish stock in its large natural and manmade lakes are crucial to the food and freshwater security of the Nile basin countries. Fish industry and fishing activities on the major water bodies of the Nile basin are not only a source of affordable animal protein, especially in poor countries, but also provide employment, income and export earnings to the riparian communities.

The principal objective of the present review is to highlight the main freshwater fishes and the major fisheries of the Nile River basin. The ecological status of the Nile fisheries and their economic importance to populations living in the countries of the Nile basin are also reviewed. The main lakes and river systems of the Nile basin will be compared on the basis of their fish species, fisheries potentiality and economic importance to the countries that host them.

2. Major lakes of the Nile basin

The lengthy course of the Nile River is interrupted by several important lakes. Some of them are natural, while others were engineered by dam constructions and artificial reservoirs projects. Within the Nile basin, there are five major natural lakes with a surface area totaling more than 100,000 km^2 (Victoria, Edward, Albert, Kyoga and Tana). These are vast areas of permanent wetlands and seasonal flooding (the Sudd, Bahr Al-Ghazal, and Machar marshes). There are also five major reservoir dams (Aswan High Dam, Roseires, Khasham El-Girba, Sennar and Jebel Aulia) in the basin. Before discharging its water into the Mediterranean Sea, the Nile fills four coastal lakes within its Delta (Lake Mariut, Lake Edku, Lake Burullus and Lake Manzala) with a total area of 1100 km^2 [2].

The share of the Nile basin area for each of the ten countries hosting the river may represent a major or minor fraction of the total area of the country (Table 1). However, the economic significance of the basin may be quite disproportionate to its area, such as it is for Kenya and Tanzania [3].

Nile waters are the mainstay for freshwater supply for agriculture irrigation, navigation, water for human use ("drinking, industrial, domestic uses"), hydroelectric power generation, and of course for exploitation of natural fish stocks and aquaculture fish farming projects.

3. Fish and fisheries of tropical lakes (L. Victoria and L. Tana)

Fishes and fisheries of Lake Victoria and Lake Tana will be reported here, as natural examples of tropical large water bodies of the Nile basin. The combined surface area of these lakes is

Country	Area in basin (km³)	Percentage of total country area in Nile Basin
Burundi	13 000	46%
D R Congo	22 300	1%
Eritrea	25 700	21%
Ethiopia	366 000	32%
Egypt	307 900	33%
Kenya	52 100	9%
Rwanda	20 400	83%
Sudan	1943 100	78%
Tanzania	118 400	13%
Uganda	238 700	98%

Table 1. National areas within the Nile Basin "Source [14]"

75% of the total of the five tropical lakes of the Nile basin. They are also important because they lie at the source of both the White Nile and the Blue Nile, respectively. Their main fishes are representative of the fish species living in the other lakes (Edward, Albert and Kyoga). Details on the other tropical lakes (Edward, Albert, Kyoga, and George) fish fauna and its fisheries are found elsewhere [4], [5].

3.1. Lake Victoria (Kenya, Tanzania and Uganda)

Lake Victoria is the largest natural lake in Africa (Fig. 1) and the second largest lake in the world after Lake Superior (USA-Canada). It has a surface area of 68,800 km² and a very large catchment area of 193,000 km². The shoreline measures 3450 km in length and the lake has a mean depth of 40 m and a maximum depth of 80 m. The lake is shared by Kenya (6%), Uganda (43%), and Tanzania (51%). The lake basin is estimated to have a population of 30 million individuals which is growing at a rate of >3% per annum. Three major cities (Kampala, Kisumu and Mwanza) with a combined population of about six million inhabitants depend on the lake for domestic and municipal water supply and waste disposal. Recent data has shown it is becoming eutrophic [6].

Two introduced species, Nile perch (*Lates niloticus*) and Nile tilapia (*Oreochromus niloticus*), and one native cyprinid (*Rastrineobola argentea*) dominate the lake's fisheries. The introduction of the Nile perch (*L. niloticus*), in the early 1960's had major ecological consequences. It is believed that some 200 endemic halochromine species (previously comprising 90% of the lake fish biomass) have become extinct in the lake due to predation by the Nile perch [7]. Most of these species had no commercial value, but are now considered relevant to the sustainability of the Lake Victoria ecosystem [8]. Lake Victoria fisheries production has increased markedly since the introduction of the Nile perch. Despite the general unreliability of published data, there is agreement that between the late 1980's and early 1990's there was an increase in the quantity of fish landed in Lake Victoria fisheries reaching a total of 500,000 tons per year. This figure was in decline by the late 1990's and early 2000's when it reached a total of 340,000 ton per year (Fig. 2). This reduction was mainly due to the overexploitation of the fisheries resources, and to the increase in the number of motorized boats, and to the Nile perch processing industries. Reportedly the number of fishing boats in the lake increased from 12,000 boats during 1982 to 22,700 boats during 1990 and more than 52,000 boats in 2002 [6]. In recent years, as fishing pressure on the Nile perch has intensified, there are signs of recovery in at least some of the former haplochromine species that were thought to be extinct from the lake environment [9], [10], [5].

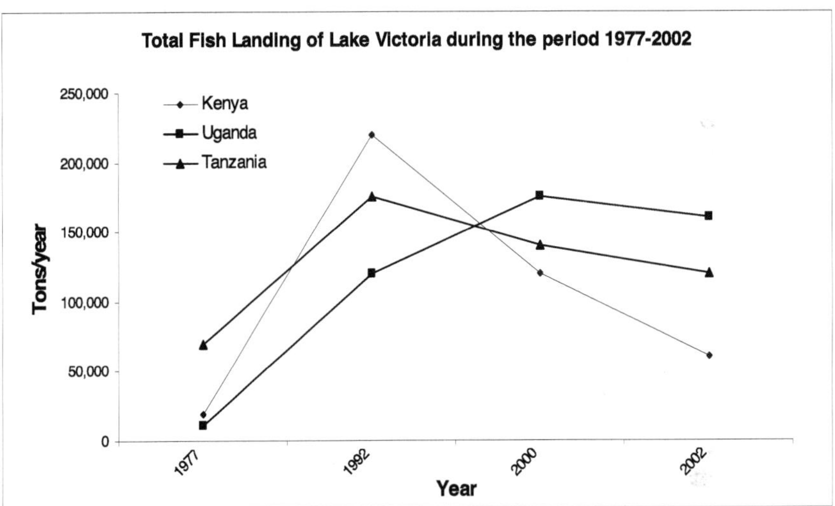

Figure 2. Variation of the total fish landing of Lake Victoria during the period 1977-2002 [8]

Extracted data from the results obtained by [6], from the ECOPATH prediction model have revealed that the stock biomass (tons km^{-2}) of the fishes (Nile perch, tilapia and sardine) are different in the waters of Lake Victoria shared by different countries. The same is also true for the ratios of production per biomass except for the sardine species (Fig. 3). The data also shows that the Ugandan lake areas have lowest stock of Nile perch biomass, while the highest stocks are found in the Kenyan lake areas. Interestingly, the Ugandan stock biomass of tilapia is the highest while the Tanzanian stocks are the lowest. The sardine stock biomass is higher in the Kenyan lake areas followed by the Tanzanian, with the lowest recorded in the Ugandan lake areas (Fig. 3). An overall estimate shows that the average stock biomass for the whole of Lake Victoria is 4.59 tons km^{-2} for the Nile perch, 3.27 tons km^{-2} for the sardine and 2.36 tons km^{-2} for the tilapia. The same trend is observed for the estimate of production/biomass $year^{-1}$, giving the ratios 3.75, 3.0 and 1.75 for the three fishes, respectively.

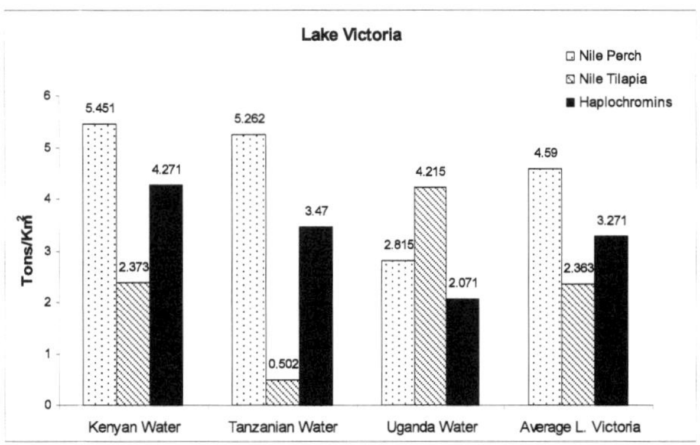

Figure 3. Biomass potentials of different water sectors of Lake Victoria for the main three fish species (tons km^{-2}) [8]

The artisanal fishing gear was in use on traditional fishing boats in Lake Victoria fisheries until the industrial exploitation of the Nile perch for international commercialization. At that time, motorized boats and improved fish gear "even illegal" was used. The most common gear used in Lake Victoria fisheries are gill nets and long-line hooks. During the period of 2000 and 2004, the numbers of gill nets in use increased by 50% reaching 980,000 gill nets, while the numbers of long-line hooks is estimated at 8 million in 2001. This number was adjusted to only 6 million during 2006 [6]. The current haul is estimated at 45 kg boat $^{-1}$ day $^{-1}$ compared with the 80 kg boat $^{-1}$ day $^{-1}$ figure during 1990's. This indicates over-intensive exploitation of the fisheries of Lake Victoria fish resources and argues for sustainable management intervention.

3.2. Lake Tana *(Ethiopia)*

Lake Tana is the largest lake in Ethiopia (surface area 3,050 km^2). It lies at an altitude of approximately 1800 m a.s.l. in the north-western highlands of Ethiopia, 500 km north of the capital Addis Ababa (Fig. 1). The lake is shallow (average depth 8 m, maximum depth 14 m), and its trophic status is described as oligo-mesotrophic [11]. The lake's fishery is 30 km downstream from the Blue Nile outflows, and is isolated from the lower Nile basin by 40m high waterfalls. Physical and hydrological features of the lake water follow a seasonal pattern. Rainfall peaks in July-August are followed by a rise in the lake water level by 1.5 m, reaching highest levels in September-October [12].

In their study, DeGraaf et al. [12] identified the three endemic fish species groups Barbs or yellow fish (*Labeobarbus* spp.), African catfish (*Clarias gariepinus*), and Nile tilapia (*Oriochromus niloticus*), as the main representatives of Lake Tana fisheries. They noted that before 1986 Lake Tana was a subsistence fishery exploited predominantly by reed boat fishing vessels. This type of fishery was limited to the shore areas and targeted the native Nile tilapia using locally made fish traps and small gill nets.

In 1992-1993 about 113 reed boats were counted on the lake and about 374 gill nets were in use, with an overall daily haul averaging 12.3 kg (7.8 kg Nile tilapia, 4.3 kg *Labeobarbus* spp. and 0.2 kg African catfish). Here it is important to indicate that *Labeobarbus* spp. in this lake is represented by nearly 15 species. It was also reported at that time that the fishermen in the lake numbered about 400 individuals. This meant that each reed boat had only one fisherman [12]. The introduction of motorboats to Lake Tana fisheries was a consequence of the increasing demand for fish from the capital Addis Ababa. This introduced an additional 130 professional fishermen, which markedly negatively influenced the fish stock in the lake. In fact, the annual total catch fell to 255 metric tons in 2001 from a value of 360 metric tons in 1997. Not only was there a reduction in total catch but also the percentage composition of the individual fish species yield varied between 1993 and 2001, so that the catch of *Labeobarbus* spp. declined by almost third in 2001 compared to 1993. Notably, the catch of Nile tilapia increased 50% in 2001 compared to 1993 (Fig. 4). This may reflect the intensive exploitation of *Labeobarbus* spp. and the superior ability of the Nile tilapia to multiply in the lake.

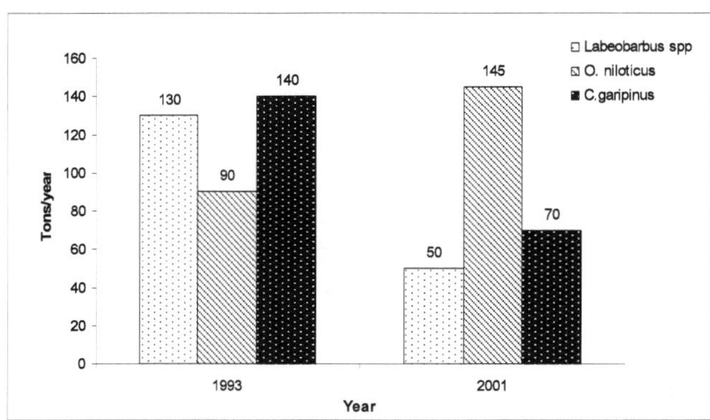

Figure 4. Total catch of the three main species from Lake Tana during the years 1993 and 2001 [13]

4. Fish and fisheries of Mid Nile (Sudan freshwater fish and fisheries)

Although Sudan is a country with a very large surface area (2,505,825 km^2), its inland water bodies occupy only 114,000 km^2 during periods of high water level. The inland fisheries of Sudan are based on the Nile river tributaries, contributing over 90% of the estimated production potential of the country. The Sudd swamps in the south (Fig. 1), and the man-made lakes on the White Nile, the Blue Nile, Atbara river and the main Nile river (Lake Nasser portion in Sudan "Nubia") count as the major fishing localities with respect to fish resource magnitude and exploitation thrust [13]. The total fish production of the various inland water bodies and wetlands are estimated at 50,000 tons year $^{-1}$. This figure is not representative of the full potential of the various fisheries (Table 2), as in some areas civil war disturbances and dense cover of aquatic macrophytes, and inadequate fishing gear have had a negative effect on local fish production.

Area	Fish Catch Potential Tons year^{-1}	Actual Yields Tons year^{-1}	Percentages of Exploitations
Sudd Region	75,000	30,00	43%
White Nile Reservoir	15,000	13,000	86%
Blue Nile Reservoir	1,700	1,500	88%
Senner Reservoir	1,100	1,100	91%
Lake Nubia Portion	5,100	1,000	19.6%
Others	4,000	4,000	100%
Total	101,900	50,500	≈ 50%

Table 2. Calculated Potential and Effective Fish Catch from Different Inland Water Bodies of Sudan

The predominant fishing gear includes active and passive gill nets, seine nets, trammel nets, long line hooks, cast nets and baskets. The commercially important fish are Nile perch (*Lates niloticus*), Bagrid catfish (*Bagrus bayad*), Silver catfish (*Bagruc docmac*), Nile tilapia (*Oreochromus niloticus*), Carp fish (*Labeo* spp.), Barbs fish (*Barbus binny*), Mormyrus fish (*Mormyrus* spp.), Nile Distichodus (*Distichodus* spp.), Tiger fish (*Hydrocyon* spp.) and Nile robber (*Alestes* spp.). There are many other species, though they have no commercial value [13].

It is important to state that, apart from the FAO reports, there is almost no scientific literature available on the ecological features and dynamics of fish populations in the inland waters of Sudan. This underlines the importance of organizing a research program on inland fisheries in this area in order to fill the gap of knowledge about the fisheries status in this important part of the Nile basin. This is especially important for the wet lands which are not limited to Sudan but also exist in the ten countries of the Nile basin, and account for an area of almost 200,000 km². A characteristic of wetlands fisheries is the variety of traps that are used to catch fish in dominantly submerged or emergent vegetation habitats. Many of these traps have traditional designs adapted to local conditions and most are made from local plant materials - often from wetlands themselves. One type of fish requiring a special fishing technique is the African lungfish (*Protopterus aethiopicus*), which is native to Ethiopian wetlands and inhabits the seasonal wetlands where it aestivates in the dried soil and is "hunted" during the dry season.

5. Downstream fishes and fisheries (Egyptian freshwater fisheries)

Egypt is the downstream country hosting the last 1530 km of the Nile river channel. The Nile enters Egypt through the Nubian Lake which continues as Lake Nasser behind the Aswan High Dam (AHD). The water allowed to pass through the AHD is the Egyptian quota of Nile water (55.5 Km³ year $^{-1}$) as outlined in an agreement with Sudan in 1959. This quantity satisfies the freshwater demands of the Egyptian population for use in agricultural irrigation, industry, domestic and navigation purposes. On the way to its Mediterranean destination, the Nile water fills many canals within the irrigation network in the western desert and the Nile delta. It also contributes to the water budget of the Northern Delta lakes [2]. Thus the freshwater inland water fisheries in Egypt include Lake Nasser (the second largest man made reservoir in Africa, after Lake Volta-Ghana), the Nile main river channel, irrigation channels, some water bodies in the Western Desert, the Nile branches (Rosetta and Damietta), and the Northern Delta Lakes (Manzala, Burullus, Idko and Mariut). The excess water is then discharged into the Mediterranean Sea through the lake connections with the sea via direct and indirect inlet openings [2]

The annual fish yield from freshwater fisheries of the Egyptian Nile basin has increased annually from 157,888 metric tons in 1990 to 224,940 metric tons in 2000 [14]. The catch increased for a number of reasons, such as the nutritive effect of sewage and fertilizer discharge in drainage channels and lakes, the intensified fishing activities and adjustment of incensement statistical techniques. Apart from the intensive aquaculture that has developed during the last twenty years in Egypt, the major contributors to the freshwater fish catch could be listed as Lake Nasser, Nile branches and irrigation channels as well as the northern Delta lakes. Their contribution may be summarized as follows:

5.1. Lake Nasser fishes and Fisheries

Lake Nasser lies behind the AHD and its extension in the northern Sudan is referred to as the Nubia reservoir. This lake has a depth reaching 180m and covers an area of 6216 km², 5248 km² of which are in Egypt and the rest in Sudan (Fig. 1). The total fish catch from Lake Nasser has been estimated as 28,153 tons year $^{-1}$ [15]. Tilapia species, mainly the Nile tilapia represent 90% of the total fish catch, while Nile perch and *Barbus* spp. cover the remaining 10%. Despite the large area of the lake, it contributes only about 10% of the total freshwater fish landing in the Egyptian Nile basin. This is probably due to its deep waters and the low number of motorboats with suitable fishing gear adapted to its bathymetry.

5.2. Nile branches and irrigation channels fishes and fisheries

There are more than 100 sites recorded along the Nile branches and major irrigation channels at which freshwater fishes are collected for marketing. Most fishermen in these areas are not registered. The fishing fleets at these sites comprise hundreds of small wooden boats (4-6 m in length). The fishing gear used is mostly primitive, though trammel nets are in use in some areas. The common fish species caught from these channels are tilapia species (*O. niloticus, O. aureus, Sarotherodon galilaeus* and *Tilapia zilli*) as well as *Clarias* spp.

Although these fishing sites are sparsely distributed and reliable statistics are in short supply, it has been estimated that 34% of the total Egyptian freshwater fish catch is slated to the two main Nile branches (Rosetta and Damietta) and the major irrigation channels [16].

5.3. Northern Delta lakes (NDL)

As the final reservoirs of Nile river water before it flows into the Mediterranean, the four lakes (Mariut, Edku, Burullus and Manzalah) are the last opportunity for Egyptians to use the Nile water within the Delta area (Fig. 5). The four lakes occupy an area of about 1100 km². They are commonly shallow (average depth 1.10m), and their water salinity is known to change from fresh to brackish in the seaward direction. They are connected to the Mediterranean Sea either directly or indirectly.

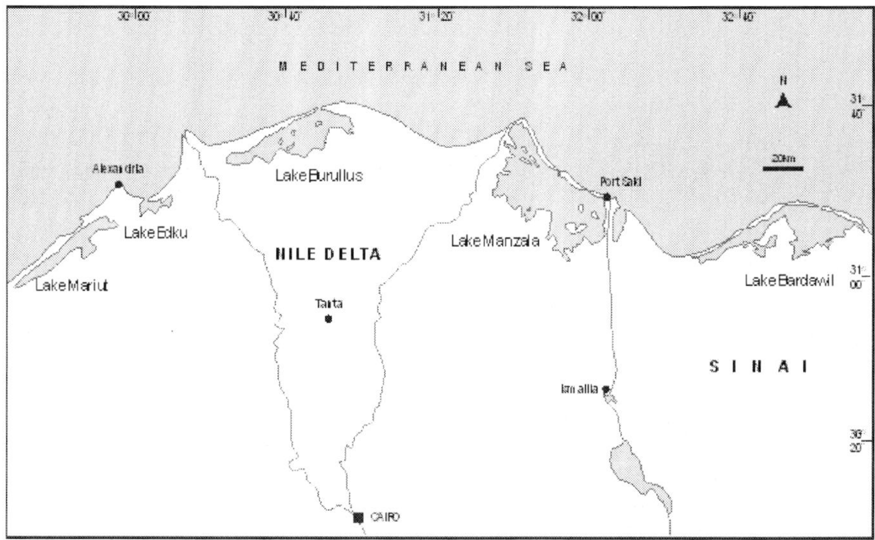

Figure 5. The Northern Nile Delta lakes

During the last 50 years, the surface area of the four lakes has been reduced to as little as 50% of the area they covered during the 1950's (Table 3). The lake areas lost have been used for highway construction, reclamation of new lands for agriculture, and in some cases, for residential constructions. Such unnatural use of the lake basins has certainly modified their ecosystems as a whole, beginning with hydrological changes and resulting in the reduction of their economic and natural value. Due to their location along the coastal area of the Egyptian Mediterranean, the lakes are surrounded by or included within the highly populated coastal cities of the country. This has become more pronounced, particularly during the last 20 years, in the various forms of human impact on these water bodies, such as industrialization and the construction of new infrastructures in response to population density increases. These impacts are manifested in the deterioration of the water quality and reduction in lake surface area [17]. The consequent negative effect of human impact on the NDL ecosystems became evident in the decline in fish production. Up to 1985 the annual fish production of these lakes accounted for >50% of the total annual yield for the country.

Year's	L. Mariut	Edku L.	L. Burullus	L. Mazalah
1950's	136.9	150.1	571.8	1274.2
1970's	68.4	129.2	534.4	997.3
1980's	66.1	115.3	481.6	904.1
1990's	62.3	109.7	350.8	650.3

Table 3. Gradual changes in the surface areas (Km^2), of the Northern Delta Lakes during the last 50 years "After, [17]"

According to the year 2001 records, the lakes fisheries had a total fish production of about 140,000 tons year $^{-1}$, which represented about 60% of the freshwater fish catch and about 30% of the total annual yield of Egypt [16]. The main freshwater fish species living in the NDL are Nile tilapia (*Oreochromis niloticus*), Blue tilapia (*O. aureus*), Mango tilapia (*Sartherodon galilaeus*) and *Tilapia zilli* also known as *Tilapia melanopleura*, Nile perch (*Lates niloticus*), Bagrid catfish (*Bagrus bayad*), Nile carp (*Labeo niloticus*), Yellow fish (*Barbus prince*), Barbus fish (*B. bynni bynni*), and African catfish (*Calrias gariepinus*). Other species, such as Grey mullet (*Mugil cephalus*),Thinlip grey mullet (*Liza ramada*), Saline mullet (*Liza salinus*) and Eel fish *Anguilla anguilla*, belonging to the brackish-marine sectors of the lakes, represent a minor part of the total catch of these lakes. The fishing fleets and gear used in the lakes are generally primitive and they are adapted to the shallow water depth and the presence of abundant emergent and submerged macrophytes. Thus wooden boats varying in size from 4 to >20 m in length and a number of net types and sizes are used.

6. Water quality in Major Nile basin fisheries and its impact on fish production

6.1. Water quality in the upstream basins

Studies of water quality of the Nile basin and its channels and lakes have demonstrated a trend of lower water quality in the downstream basins compared with those located upstream. This is mainly due to the following:

The all-year round continuous flow of the river from south to north which dilutes existing pollutants and contaminants and carries them downstream.

Less industrialization of upstream countries compared to the downstream ones.

Neither water multi-use nor recycling processes are common practice in upstream countries compared with the downstream ones.

Low population densities and less modernized sewage systems at upstream countries compared with the downstream ones.

In addition to the above, the main impact of human activities in the areas surrounding the upstream Nile river water bodies may result from the uncontrolled drainage and discharge of agriculture fertilizers, insecticides, herbicides and untreated sewage into the Nile basin. This is probably true for Lake Victoria and Lake Tana, since industrial pollution and discharge of inorganic pollutants are more limited. However, in these upper parts of the Nile basin, the existing civil wars and the absence of hygiene have resulted in periodic devastating epidemics such as cholera and other infectious diseases that may also negatively influence fisheries production either directly or indirectly.

Studies of Lake Victoria water quality have indicated that notable changes in physical, chemical and biological features of the lake ecosystem have occurred during the last three decades. These changes have resulted in increasing eutrophication and decreasing water transparency and decreasing dissolved oxygen concentrations at the hypolimnion layer during periods of stratification [18], [19]. These studies indicated also that water eutrophication in Lake Victoria is a direct result of nutrient enrichment from human activities in the catchments area as well as industrial and domestic sewage discharge from combustion processes. That is in addition to soil erosion and fertilizers' washout through drainage since the introduction of tea, cotton and maize farming.

Recent studies of pesticide residues in Nile tilapia and Nile perch from southern Lake Victoria has implications of recent exposure of these two species to DDT and endosulfan isomers. The study concluded that most of the analyzed samples contained residue levels higher than the method detection limits, though below or within the Accepted Daily Intake (ADI) limits [20]. Other studies have indicated that spraying of endosulfan in cotton fields near Tana River (Kenya) result in levels of fish contamination [21], [22].

The most surprising finding of these studies is the detection of these compounds in fish but not in local sediments and water. It has been found that the high demand for fish has pushed some Lake Victoria fishermen to use DDT as a mean to catch fish in certain areas of Lake Victoria. The reason of the absence or the inability to detect such compounds in the lake sediments could be due to dilution effect of the running water in these basins which made their residuals below the detection limits.

6.2. Water quality in the Mid-River basins

In Sudan the water quality of the southern wetlands (Sudd area) depends on how much water is supplied annually to its basins. The continuous alteration of wetlands by human activities and their utilization for other purposes have negatively influenced both the quality and the quantity of water reaching this area. The redirection of water away from the Sudd for human activities has resulted in the formation of isolated ponds, promoting water stagnation and growth of macrophytes, and weeds. These practices have negatively influenced the capacity of fisheries in this part of the Nile basin.

At Lake Nasser, where both the White Nile and the Blue Nile water mix in different portions, the residues of Nile water contaminants from the upstream countries via Sudan reach this huge man-made reservoir shared by Sudan and Egypt. Lake Nasser water quality is affected also by the continuous fluctuation of its water level. In this regard, most of the studies focused on detecting contaminant levels in the lake water and its effect through bioassay experiments of living biota. Unfortunately, these types of studies did not show the full effect of these pollutants on fish and the degree of pollution of the lake water. Recently, a monitoring study of a range of heavy metals (Co, Cr, Cu, Fe, Mn, Ni, Sr, and Zn) in Nile tilapia fish of different ages (1, 1.5, 2, 2.5 and 3 years) from Lake Nasser was undertaken by [23]. This study examined the trace element concentrations in water, sediments, and aquatic plants, as well as different edible and non-edible fish organs at different ages. The study showed that both water samples and aquatic plants have higher levels of Fe compared with concentrations detected in fish edible parts (muscles). On the other hand, the concentrations of the various metals in sediments were higher than their concentration in fish muscles. The author concluded that the concentrations of Co, Cr, Cu, Fe, Mn, Ni, Sr, and Zn in the edible parts of the examined Tilapia fish were within the recommended permissible levels for human consumption (Table 4).

Elements	Zn	Sr	Ni	Mn	Fe	Cu	Cr	Co
Aquatic plantMean (µg/g DW)	175 (±6)	240 (±9)	19(±1.2)	740 (±8)	1720 (±14)	68 (±6)	29 (±1.1)	35(±1.6)
SedimentMean (µg/g DW)	143 (±2.9)	455 (±12)	122(±3.2)	1000 (±19)	51500 (±24)	109 (±2.6)	79 (±2.0)	89.5 (±2.6)
WaterMean (µg/g DW)	230 (±0.84)	1852 (±18)	145 (±1.02)	186 (±0.9)	1420 (±8)	220 (±1.1)	240 (±0.58)	185 (±0.8)
FishMean (µg/g DW)	1.55 (±0.08)	1.92 (±0.99)	0.19 (±0.25)	0.5 (±0.066)	6.45 (±1.1)	0.27 (±0.08)	0.29(±0.22)	0.25 (±0.05)

Table 4. Heavy metal concentrations in fish (*T. nilotica*), aquatic plant (*N. armeta*), sediment and water samples from Lake Nasser. "Modified from, [23]". DW, dry weight; (±), Standard deviation

6.3. Water quality in the downstream basins

In their location farthest downstream in Nile basin, the Nile Delta branches (Damietta and Rosetta) and the Northern Delta Lakes (NDL) are, not surprisingly, the most polluted and have the lowest water quality. The trophic status in these water bodies has been described in the scientific literature as varying between eutrophic and hypertrophic conditions [17], [2]. The continuous aggressive human impacts on the water bodies of the NDL since the 1950's until now have resulted in various negative consequences, beginning with the reduction of their surface area (Table, 3), and increase of nutrients discharge through agricultural, industrial drainages and domestic sewage; and leading to the dramatic decline of their fisheries [24]. Both domestic and industrial sewage represent a major source of nutrient enrichment of water bodies of the Nile delta. These contribute significantly to the development of the eutrophication phenomenon and the consequent degradation of water quality. In a number of cases, municipal and rural domestic wastewater is discharged directly into waterways. The constituents of domestic and industrial input to water resources are pathogens, nutrients, trace metals, suspended solids, salts and oxygen demanding materials [25]. Siegel [26] reported that the nutrient base for aquaculture in Lake Manzalah is sewage carried by drains from as far away as Cairo, 140 km to the south. He added that untreated or poorly treated industrial wastes, heavy metals and other pollutants have been released into the Nile Delta drainage network and have been discharged, along with sewage and agriculture wastes, into the northern delta lakes and their associated wetlands. Studies of sedimentary deposits in the southeastern sector of Lake Manzalah have detected different trace metal concentration. They have detected Hg (up to 822 ppb), Pb (up to 110 ppm), Zn (up to 635 ppm), Cu (up to 275 ppm), Cr (up to 215 ppm), Sn (up to 14 ppm) and Ag (up to 4.7 ppm). The results suggest that high concentrations of heavy metals in lake sediments may cause contamination in fish, especially bottom feeders in such environments [26]. This may be the case in the northern lakes, since all of them receive the sewage discharges of major cities.

Location	Sampling	HCB	DDTs	PCBs	Γ-HCBs
River Nile					
Cairo	1993	<0.001	0.08-0.12	3-640	0.05-0.07
Rosetta Branch	1998	1-77	0.2-99	5-161	
Damietta Branch	1988	<DL-93	90-102	25-53 7-21	<DL-126
Nile Estuaries					
Rosetta	95-97	197-217	83-97	166-181 390-430	286-310
Damietta	95-97	195-240	109-128	270-330 581-700	312-352
Coastal Lakes					
Lake Manzala	1993	0.03-0.18	0.10-0.56	18-48	0.37-1.57
	92-93	0.84-2.28	26-55 3-26	<DL-20	8.6-12.1 1-11

Table 5. Concentrations (ng/l) of HCB, DDT's, PCB's and lindane in freshwaters from different locations of the Nile Delta- Egypt "Modified from [27]" <DL: below detection limit

Although both agricultural drainages and sewage discharges into the lake environments may be considered as sources of nutrient that can compensate the nutrients reduction in the Nile water after the construction of Aswan High Dam (AHD), these discharges have different pollutants, such as organochlorine compounds, which are not only harmful to the fisheries but also to the consumers of fish. In his study, [27] noted that the use of organochlorine insecticides in Egypt began in the 1950's and was extensively used until 1981 to protect crops from insects, disease fungi and weeds, to remove undesired vegetation and for domestic household use in the control of insects. The reported active ingredients of major organochlorine pesticides during a 30-year period include: toxaphene 45,000 Mt (1955-1961), endrin 10,500 Mt (1961-1981), DDT 13,500 Mt (1952-1971) and lindane 21,000 Mt (1952-1978) [28]. DDT is still in limited use in the country as a rodenticide and termiticide. These pollutants are released into drainage canals and reach the delta branches and NDL. According to the data provided by the Ministry of Agriculture, the application of pesticides decreased in the country, from 20,500 tons in 1980 to 16,435 tons in 1995. Since chlorinated pesticides have been banned, the majority of the pesticides currently in use are organophosphorus compounds [27]. Table 5 shows concentrations of PCB's, DDT's, HCB and Lindane present in the water bodies located in the Delta region.

7. Freshwater aquaculture in the Nile basin

Freshwater aquaculture activities in the Nile basin countries depend on the following main factors:

1.   The status of fish as a main food source according to the customs of the local populations

The existence of a market demand for specific fish living in the local river and lake waters

The economic value of these activities for the private sector compared with other projects

This is clearly demonstrated by the estimation of aquaculture production within the Nile basin countries (Table 6). Although Lake Victoria produces almost 25% of the freshwater fish catch in Africa, the aquaculture productivity of countries sharing the basin of Lake Victoria is very limited. This could be due to the lack of interest of the private sector in developing aquaculture activities, since the Lake production easily satisfies local and international market demand. In Uganda the number of ponds totals 6200, covering a total area of 124.6 ha. In some of these ponds yields of about 10,000 kg ha^{-1} y^{-1} were recorded. The most widely distributed species are mirror carp and various tilapia species. However, since the Nile perch is the most popular fish for export, there is little interest in developing aquaculture. It is expected that future expansion in freshwater farming in this part of the Nile basin is promising, especially when the fish stock at Lake Victoria become overexploited.

Although freshwater aquaculture was not commercially practiced in Sudan until 1990, the population density increase and local market demands have encouraged large scale aquaculture of freshwater fish, especially Tilapia species. The main demand comes from the capital Khartoum, where population density is rising, with growing interest in tilapia fish as an alternative source of animal protein [3].

In Egypt, where water conservation is becoming the highest priority, the reutilization of drainage water from other activities has assisted private sector investment at aquaculture activities in general and freshwater fish farms in particular. This has been spurred by the Egyptian tradition of consuming fish as an essential food item. The local market demand for fish has increased tremendously during the last three decades due to population growth.

Country	1987	1988	1989	1990	1991	1992	1993	1994	1995	1996 -	2007
Burundi	25	24	25	30	50	50	55	55	50	10	200
DR Congo	723	759	760	700	700	730	700	750	750	750	2970
Eritrea	-	-	-	-	-	-	-	-	-	-	-
Ethiopia	-	1	33	36	36	22	28	33	55	85	-
Egypt	45000	50000	55000	55916	877	54195	45380	46887	45473	53302	635516
Kenya	126	251	603	698	722	749	763	813	893	500	4240
Rwanda	64	38	44	164	58	53	53	50	49	50	4038
Sudan	43	45	100	234	203	200	200	200	1000	1000	1950
Tanzania	35	37	375	375	400	350	200	150	200	200	10
Uganda	38	34	42	52	63	77	87	179	194	210	51110

Table 6. National aquaculture production in the countries of the Nile Basin [29], [30]

Consequently, the price of captured fish has risen, assisted further by the degradation of the natural fisheries. The result was that fish moved out of the affordable range for poor people. The high price of fish was remedied by freshwater aquaculture, which is a very well-known practice in Egypt since 1930. Although records were not available at that time on the size of production, the FAO statistical services sector was able to establish records of the Egyptian fisheries and aquaculture beginning in 1950. These records indicate that the total fish catch from aquaculture has linearly increased up to 539,748 tons year $^{-1}$ in 2005 (Figure 6). This quantity represents not only aquaculture freshwater fish but also brackish and marine cultured ones. Data suggests that >60% of this quantity consists of freshwater fishes, mainly Tilapia species.

During the last two decades, and due to the reutilization of drainage water, the salinity in these water bodies has increased, leaving them brackish. These changes in water salinity are paralleled by changes in the cultured fish species. In fact, most of the fish farms responded to changes in water salinity by turning from Tilapia to carp fish farms and then to grey mullet. The total income of produced fish from aquaculture in Egypt is estimated at 971,846 * 10^3 US $ per year (Figure 6), according to the available estimate of 2005 [14]. Based on these figures it is estimated that the supply of fish per capita has increased from 7 kg year $^{-1}$ in 1990 to 15 kg year^{-1} in 2003. In fact, Egypt is not only the leading country in the Nile River basin in freshwater aquaculture (Table 6), but also the first amongst the Mediterranean countries and the Arab world [31], [29], [3], [13], [14].

8. Economic value of freshwater fisheries of the Nile basin

There is no doubt that freshwater fisheries provide an essential supply of animal protein to large parts of the developing world. At national, community and family levels these systems

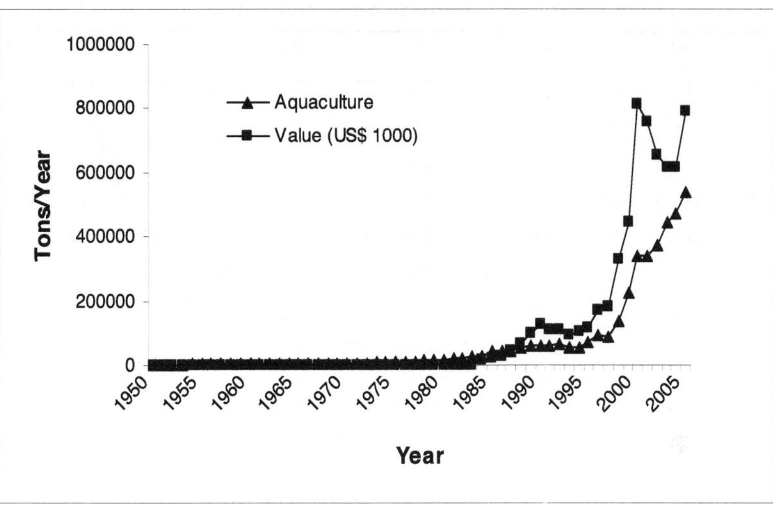

Figure 6. Aquaculture fish production in Egypt and its Economic value during the period 1950-2005 [14]

are critically important in sustaining food security. However, this is not the case for the upstream countries of the Nile River basin, where, although the great lakes have abundant fish production, there is a lot of concern about malnutrition of the populations surrounding the basin. An example is provided by Lake Victoria which supports Africa's largest fishery. Its most valuable product is the Nile perch, much of which is exported. This has given rise to arguments claiming a direct linear relationship between perch exports and disturbingly high rate of malnutrition along the lake's shore [32]. This state of affairs is mainly due to the foreign fish processing companies that control the fish exports, and the high demand on the Nile perch especially in the EU countries. With such high demands for Nile perch, the value of the fisheries has risen considerably. In 1983, there were 12,041 boats operating on the lake. By 2004, there were 51,712 boats and 153,066 fishermen. Of the 1433 landing sites identified in 2004 along the lake shores, just 20% of the population had communal lavatory facilities, 4% were served by electricity and 6% were served by potable water supply [32]. The authors have concluded that, Policy options can include targeting health care specifically at the lake's communities. They also added that local organizations such as LVFO (Lake Victoria Fishery Organization), should translate the largest inland fishery benefit into discernible positive impact on the ground. That means positive economic return has to be reflected on fishermen families.

The Ugandan portion of Lake Victoria fish is among the highest revenue foreign exchange earner for non-traditional exports. In fact, the fish exports registered unprecedented growth in earnings, coming second after coffee exports in 2001. Its proportional contribution to export earnings rose from 5% in 1994 to 17% in 2001. It is estimated that fish exports in 2001 were in the order of 28,000 metric tons, with a value of approximately US$ 78 million. That estimate is

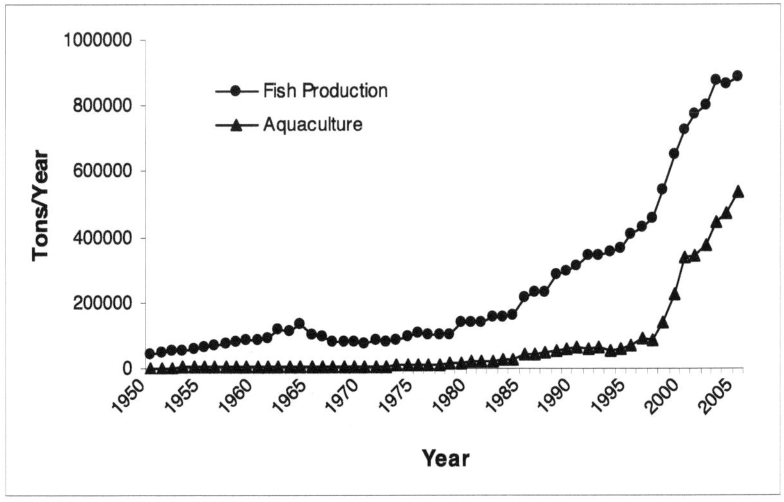

Figure 7. Growth trends in total fish production in Egypt (marine and freshwater catch) compared with aquaculture during the period 1950-2005 [14].

only for the Nile perch. Other estimates included Nile tilapia from other countries bordering the lake (Kenya and Tanzania). These account for a total catch of 110 metric tons. However, the difference in value is marked, since the Nile perch is exported to EU countries, while the other species and small size Nile perch are locally consumed or regionally exported at lower prices. Recent statistics on the annual total fish production of the three countries (Kenya, Tanzania and Uganda) from Lake Victoria is amounted to 960, 500 tones [30]

In Sudan freshwater fisheries contribute very little to the national economy, especially as the estimated fish consumption per capita is only 1.7 kg year^{-1}. In general, the fishing industry in 2003 is estimated to have exported 1,629 tons of fish with a value of US$ 533,000, while a quantity of 157 tons was imported for local consumption at a cost of US$ 324,000. This means that the total earnings from fish production did not exceed US$ 200,000 representing only 0.4% of the national GDP [13].

As previously highlighted, Egyptian fisheries, including both natural fisheries and aquaculture represent the largest producers of freshwater fishes within the Nile basin countries. This is mainly due to the rapid growth in freshwater aquaculture production, amounting to about 60% of the national total fish catch (Figure 7). Despite high fish production, the fishing industry plays a relatively minor direct role in the economy of Egypt. Nevertheless, domestic fish production makes a valuable contribution to the national food supply and to the traditional way of life, for which fish eating remains an important part. In addition, it is a significant source of food for the tourist industry.

Fishing industry is also important for the livelihoods of over 65,000 fishermen and other people employed full-time in related activities (estimated at some 300,000 men). The fish industry in Egypt also produces fish powder (from non-edible fish) of about 543 tons to be used as nutritive food additives in poultry farms [13].

9. Future of freshwater fish production of the Nile basin

Without exception, the freshwater fisheries in the Nile basin countries are under severe and constantly stressful conditions. The Nile fisheries have continued to deteriorate despite its impressive freshwater resources and the many efforts by national and regional institutions with national and international support. This decline is mainly due to the rapid increase in population densities surrounding the water bodies of the Nile basin, which has negatively influenced the water quality of these basins and their fisheries. In addition, overexploitation of the fish resources has advantaged non-commercial and invasive species in their competition with native and economically valuable ones. The lack of regulatory fishing plans and management policy in most of the Nile basin fisheries has also added complexity to any future strategic plans to safeguard such valuable resources. In order to manage this important sector along the Nile basin there is a need to:

- Make appropriate information data available to guide decision makers

- Enforce laws and sustainable management of fisheries resources and the fish habitat.

- Control the use and dumping of Pesticides residuals in the Nile basin

- Promote sustainable practices with the help of scientific knowledge.

- Provide adequate financial resources and human capacity to implement fisheries program.

- Investigate and improve the socio-economic status of fishermen.

- Regulate fishing gear and practices that may otherwise damage fish stocks and habitat.

The poverty of many of the diverse populations living along the Nile basin remains a barrier to the implementation of concepts of natural resources conservation, despite the many efforts of governments and NGO's to disseminate this information.

Faced with a continuing large gap between global supply and demand for fish protein, with critical shortage in some regions, aquaculture is widely regarded as having a crucial role to play in meeting global and regional food requirements over the next 20 years. Aquaculture can be a water efficient means of food production, and also brings wider resources management benefits. To this end, it is advisable that the Nile basin countries take advantage of the existing water resources and encourage the private sector in implementing aquaculture activities as an actual and future alternative of animal protein and as a substitute for the degraded natural fisheries of the Nile basin.

Author details

Waleed Hamza*

Address all correspondence to: w.hamza@uaeu.ac.ae

Biology Department, Faculty of science - United Arab Emirates University, Al-Ain, UAE

References

[1] Roskar, J., 2000. Assessing the water resources potential of the Nile river based on data, available at the Nile forcasting center in Cairo. Geografski zbornik. 53:32-65.

[2] Hamza, W., 2006. Estuary of the Nile. In: Wangersky P. (Ed.), *Estuaries; Hdb. Env. Chem. Springer Verlag Publisher*, Vol. 5 H: 149-173 pp.

[3] FAO, 2000. Nile basin initiative report to ICCON "Water and Agriculture in the Nile Basin", Land and water development division AGL/MISC/29/2000 Rome 1-68pp.

[4] Witte, F., van Oijen and N. Sibbing, 2009a. Fish fauna of the Nile. In: Dumont H.J. (ed.), The Nile. Monographiae Biologicae, Vol. 89:647-675. Springer, Dordrecht.

[5] Witte, F., M. deGraaf, O.Mkumbo, A.I. El-Moghraby and F. Sibbing, 2009b. Fish fauna of the Nile. In: Dumont H.J. (ed.), The Nile. Monographiae Biologicae, Vol. 89:647-675. Springer, Dordrecht.

[6] Matsuishi, T., L. Muhoozi, O. Mkumbo, Y. Budeba, M. Najiru, A. Asila, A. Othina and I. G. Cowx, 2006. Are the exploitation pressure on the Nile Perch fisheries resources of Lake Victoria a cause of concern?. Fisheries Management and Ecology 13 (1) 53-71.

[7] Witte, F., T. Goldscmidt, P. Goudswaard, W. Ligtvoet, M. van Oijen and J.H. Wanink, 1992. Species extinction and concomitant changes in Lake Victoria. Netherlands Journal of Zoology 42: 214- 232.

[8] Balirwa, J.S., C.A Chapman, L.J. Chapman, I.G. Cowx, K. Geheb, L. Kaufman, R.H. Lowe-McConnell, O. Seehausen, J.H. Wanink, R.L. Welcomme and F. Witte, 2003. Biodiversity and fisheries sustainability in the Lake Victoria basin: an unexpected marriage? Bioscience 53: 703-716.

[9] Cowx, I.G., M. van der Knaap, L. I. Muhoozi and A. Othina, 2003. Improving fishery catch statistics for Lake Victoria. Aquatic Ecosystem Health & Management, 6(3): 299-310.

[10] Njiru, M., J. Ojuok, A. Getabu, T. Jembe, M. Owili and C. Ngugi, 2008. Increasing dominance of Nile tilapia, *Oreochromis niloticus* (L) in Lake Victoria, Kenya: Conse-

quences for the Nile perch *Lates niloticus* (L) fishery. Aquatic Ecosystem Health & Management, 11(1): 42-49.

[11] Dejen, E., J. Vijverberg, and F.A. Sibbing, 2004. Temporal and special distribution of microcrustacean zooplankton in relation to turbidity and other environmental factors in a large tropical lake (L. Tana, Ethiopia), Hydrobiologia 513: 39-49.

[12] DeGraaf, M, P.A.M. van-Zweiten, M.A.M. Machiels, E. Lemma, E. Dejen, and F.A. Sibbing, 2006. Vulnerability to a small scale commercial fishery of Lake Tana's (Ethiopia) endemic Labeobarbus compared with African catfish and Nile tilapia: an example of recruitment-overfishing? Fisheries Research 82: 304-318.

[13] FAO, 2002. Fishery country profile, FID/CP/SUD. February, 2002.

[14] FAO, 2008. FAO Fisheries and Aquaculture Information and Statistics Service (Egypt). http://www.fao.org/fishery/countrysector/FI-CP_EG/1

[15] Earth Trend, 2003. Water resources and freshwater ecosystems-Egypt. Available online (www.earthtrends.wri.org).

[16] GAFRD. 1995-2002. Annual fishery statistics reports (General Authority for Fish Resources Development), Cairo-Egypt.

[17] Hamza, W., 1999. Differentiation of phytoplankton communities of Lake Mariut: A consequence of human impact. Bull. Fac. Sci. Alex. Uni. 39 (1-2): 159-168.

[18] Mugidde, R., 1993. The increase in phytoplankton production and biomass in Lake Victoria (Uganda). Ver. Internat. Verein. Limnol. 25: 846-849.

[19] Bootsma, H., & R.E. Hecky, 1993. Conservation of the African Great lakes: a limnological perspective. Conservation Biology 7: 644-656.

[20] Henry, L., M.A. Kishimba, 2006. Pesticide residue in Nile tilapia (Oreochromis niloticus) and Nile Perch (Lates niloticus) from Southern Lake Victoria, Tanzania. Environmental Pollution 140: 348-354.

[21] Munga, D., 1995. DDT and endosulfan in the fish from Holla irrigation scheme. M.Sc. Thesis. University of Nairobi, Tana River, Kenya.

[22] Wadinga, S.o., 2001. Use and distribution of organochlorine pesticides; The future in Africa. Pure and Applied Chemistry 73:1147-1155.

[23] Rashed, M.N., 2001. Monitoring of environmental heavy metals in fish from Nasser Lake. Environment International 27: 27-33.

[24] Oczkowwski, A and S. Nixon, 2008. Increasing nutrient concentrations and the rise and fall of a coastal fishery; A review of data from the Nile Delta-Egypt. Estuarine, Coastal and Shelf Science. 23: 189-219.

[25] Hamza, W. & Mason, S., 2004. Water availability and food security challenges in Egypt. Proceedings International forum on Food security under water scarcity in the Middle East: Problems and Solutions. Como- Italy 24-27 Nov., 2004.

[26] Siegel, F., 1995. Environmental geochemistry in development planning: an example from the Nile delta, Egypt. Journal of Geochemical Exploration. 55: 265-273.

[27] Barakat, A., 2004. Assessment of persistent toxic substances in the environment of Egypt. Environment International. 30: 309-322.

[28] El-Sebae, A.H, Abou Zeid, M., & Saleh, M.A., 1993. Status and environmental impact of Toxaphine in the third world- a case study of African agriculture. Chemoshere. 27:2063-2072.

[29] FAO, 1998. Aquaculture production statistics-1987-1996. FAO Fisheries Circular, No. 815, Revision 10 [FIDI/C815] Rome.

[30] FAO, 2009. Fisheries and Aquaculture statistics. http://www.fao.org/statistics

[31] FAO, 1994. Seminar on freshwater culture, Damascus, May 2-4, 1994; Edited by ME-DRAP II Regional Center Tunis-Tunisia.

[32] Geheb, K., S. Kalloch, M. Medard, A. Nyapendi, C. Lwenya and M. Kyangwa, 2007. Nile perch and the hungry of Lake Victorai: Gender, status and food in an East African fishery. Food Policy- Elsiver (in press).